Technical Drawing

Fourth Edition

WORKBOOK

William S. Chalk

Delmar Publishers

an International Thomson Publishing company I(T)P®

Albany • Bonn • Boston • Cincinnati • Detroit • London • Madrid
Melbourne • Mexico City • New York • Pacific Grove • Paris • San Francisco
Singapore • Tokyo • Toronto • Washington

NOTICE TO THE READER

Publisher does not warrant or guarantee any of the products described herein or perform any independent analysis in connection with any of the product information contained herein. Publisher does not assume, and expressly disclaims, any obligation to obtain and include information other than that provided to it by the manufacturer.

The reader is expressly warned to consider and adopt all safety precautions that might be indicated by the activities herein and to avoid all potential hazards. By following the instructions contained herein, the reader willingly assumes all risks in connection with such instructions.

The Publisher makes no representation or warranties of any kind, including but not limited to, the warranties of fitness for particular purpose or merchantability, nor are any such representations implied with respect to the material set forth herein, and the publisher takes no responsibility with respect to such material. The publisher shall not be liable for any special, consequential, or exemplary damages resulting, in whole or part, from the readers' use of, or reliance upon, this material.

Delmar Staff:
Publisher: Michael McDermott
Acquisitions Editor: Sandy Clark
Developmental Editor: John Fisher
Production Coordinator: Jennifer Gaines
Project Editor: Christopher Chien
Art and Design Coordinator: Mary Beth Vought
Editorial Assistant: Allyson Powell
Marketing Manager: Kitty Kelly
Marketing Coordinator: Paula Collins

Printed in the United States of America
5 6 7 8 9 10 XXX 05 04 03

For more information, contact Delmar Learning, Executive Woods, 5 Maxwell Drive, Clifton Park, NY 12065; or find us on the World Wide Web at http://www.delmar.com

International Division List

Japan:
Thomson Learning
Palaceside Building 5F
1-1-1 Hitotsubashi, Chiyoda-ku
Tokyo 100 0003 Japan
Tel: 813 5218 6544
Fax: 813 5218 6551

Australia/New Zealand
Nelson/Thomson Learning
102 Dodds Street
South Melbourne, Victoria 3205
Australia
Tel: 61 39 685 4111
Fax: 61 39 685 4199

UK/Europe/Middle East:
Thomson Learning
Berkshire House
168-173 High Holborn
London
WC1V 7AA United Kingdom
Tel: 44 171 497 1422
Fax: 44 171 497 1426

Latin America:
Thomson Learning
Seneca, 53
Colonia Polanco
11560 Mexico D.F. Mexico
Tel: 525-281-2906
Fax: 525-281-2656

Canada:
Nelson/Thomson Learning
1120 Birchmount Road
Scarborough, Ontario
Canada M1K 5G4
Tel: 416-752-9100
Fax: 416-752-8102

Asia:
Thomson Learning
60 Albert Street, #15-01
Albert Complex
Singapore 189969
Tel: 65 336 6411
Fax: 65 336 7411

South Africa:
Thomson Learning
Zonnebloem Building
Constantia Square
526 Sixteenth Road
P.O. Box 2459
Halfway House, 1685
South Africa
Tel: 27 11 805 4819
Fax: 27 11 805 3648

Library of Congress Cataloging-in-Publication Data: 98-42472
ISBN 0-7668-0532-8

PREFACE

To the student:

The **primary objective** of this workbook is to assist you in developing your skills in making and reading technical drawings. In your career as a drafter, designer, or engineer you will need a working technical vocabulary, as well as skills with which to use this vocabulary to read and prepare technical drawings, whether the drawings are done in pencil, in ink, or on a computer. Exercises in the first three sections of the workbook focus on these skills and on expanding your technical vocabulary.

Thus, for a brief introduction to "reading" drawings, two example technical drawings are included in introductory problems in the first section with questions related to the drawings for you to consider. The questions are intended to direct your attention to the type of vocabulary information you would be looking for in a technical drawing. Answers to these preliminary questions are on a page immediately following the two drawings.

The **secondary objective** of the workbook is to take you beyond the fundamentals to such areas as: advanced dimensioning and tolerancing; commercial components such as fasteners cams, gears, and springs; graphics employed by different drafting disciplines such as architectural, structural, civil, and electrical; the design process; and charts and graphs. The workbook also includes a special section on mechanical drafting mathematics, and involves you in CAD applications in appropriate chapters.

To the Instructor:

The topics of the workbook problems parallel the order of topics in Goetsch, Nelson and Chalk, TECHNICAL DRAWING, 4th Edition, by Delmar Publishers; however, the workbook can be used with any beginning technical drawing or engineering graphics text that covers the fundamentals of technical drawing. The problems are grouped in sections that follow the sections and chapters in our text.

Primary emphasis of the workbook: Sections 1,2, & 3

- **Section 1** Basics
- **Chapters:** Introduction – Graphic Communications and Technical Drawing; **2** – Instruments; **3** – Lettering, Sketching, & Line Techniques; **4** – Geometric Construction
- **Section 2** Technical Drawing Fundamentals
- **Chapters: 5** – Multiview Drawings; **6** – Sectional Views; **7** – Auxiliary Views; **8** – Descriptive Geometry; **9** – Patterns and Developments; **10** – Dimensioning and Notation
- **Section 3** Design Drafting Applications
- **Chapters: 12** – Geometric Dimensioning and Tolerancing; **13** – Fasteners; **14** – Springs; **15** – Cams; **16** – Gears; **17** – Assembly and Detail Drawings for Design; **18** – Pictorials
- Selected chapters have **CAD** problems to give the students additional challenges.

Secondary emphasis of the workbook: Section 4 and Mechanical Drafting Mathematics

- **Section 4** Related Technologies, Applications, and Processes
- **Chapters: 19** – Welding; **20** – Modern Manufacturing: Materials, Processes, and Automation; **21** – Drafting Applications: Pipe, Structural, and Architecture; **22** – Drafting Applications: Electronics and Printed Circuit Boards; **23** – Charts and Graphs; **24** – The Design Process
- **Mechanical Drafting Mathematics**: Basic mathematics, geometry, trigonometry, and applications

(Note: since Chapters 1, 11, 25, and 26 in our text cover descriptive materials related to CAD, Engineering Visualization, and Employability Skills for finding and keeping a job, the chapters are not covered in this workbook.)

Solutions to the workbook problems appear in the instructor's guide for our text.

General Comments

Title blocks for the problems contain information to assist you in selecting appropriate exercises to accompany your lectures. For example:

Ch4 Geometric Construction

Section-Chapter-Problem **1 4 11**	Line types Figure 3-17 Ogee curves Figure 4-46	Course_______Section____ **Student**_________Date___	**Problem** **1-4-11**

The chapter and chapter heading occur at the beginning of a set of problems for the chapter. The problem number **1-4-11** refers to **Section 1** and **Chapter 4** of the text, and problem number **11** in the set for the chapter. Also, each problem page usually contains a specific reference to a figure (s) in our text. A separate space is provided for the student's name, the date, and the course. Finally, the page number is the same as the problem number.

Each section is preceded by objective statements that tell students what they should be able to achieve by studying and doing the problems. The Objectives are followed by supplementary comments on the problems.

Industrial topics are emphasized throughout the workbook. A significant number of the problems focus on the Detail and Assembly drawings employed in industry.

Two new features appear in our workbook:

1. **Mechanical Drafting Mathematics**. Basic mathematics, geometry, trigonometry, and related applications are discussed and included in problems.
2. **Computer Aided Drafting** problems are offered in appropriate chapters.

TABLE OF CONTENTS

Preface iii

Section 1 • Basics**

Objectives: Section 1 1
Supplemental Comments: Section 1 1

Introduction: Graphic Communications and Technical Drawing

Section	Chapter	Problem	Topic	Page/Problem No.
1	Intro	1	Reading drawings Figure I-29	1-Intro-1
1	Intro	2	Reference Chapter 10: Dimensioning and Notation	1-Intro-2
1	Intro	3	Reference Chapter 10: Dimensioning and Notation	1-Intro-3
1	Intro	4	Reading drawings Figure I-29 (answers) Reference Chapter 10: Dimensioning and Notation	1-Intro-4

Chapter 2 Technical Drawing with Instruments

Section	Chapter	Problem	Topic	Page/Problem No.
1	2	1	Scales (Half size & Quarter size)	1-2-1
1	2	2	Scales (Full size, 10X, Metric full)	1-2-2
1	2	3	Scales (Metric, Engineer's)	1-2-3
1	2	4	Scales (Architect's: 1/8 & 1/48 size)	1-2-4
1	2	5	Scales (Architect's, Engineer's & Metric)	1-2-5
1	2	6	Layout to scale from freehand sketch	1-2-6
1	2	7	Layouts to scale from freehand sketches	1-2-7
1	2	8	ASME Y14.5-1994M standard symbols	1-2-8
1	2	9	French curves	1-2-9
1	2	10	Protractor	1-2-10
1	2	11	Micrometer	1-2-11
1	2	12	Geometric Patterns	1-2-12
1	2	13	French curves	1-2-13

Chapter 3 Lettering, Sketching, and Line Techniques

Section	Chapter	Problem	Topic	Page/Problem No.
1	3	1	Lettering Uppercase Gothic	1-3-1
1	3	2	Lettering Spur Gears	1-3-2
1	3	3	Lettering Line Types	1-3-3
1	3	4	Lettering Title Block	1-3-4
1	3	5	Lettering Attributes	1-3-5

Section	Chapter	Problem	Topic	Page/Problem No.
1	3	6	Lettering Specifications	1-3-6
1	3	7	Freehand sketching Three views	1-3-7
1	3	8	Freehand sketching Isometric	1-3-8
1	3	9	Straight lines Instruments	1-3-9
1	3	10	Freehand sketching Oblique, Perspective	1-3-10
1	3	11	Freehand sketching Oblique, Perspective	1-3-11
1	3	12	Freehand sketching Two point Perspective	1-3-12
1	3	13	Freehand sketching Three view & Isometric	1-3-13
1	3	14	Freehand sketching Three view & Isometric	1-3-14
1	3	15	CAD Problems: Title Block and Drawing Changes	1-3-15

Chapter 4 Geometric Construction

Section-Chapter-Problem			Topic	Page/Problem No.
1	4	1	Line types Instruments	1-4-1
1	4	2	Line types Instruments Layout	1-4-2
1	4	3	Line types Instruments Layout	1-4-3
1	4	4	GD & T Symbols	1-4-4
1	4	5	Elemental geometrical constructions	1-4-5
1	4	6	Elemental geometrical constructions	1-4-6
1	4	7	Transfer shapes	1-4-7
1	4	8	Elemental geometrical constructions	1-4-8
1	4	9	Geometrical constructions	1-4-9
1	4	10	Arc tangents	1-4-10
1	4	11	Ogee curves	1-4-11
1	4	12	Parabolic curves	1-4-12
1	4	13	Parabola & Hyperbola curves	1-4-13
1	4	14	Helix	1-4-14
1	4	15	Involute curves	1-4-15
1	4	16	Kinematic layout	1-4-16
1	4	17	CAD Problems: Geometric Shapes; Logo Design	1-4-17

Section 2 • Technical Drawing Fundamentals

Objectives: Section 2 103

Supplemental Comments: Section 2 103

Chapter 5 Multiview Drawings

Section	Chapter	Problem	Topic	Page/Problem No.
2	5	1	Principal viewing planes: "box"	2-5-1
2	5	2	Multiview drawings	2-5-2
2	5	3	Multiview drawings: matching views	2-5-3
2	5	4	Freehand sketching	2-5-4
2	5	5	Freehand sketching	2-5-5
2	5	6	Reading drawings: inclined surfaces	2-5-6
2	5	7	Inclined & Oblique surfaces	2-5-7
2	5	8	Centering orthographic views	2-5-8
2	5	9	Centering orthographic views	2-5-9
2	5	10	Centering orthographic views	2-5-10
2	5	11	Projections: third angle & first angle	2-5-11
2	5	12	Centering orthographic views	2-5-12
2	5	13	Orthographic views: radii & holes	2-5-13
2	5	14	Orthographic views: radii & holes	2-5-14
2	5	15	Orthographic views: better practices	2-5-15
2	5	16	Orthographic views: inclined surfaces	2-5-16
2	5	17	Orthographic views: oblique, inclined, & curved	2-5-17
2	5	18	Orthographic views: oblique surfaces	2-5-18
2	5	19	Orthographic views: inclined & curved surfaces	2-5-19
2	5	20	Orthographic views: centering views	2-5-20
2	5	21	Orthographic views: centering views	2-5-21
2	5	22	Structural drafting	2-5-22
2	5	23	Visualization practice	2-5-23
2	5	24	CAD Problems: House Views; Hinge Location	2-5-24

Chapter 6 Sectional Views

Section	Chapter	Problem	Topic	Page/Problem No.
2	6	1	Full sections	2-6-1
2	6	2	Half & Broken out sections	2-6-2
2	6	3	Offset, Full & Half sections	2-6-3
2	6	4	Aligned & Removed sections	2-6-4
2	6	5	Offset sections	2-6-5
2	6	6	Full section & Detail view	2-6-6
2	6	7	Section assembly drawing	2-6-7
2	6	8	Broken out section & Parts not sectioned	2-6-8
2	6	9	Removed sections	2-6-9
2	6	10	Aligned & Enlarged sections	2-6-10

Section	Chapter	Problem	Topic	Page/Problem No.
2	6	11	Auxiliary & Thinwalls in section	2-6-11
2	6	12	CAD Problems: Full Section; Offset Section	2-6-12

Chapter 7 Auxiliary Views

Section-Chapter-Problem			Topic	Page/Problem No.
2	7	1	Auxiliary features	2-7-1
2	7	2	Reference & Fold line methods	2-7-2
2	7	3	Auxiliary: Reference line method	2-7-3
2	7	4	Auxiliary: Fold line method	2-7-4
2	7	5	Auxiliary: Fold line method	2-7-5
2	7	6	Reference & Fold line methods	2-7-6
2	7	7	Reference & Fold line methods	2-7-7
2	7	8	Reference, Fold line & Intersecting Surfaces	2-7-8
2	7	9	Reference & Fold line methods	2-7-9
2	7	10	Auxiliary: Fold line method	2-7-10
2	7	11	Auxiliary: Reference line method	2-7-11
2	7	12	CAD Problems: Paper Airplane; Auxiliary Views	2-7-12

Chapter 8 Descriptive Geometry

Section-Chapter-Problem			Topic	Page/Problem No.
2	8	1	True length of a line	2-8-1
2	8	2	Edge view and true size of a plane	2-8-2
2	8	3	Finding shortest distances	2-8-3
2	8	4	Projections and dihedral angles	2-8-4
2	8	5	Shortest distance between two non-parallel lines	2-8-5
			Shortest distance between a point and a plane	2-8-5
2	8	6	Shortest distance between two non-parallel lines	2-8-6
2	8	7	True size of planes	2-8-7
2	8	8	True size of planes	2-8-8
2	8	9	Dihedral angles	2-8-9
2	8	10	Dihedral angles	2-8-10
2	8	11	Intersections: lines, planes, cylinders, prisms	2-8-11
2	8	12	Strike & dip. Bearing, grade, and length	2-8-12
2	8	13	Revolution: true length lines	2-8-13
2	8	14	Contour lines. Cuts & fills	2-8-14
2	8	15	Vector graphics	2-8-15
2	8	16	CAD Problems: Intersections; Topography	2-8-16

Chapter 9 Patterns and Developments

Section	Chapter	Problem	Topic	Page/Problem No.
2	9	1	Development: parallel line technique (prism) (Deduce shape from development)	2-9-1
2	9	2	Development: parallel line technique (cylinder) (Deduce shape from development)	2-9-2
2	9	3	Development: parallel line technique (prism)	2-9-3
2	9	4	Developments: parallel line technique (prism)	2-9-4
2	9	5	Development: triangulation technique	2-9-5
2	9	6	Developments: parallel line technique (cylinders)	2-9-6
2	9	7	Developments: line technique and intersections	2-9-7
2	9	8	Development: radial line technique	2-9-8
2	9	9	Development: radial line technique	2-9-9
2	9	10	Developments: radial line & triangulation techniques	2-9-10
2	9	11	Development: triangulation technique (Deduce shape from development)	2-9-11
2	9	12	Development: parallel line technique	2-9-12
2	9	13	Developments: bend allowances	2-9-13
2	9	14	CAD Problems: Flat Patterns for Food Containers	2-9-14

Chapter 10 Dimensioning and Notation

Section	Chapter	Problem	Topic	Page/Problem No.
2	10	1	General Dimensioning	2-10-1
2	10	2	Dimensioning: three rectangular views	2-10-2
2	10	3	Geometric breakdown: angles & cylinders	2-10-3
2	10	4	Geometric breakdown: radii & cylinders	2-10-4
2	10	5	Geometric breakdown: cylinders, knurls & slots	2-10-5
2	10	6	Geometric breakdown: radii & irregular curves	2-10-6
2	10	7	Geometric breakdown: coordinate & bolt circle	2-10-7
2	10	8	Geometric breakdown: slots & arcs	2-10-8
2	10	9	Geometric breakdown: size & location	2-10-9
2	10	10	Notes & callouts	2-10-10
2	10	11	Holes callouts & bolt centers	2-10-11
2	10	12	Notes & callouts	2-10-12
2	10	13	RC Fits: HOLE BASIS	2-10-13
2	10	14	RC Fits: SHAFT BASIS	2-10-14
2	10	15	FN Fits: SHAFT BASIS	2-10-15
2	10	16	RC & FN Fits	2-10-16
2	10	17	Reading Drawings	2-10-17
2	10	18	CAD Problems: Dimensions; ECO	2-10-18

Section 3 • Design Drafting Applications

Objectives: Section 3 301
Supplemental Comments: Section 3 301

Chapter 12 Geometric Dimensioning and Tolerancing

Section-Chapter-Problem			Topic	Page/Problem No.
3	12	1	Unilateral and Orientation Tolerances	3-12-1
3	12	2	Geometric Dimensioning and Positioning Tolerancing	3-12-2
3	12	3	Geometric Dimensioning and Positioning Tolerancing, Datum Surfaces and MMC	3-12-3
3	12	4	Parallelism and Concentricity	3-12-4
3	12	5	Perpendicularity and Total Runout	3-12-5
3	12	6	Datum Targets	3-12-6
3	12	7	Geometric Dimensioning and Tolerancing: Reading Drawings	3-12-7
3	12	8	Geometric Dimensioning and Tolerancing: Perpendicularity and Parallelism	3-12-8
3	12	9A	Problem Statement: Control Blade Shaft*	3-12-9A*
3	12	9B	GD & T: Control Blade Shaft*	3-12-9B*
3	12	10	CAD Problems: Datums; Tolerancing	3-12-10

*Topic titles for problem numbers with letters usually indicate the assignment for that page.

Chapter 13 Fasteners

Section-Chapter-Problem			Topic	Page/Problem No.
3	13	1	Holes for Fasteners	3-13-1
3	13	2	Drawing Fasteners	3-13-2
3	13	3	Drawing Fasteners, Fasteners in Section	3-13-3
3	13	4	Fasteners in Section	3-13-4
3	13	5	Spring Pin, Retaining Ring	3-13-5
3	13	6	Thread Specifications and Standards	3-13-6
3	13	7	CAD Problems: Drawing Bolts; Dimensioning	3-13-7

Chapter 14 Springs

Section-Chapter-Problem			Topic	Page/Problem No.
3	14	1	Ground, Closed Ends Spring	3-14-1
3	14	2	Plain, Closed Ends Spring	3-14-2
3	14	3	CAD Problems: Spring Design	3-14-3

Chapter 15 Cams

Section-Chapter-Problem			Topic	Page/Problem No.
3	15	1	Cam Displacement Diagram	3-15-1

Section	Chapter	Problem	Topic	Page/Problem No.
3	15	2	Drawing and Dimensioning A Cam	3-15-2
3	15	3	CAD Problems: Displacement; Cam Design	3-15-3

Chapter 16 Gears

Section-Chapter-Problem			Topic	Page/Problem No.
3	16	1	Spur Gear Nomenclature	3-16-1
3	16	2	Spur Gear Tooth Data	3-16-2
3	16	3	CAD Problems: Gear Design; Gear Train	3-16-3

Chapter 17 Assembly and Detail Drawings for Design

Section-Chapter-Problem			Topic	Page/Problem No.
3	17	1A	Examples: Freehand Sketch & Design Layout for Anchor Bolt Puller	3-17-1A***
3	17	1B	Examples: Detail Drawings: Anchor Bolt Puller	3-17-1B***
3	17	1C	Examples: Detail Drawings: Anchor Bolt Puller	3-17-1C***
3	17	1D	Example: Assembly Drawing: Anchor Bolt Puller	3-17-1D***

***Examples only: Freehand, Layout, Assembly, and Detail Drawings

Section	Chapter	Problem	Topic	Page/Problem No.
3	17	2A	Conceptual Drawing: Toggle Clamp	3-17-2A*
3	17	2B	Design Layout: Toggle Clamp	3-17-2B*
3	17	2C	Detail Drawings: Toggle Clamp	3-17-2C*
3	17	2D	Detail Drawings: Toggle Clamp	3-17-2D*
3	17	2E	Assembly Drawing: Toggle Clamp	3-17-2E*

*Topic titles for problem numbers with letters usually indicate the assignment for that page.

Section	Chapter	Problem	Topic	Page/Problem No.
3	17	3A	Engineering Change Order (ECO): Anchor Bolt Puller	3-17-3A*
3	17	3B	ECO Design Layout: Anchor Bolt Puller	3-17-3B*
3	17	3C	ECO Side Frame Details: Anchor Bolt Puller	3-17-3C*
3	17	3D	ECO Spacer Details: Anchor Bolt Puller	3-17-3D*
3	17	3E	ECO Support Plate Details: Anchor Bolt Puller	3-17-3E*
3	17	3F	ECO Angle Stop Details: Anchor Bolt Puller	3-17-3F*
3	17	3G	ECO Assembly Drawing: Anchor Bolt Puller	3-17-3G*
3	17	3H	ECO Parts List: Anchor Bolt Puller	3-17-3H*

*Topic titles for problem numbers with letters usually indicate the assignment for that page.

Section	Chapter	Problem	Topic	Page/Problem No.
3	17	4A	Freehand Sketch: Paravane Stabilizer	3-17-4A*
3	17	4B	Design Layout: Paravane Stabilizer	3-17-4B*
3	17	4C	Vane Details: Paravane Stabilizer	3-17-4C*
3	17	4D	Weight Detail: Paravane Stabilizer	3-17-4D*
3	17	4E	Assembly with Parts List: Paravane Stabilizer	3-17-4E*

*Topic titles for problem numbers with letters usually indicate the assignment for that page.

3	17	5A	Freehand Sketch: ECO Paravane Stabilizer	3-17-5A*
3	17	5B	Design Layout: ECO Paravane Stabilizer	3-17-5B*
3	17	5C	Horizontal Vane Details: ECO Paravane Stabilizer	3-17-5C*
3	17	5D	Vertical Vane Details: ECO Paravane Stabilizer	3-17-5D*
3	17	5E	Weight Detail: ECO Paravane Stabilizer	3-17-5E*
3	17	5F	Assembly with Parts List: ECO Paravane Stabilizer	3-17-5F*
3	17	6	CAD Problems: Assembly Drawings—palette, rocket	3-17-6

*Topic titles for problem numbers with letters usually indicate the assignment for that page.

Chapter 18 Pictorial Drawings

Section	Chapter	Problem	Topic	Page/Problem No.
3	18	1	Isometric Drawing	3-18-1
3	18	2	Isometric Drawing	3-18-2
3	18	3	Isometric and Oblique Drawing	3-18-3
3	18	4A	Two Point Perspective Drawing	3-18-4A*
3	18	4B	Two Point Perspective Drawing	3-18-4B*
3	18	5A	Analyzing Orthographic Views	3-18-5A*
3	18	5B	Two Point Perspective	3-18-5B*
3	18	6	CAD Problems: Isometric and Oblique Drawings	3-18-6

*Topic titles for problem numbers with letters usually indicate the assignment for that page.

Section 4 • Related Technologies, Applications, and Processes

Objectives: Section 4 433
Supplemental Comments: Section 4 433

Chapter 19 Welding

Section	Chapter	Problem	Topic	Page/Problem No.
4	19	1	Welding Symbols, Nomenclature	4-19-1
4	19	2	Welding Symbols, Applications	4-19-2
4	19	3	Welding Symbols, Casting into Weldment	4-19-3
4	19	4	Welding Symbols, Casting into Weldment	4-19-4
4	19	5A	Welding Practices, Picnic Table	4-19-5A*
4	19	5B	Welding Practices, Steel Details, Picnic Table	4-19-5B*
4	19	5C	Welding Practices, Wood Details, Picnic Table	4-19-5C*
4	19	5D	Welding Practices, Picnic Table Assembly	4-19-5D*
4	19	6	CAD Problems: Welding Symbols	4-19-6

Chapter 20 Modern Manufacturing: Materials, Processes, and Automation

Section	Chapter	Problem	Topic	Page/Problem No.
4	20	1	Template Jig with Drill Bushings Assembly	4-20-1
4	20	2	CAD Problems: Manufacturing Fixtures	4-20-2

Chapter 21 Drafting Applications: Pipe, Structural, and Architecture

Section	Chapter	Problem	Topic	Page/Problem No.
4	21	1A	Truss Specifications	4-21-1A*
4	21	1B	Truss Layout	4-21-1B*
4	21	1C	Truss Parts List	4-21-1C*
4	21	2	Single-Line Pipe Drawing	4-21-2
4	21	3	Hydraulic Circuit Schematic	4-21-3
4	21	4	CAD Problems: Architectural Drafting	4-21-4

Chapter 22 Drafting Applications: Electronics and Printed Circuit Boards

Section	Chapter	Problem	Topic	Page/Problem No.
4	22	1	Sketch and Finished Schematic	4-22-1
4	22	2	Sketch and Finished Schematic	4-22-2
4	22	3	CAD Problems: Schematic Circuit; PC Board	4-22-3

Chapter 23 Charts and Graphs

Section	Chapter	Problem	Topic	Page/Problem No.
4	23	1	Plotting Data	4-23-1
4	23	2	CAD Problems: Bar Chart; Survey Charts	4-23-2

Chapter 24 The Design Process

Section	Chapter	Problem	Topic	Page/Problem No.
4	24	1A	Example: Motor, Gear Train, and Drum Hoist	4-24-1A*
4	24	1B	Calculations: Drive Train with Drum Hoist	4-24-1B*

*Topic titles for problem numbers with letters usually indicate the assignment for that page.

**Goetsch, Nelson and Chalk, TECHNICAL DRAWING, 4th Edition

Appendix A: Mechanical Drafting Mathematics

Section-Problem No.	Topic	Page/Problem No.
Appendix A 1	Rounding Off; Decimal Equivalents; Evaluating Formulas	A-1
Appendix A 2	Evaluating Formulas; Ratio and Proportion	A-2
Appendix A 3	Gear Ratios; Adding and Subtracting Angles	A-3
Appendix A 4	Dividing Angles; Converting Angle Units	A-4
Appendix A 5	Converting Angle Units; Types of Angles & Triangles	A-5
Appendix A 6	Right Triangle Applications	A-6
Appendix A 7	Areas of Polygons	A-7
Appendix A 8	Multi-sided Polygons; Properties of Circles	A-8
Appendix A 9	Properties of Circles; Trigonometric Functions	A-9
Appendix A 10	Trigonometry; Basic Calculations	A-10
Appendix A 11	Trigonometry; Drafting Applications	A-11
Appendix A 12	Triangle Application; Dovetail	A-12
Appendix A 13	Law of Sines; Applications	A-13
Appendix A 14	Law of Sines; Applications	A-14
Appendix A 15	Law of Cosines; Applications	A-15
Appendix A 16	Oblique Triangle	A-16
Appendix A 17	Law of Cosines; CUT and FILL	A-17
Appendix A 18	Rise and Run; Stairway	A-18
Appendix A 19	ADDA; Sample Quiz	A-19
Appendix A 20	Proportional Equations; Heights	A-20
Appendix A 21	Concrete Bulkhead; Volume	A-21
Appendix A 22	Trigonometry; Application	A-22
Appendix A 23A	Graphical Technique; Areas	A-23A
Appendix A 23B	Graphical Technique; Areas	A-23B
Appendix A 24	Dovetail Calculations	A-24
Appendix A 25A	Taper Calculations	A-25A
Appendix A 25B	Taper Calculations	A-25B
Appendix A 26	Triangulation; Law of Sines	A-26
Appendix A 27	Mechanical Advantage	A-27
Appendix A 28	Bicycle Speed; Gear Ratio	A-28
Appendix A 29	Cutting Speed; Lathes	A-29
Appendix A 30	Highway Curve; Banking Calculation	A-30
Appendix A 31	Drive and Drive Base; Weight	A-31

Section 1
Objectives:

After doing and studying the exercises and problems in Section 1 of this workbook you should be able to:

- recognize the importance of having a technical vocabulary for reading technical drawings.
- do and recognize standard lettering and line work for technical drawings.
- use technical drawing instruments: scales, triangles, protractors, French curves, etc.
- prepare freehand sketches using the blocking-in technique.
- sketch objects in several formats such as orthographic, isometric, oblique, and 1 and 2 vanishing point perspectives.
- do layout drawings to determine dimensions graphically.
- do basic geometrical constructions such as bisecting a line, bisecting an angle, divide a line into equal segments, and enlarge the drawing of an object.
- do challenging geometrical constructions such as arc tangents, ogee curves, parabolas, hyperbolas, involutes, helixes, and kinematic layouts.

Supplemental Comments: Section 1

Introduction: Graphic Communications and Technical Drawing
Problem 1-Intro-1: Only a portion of Figure I-29 is used for this problem. Observe some additional technical drawing practices shown in this figure such as:

a. Break lines to isolate Sections and Details
b. Scales specified at each Section or Detail
c. Irregular leader lines

Problem 1-Intro-2: Letters in the circles are for instructional purposes only; however, the letters in the diamonds would appear in industry to indicate locations of revisions. If the MOUNTING GUIDE were used in an assembly of components it could appear in a parts list as follows:

Part	Qty	Description	Material
12	2	Mounting Guide Dwg 392450-B	B-158

An outline of the Mounting Guide would appear on the assembly drawing.

Note the different types of lines used on the MOUNTING GUIDE drawing.

Chapter 1 Technical Drawing with CAD
Problems related to this chapter are not included in the workbook because the chapter is more descriptive than problem oriented.

Chapter 2 Technical Drawing with Instruments
Problems 1-2-1 through **1-2-5** are basic scale usage problems. Later in school and in industry you will discover that each industry has its scale preferences (for example, the machine industry prefers decimal dimensions and the structural industry prefers feet and inches). Metric units are becoming more common, especially on drawings involved with foreign trade.

Problems 1-2-6 and **1-2-7** illustrate the initial steps in the development of technical ideas—freehand sketches redrawn to scale with instruments.

Problem 1-2-8 introduces some of the symbols used in more advanced dimensioning such as Geometric Dimensioning and Tolerancing. (Refer to Chapter 12.)

Problems 1-2-9 and **1-2-13** require a French curve (also called an irregular curve) to draw a best-fit line through a series of points. "Best-fit" implies that the line may not go through the center of each point.

Problem 1-2-10 illustrates the use of a protractor. Angles can be located more precisely with drafting machines fitted with vernier scales or digital readouts.

Problem 1-2-11 involves reading micrometer scales. Micrometers are a shop tool, but they are also an excellent device to give you an insight into tolerancing. For example, ±.003" is approximately ± the thickness of the pages in this workbook.

Problem 1-2-12 provides practice with standard drawing instruments.

Chapter 3 Lettering, Sketching, and Line Techniques
Problems 1-3-1 through **1-3-6:** The first problem shows the recommended strokes to make vertical Gothic letters. The next five problems provide an introduction to various facets of technical drawing and technology:

1-3-2 Spur Gearing nomenclature
1-3-3 "vocabulary" of **Lines** on technical drawings
1-3-4 typical **Title Block** information
1-3-5 Attribute listing
1-3-6 Specification writing

Problem 1-3-7 requires sketching freehand three views of an object shown in an isometric drawing. A major challenge to a drafter's imagination is to read two-dimensional drawings and visualize a three-dimensional picture, and to visualize a three-dimensional object and construct views of it in a two-dimensional drawing surface.

Problem 1-3-8 challenges your imagination to visualize how a flat pattern of a commercial packaging box would appear when "folded up." If needed, the pattern could be cut out, folded up, and assembled.

Problem 1-3-9 gives practice in drawing parallel lines—a useful skill for doing orthographic projection because lines of sight are parallel. (CAD programs use grids to facilitate drawing parallel and perpendicular lines.)

Problems 1-3-10 through **1-3-12** provide practice in sketching in three popular formats: isometric, oblique, and perspective. Combinations of these are used by technical personnel for "talking sketches." Many ideas have been sketched on cafeteria napkins and on the backs of envelopes.

Problems 1-3-13 and **1-3-14** present an opportunity to use visualization skills to add missing lines on objects in three-view drawings and then do isometric sketches of the objects.

Problem 1-3-15 involves using a CAD program to do a title block and make some elementary drawing changes.

Chapter 4 Geometric Construction
Problem 1-4-1 involves using scales and drawing instruments to convert a freehand sketch into a drawing.

Problems 1-4-2 and **1-4-3** illustrate the use of layout drawings (drawings drawn to scale) to determine dimensions graphically that would be challenging to find mathematically.

Problem 1-4-4 introduces more symbols used in Geometric Dimensioning and Tolerancing. (Refer to page 1-2-8.)

Problems 1-4-5 through **1-4-8** require the use of drawing instruments to solve fundamental geometric construction problems such as bisecting lines and angles, drawing arc tangents, dividing lines into equal segments, and transferring complicated shapes.

Problems 1-4-9 through **1-4-11** involve more challenging geometric constructions using arc tangents and ogee curves. Two industrial applications of ogee curves are required.

Problems 1-4-12 through **1-4-15** contain problems to perform geometrical constructions of parabolas, one hyperbola, one helix, and several involutes. Aids are provided to help you get started.

Problem 1-4-16 gives you an opportunity to do a kinematic layout, i.e., solve a motion problem graphically.

Problem 1-4-17 is a CAD problem requiring geometric shapes and creating logos.

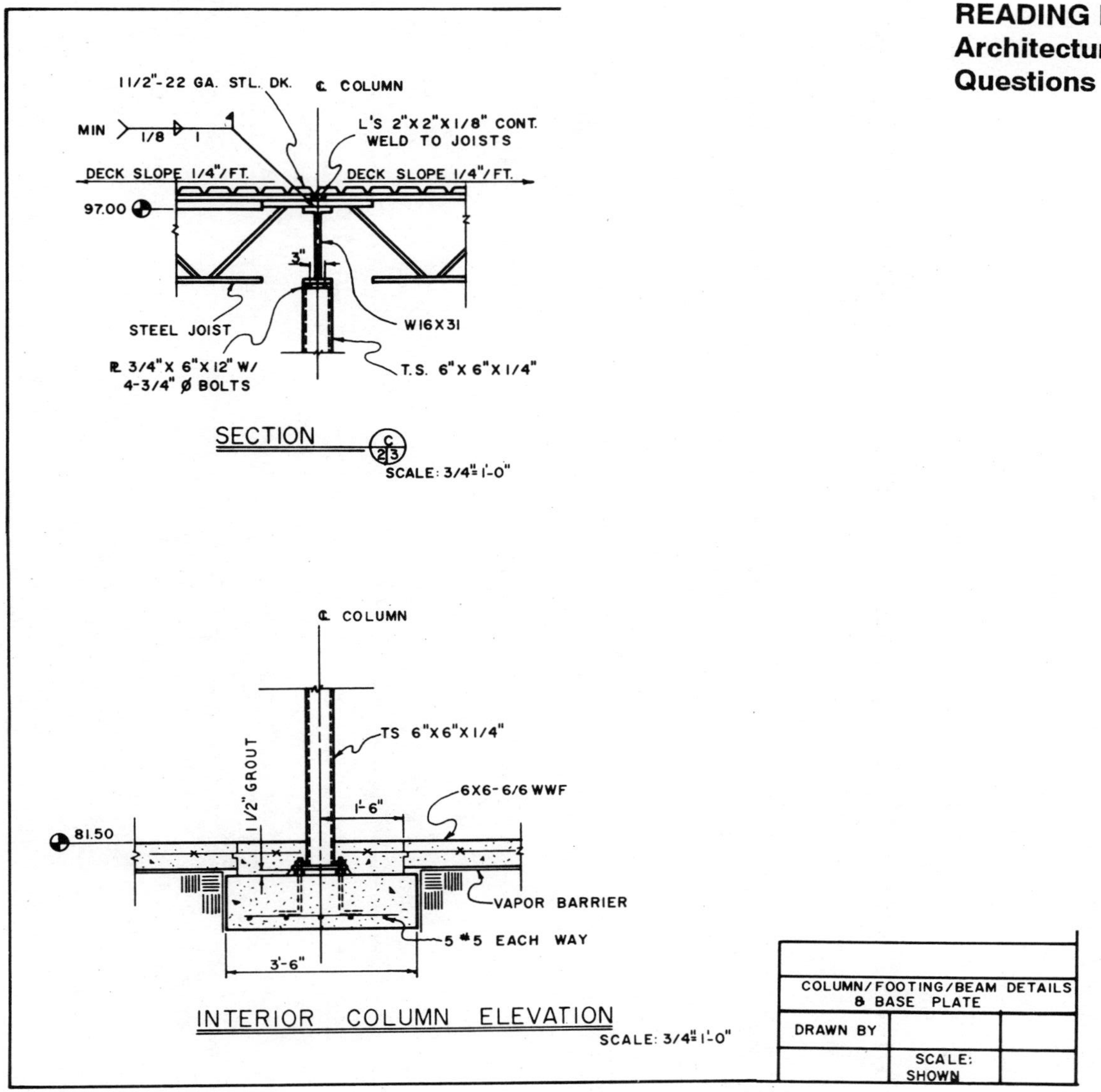

READING DRAWINGS EXERCISES
Architectural/Engineering Drawing (from Figure I-29)
Questions

(1) What is the title of the drawing?
(2) What is the scale of Section C-2-3?
(3) What does TS 6X6X1/4 mean?

(4) What is the approximate elevation of the deck (to the nearest foot)?
(5) What is the approximate clearance between the floor and the ceiling? (refer to INTERIOR COLUMN ELEVATION)
(6) What size beam is W16X31?

(7) How many and what size bolts are used to fasten the W16X31 beam to the 6X6X1/4 column?
(8) What is the slope of the deck?

(9) The floor (@81.50') is made of what material?

(10) What does this symbol mean?

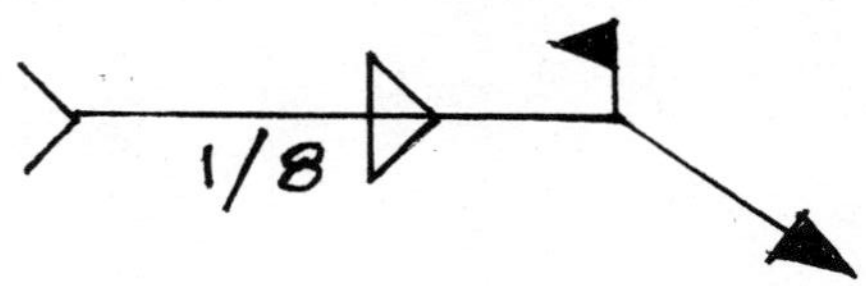

LTR.	REVISIONS	DATE	BY
A	.422 DIA WAS .53	2-23-75	KAI
B	.750 WAS .563	4-14-78	[illegible]

NOTES:

1 (in a triangle) STRESS RELIEVE BEFORE MACHINING
2 FINISH: 250 μ UNLESS SPECIFIED
3 BREAK AND DEBURR ALL EDGES
4 UNLESS OTHERWISE SPECIFIED ALL FILLETS AND ROUNDS ARE .09R

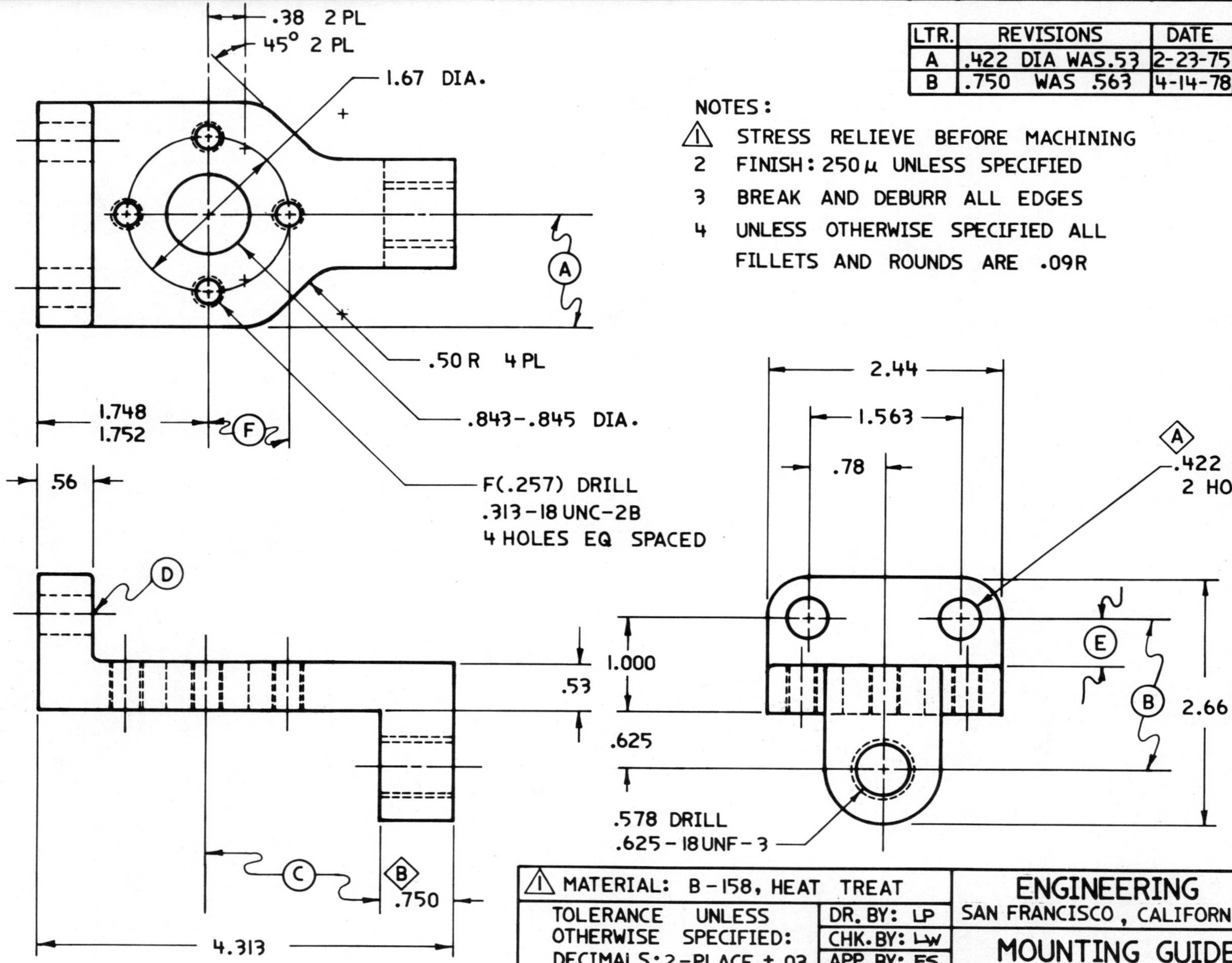

1 (in a triangle) MATERIAL: B-158, HEAT TREAT		ENGINEERING SAN FRANCISCO, CALIFORNIA	
TOLERANCE UNLESS OTHERWISE SPECIFIED:	DR. BY: LP	MOUNTING GUIDE	
DECIMALS: 2-PLACE ±.03	CHK. BY: LW		
3-PLACE ±.010	APP. BY: FS		
	DATE: 12-5-74	DWG. NO. 392450	REV. B
ANGLES: ±.5°	SCALE: FULL		

Introduction: Graphic Communication and Technical Drawing

Section-Introduction 1 Intro 2	Reference: Chapter 10 Dimensioning and Notation	Course______ Section____ Student________ Date____	Problem 1-Intro-2

The following questions apply to Engineering Dwg. No. 393450. Study the drawing and try to answer each question in the blank pace at the left . Then compare your answers to those on page 1-Intro-4.

________________ 1. What is the part name?

________________ 2. What is the drawing number?

________________ 3.What is the scale of the drawing?

________________ 4.What is the overall height of the object?

________________ 5.What is dimension (A)?

________________ 6.What tolerance is expected for these 2-place dimensions?

________________ 7.What are the dimension limits (max and min) of (B)?

________________ 8.What size tap hole is specified for the 5/8" threaded hole?

________________ 9.What threads series and threads per inch are specified for the 5/8" hole?

________________ 10.When was the drawing last revised?

________________ 11.What is the maximum dimension of (F)?

________________ 12.What is the material of the object?

________________ 13.What is the dimension of B in the "diamond"?

.What was the dimension of B in the "diamond" before the revision?

________________ 14.How many holes are in the object?

READING DRAWINGS EXERCISES
Architectural/Engineering Drawing (from Figure I-29)
Answers

(1) COLUMN/FOOTING/BEAM DETAILS

(2) 3/4"=1'-0"

(3) square tubing, steel, 6" square, 1/4" thick

(4) 97.00 feet

(5) 97.00'-depth of the steel joist-81.50'=97-1.33' (Steel joist appears to be same depth as the W16X31 beam)-81.50'=14.67'

(6) wide flange beam, 16" deep, weighs 31 pounds per foot

(7) 4-3/4" dia ϕ bolts (the column has a 6"X12"X3/4" plate welded to it)

(8) 1/4" per foot

(9) concrete

(10) the weld is 1/8" on each leg; the flag means field welding; the double triangle means weld both sides.

READING DRAWING EXERCISES
Answers to MOUNTING GUIDE drawing

MOUNTING GUIDE__________ 1. What is the part name?

392450-B__________ 2. What is the drawing number?

FULL__________ 3. What is the scale of the drawing?

2.66" __________ 4. What is the overall height of the object?

1.22"=2.44/2__________ 5. What is dimension (A)?

±.03" __________ 6. What tolerance is expected for these 2-place dimensions?

1.625 +.010+.010=1.645" and 1.625-.010-.010=1.605"__________ 7. What are the dimension limits (max and min) of (B)?

.578"ϕ__________ 8. What size tap hole is specified for the 5/8" threaded hole?

UNF, 18 __________ 9. What threads series and threads per inch are specified for the 5/8" hole?

4-14-78 __________ 10. When was the drawing last revised?

(1.67+.03)/2=.85"__________ 11. What is the maximum dimension of (F)?

B-158__________ 12. What is the material of the object?

.750", .563" __________ 13. What *is* the dimension of B in the "diamond"? What *was* the dimension of B in the "diamond" before the revision?

8 __________ 14. How many holes are in the object?

Fill-in the missing equivalents. Use a conversion chart.

FRACTION	DECIMAL .XXX	METRIC .XX
	.078	
		19.05
27/32		
		8.34
	.047	
15/32		
	.953	
1 1/2		
	5.375	
		3.18
	.438	
		15.88
1/64		
		49.21
3/8		
	.453	

Carefully measure each line using the indicated scale.
Enter each length in the spaces provided to the left.

ANSWER

SCALE: HALF

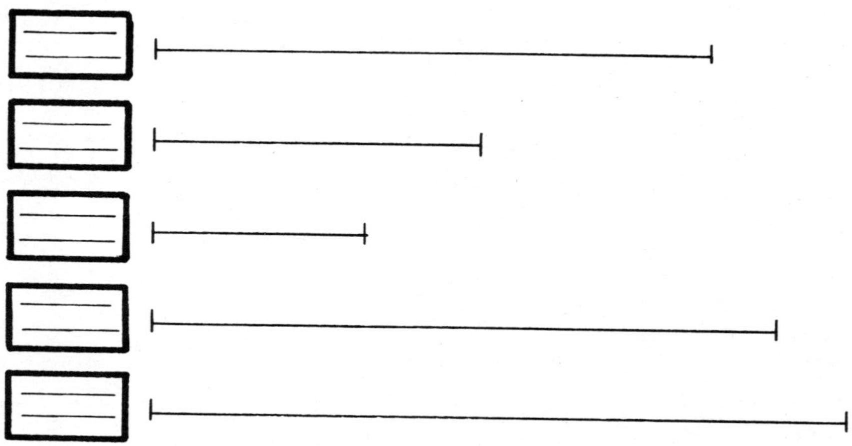

ANSWER

SCALE: QUARTER

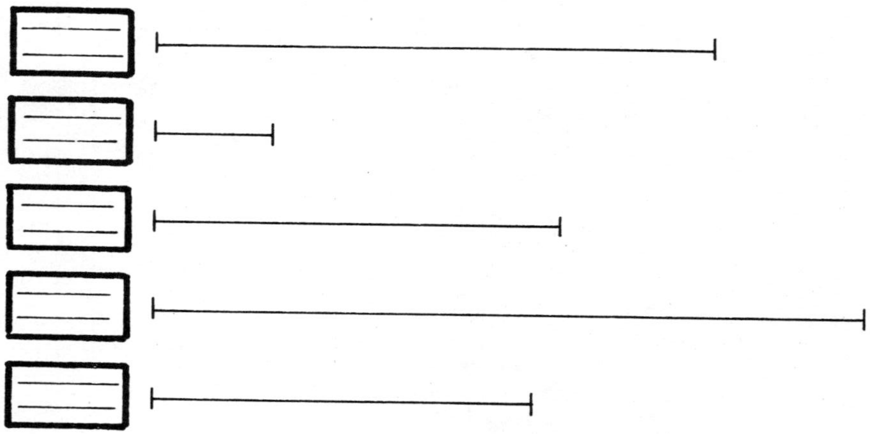

Ch2 Technical Drawing with Instruments

Section-Chapter-Problem
1 2 1

Scales: Half Size & Quarter Size
Figures 2-53 & 54

Course______ Section______
Student______ Date______

Problem
1-2-1

Carefully measure each line using the indicated scale.
Enter each length in the spaces provided to the left.

SCALE: FULL

ANSWER

SCALE: 10 X

ANSWER

METRIC FULL SIZE

ANSWER

Ch2 Technical Drawing with Instruments

Section-Chapter-Problem	Scales: Full Size, 10X & Metric Full Size Figures 2-57 & 61	Course______ Section______ Student______ Date______	Problem
1 2 2			1-2-2

All the scales on the Metric and Engineer's can be changed by a factor of 10 by simply moving the decimal point. The Metric scale is used primarily for engineering drawings. The Engineer's scale is used primarily for engineering drawings, but is also used for force diagrams, velocity and acceleration polygons, graphs, and topographic maps.

Using the example measurements below (a) determine the unknown dimensions and record the values where indicated, and (b) lay off the given distances or quantities and record the values over the dimension lines.

METRIC SCALES

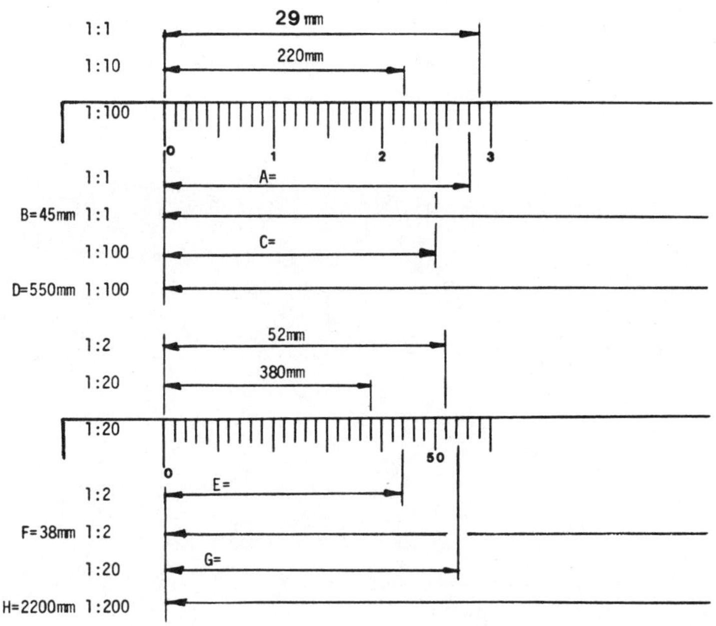

ENGINEER'S SCALE

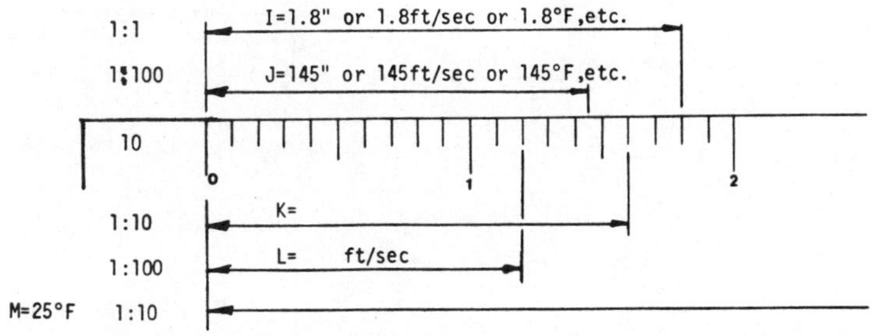

Ch2 Technical Drawing with Instruments			Problem
Section-Chapter-Problem 1 2 3	Scales: Metric & Engineer's Figures 2-57 & 61	Course__________ Section_____ Student____________________ Date_____	1-2-3

Using the example measurements given below (a) determine the unknown dimensions and record the values where indicated, and (b) lay off the given distances or quantities and record the distances over the dimension lines.

The two scales below are illustrations of the Architect's scale. The major divisons on each scale represent one foot. One of the one foot divisions is subdivided into 12 equal parts to represent the number of inches in one foot. Some are further divided into fractions of an inch. The 1-1/2"=1'-0 scale is 1/8 size and the 1/4"=1'-0 is 1/48 size. Architect's scales are used for structural drawings, commercial buildings, dwellings, topographic maps, and some machinery drawings.

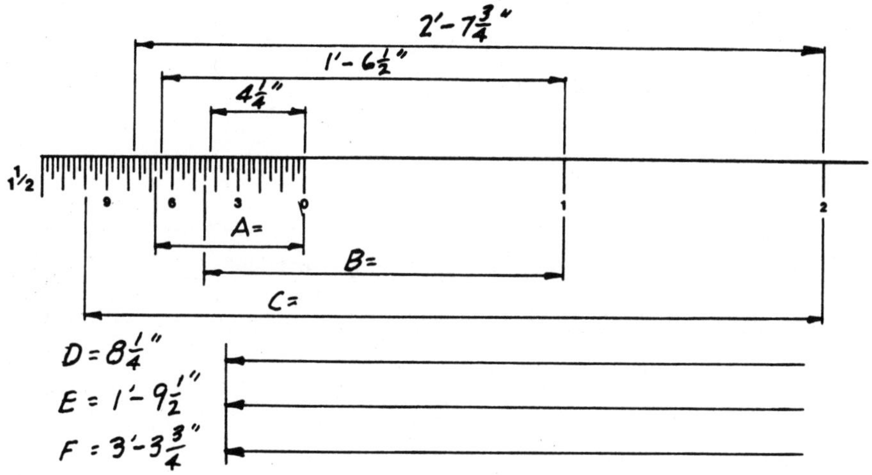

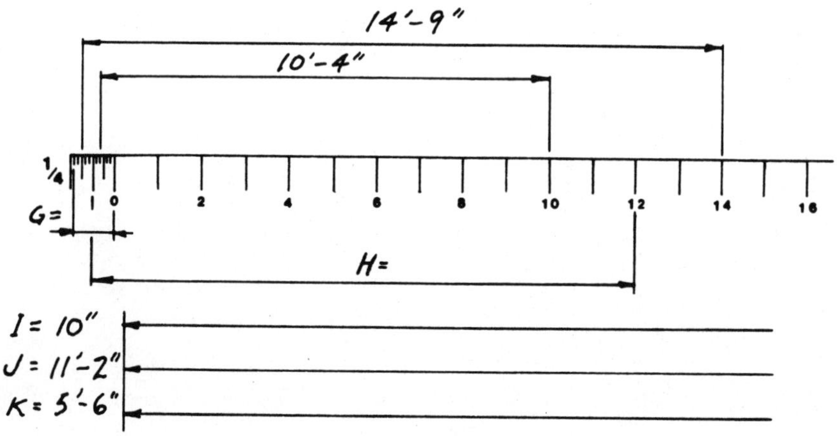

Ch2 Technical Drawing with Instruments			Problem
Section-Chapter-Problem 1 2 4	Architect's scales Figures 2-55 & 2-56	Course______ Section______ Student______ Date______	1-2-4

A. In the space at right record the measurements of the dimension lines.

1. 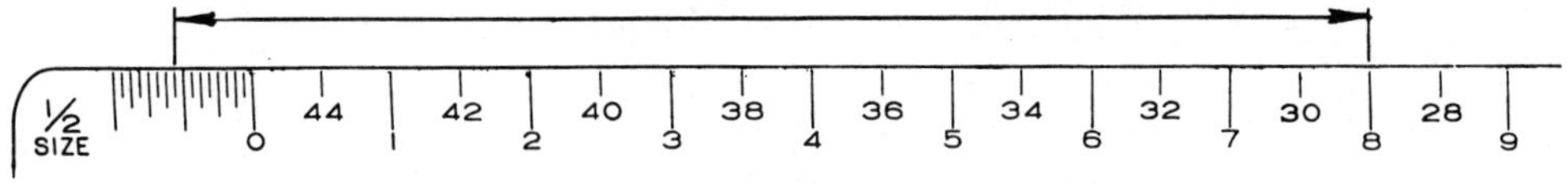

____

2. 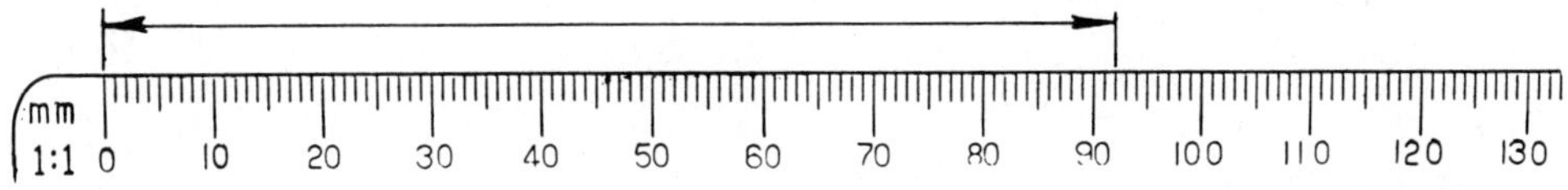

____

3. 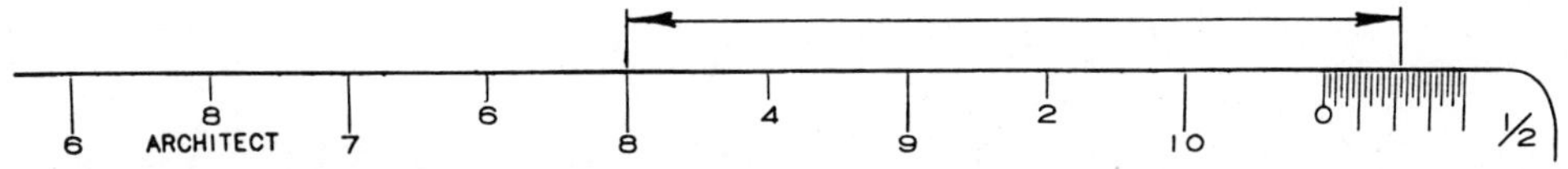

____

4. 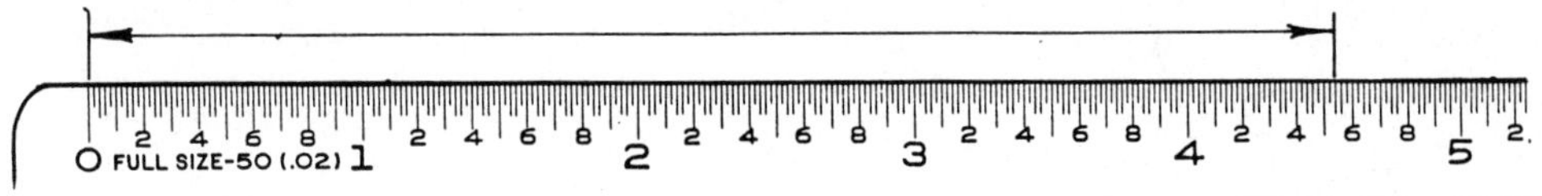

____

B. Add dimension lines to represent the measures that are listed at the right.

5. 

13 1/4 inches

6. 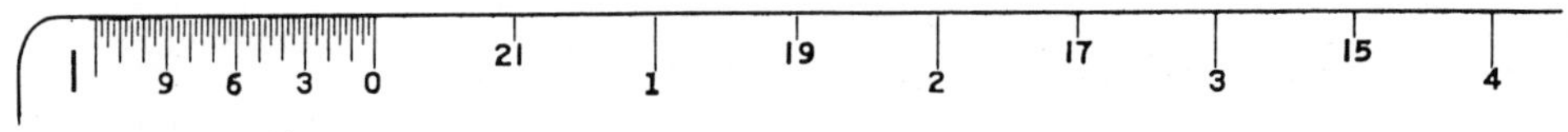
3' - 5 3/4"

7. 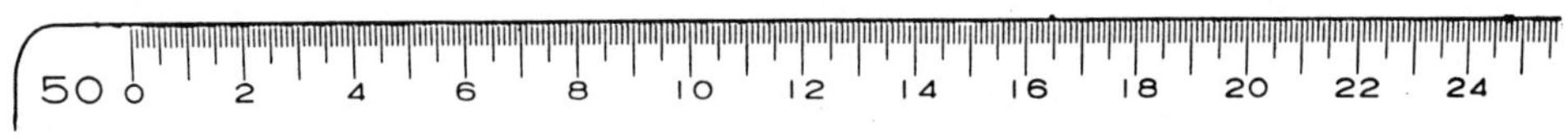

1440' @ 1" = 500'

8. 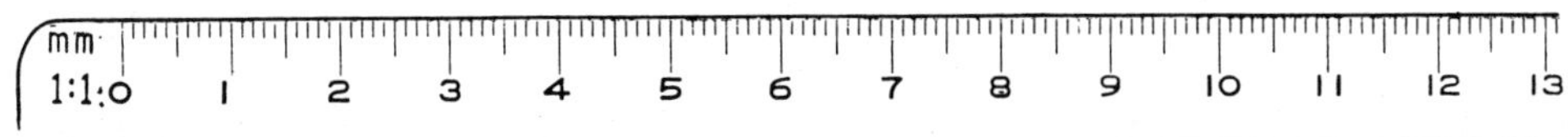

107 mm

C Measure from the left end of the long line, and mark the distance with a sharp, dark, vertical line. Use the distance and scale given at the right of eachline.

9.	⊢———	2.77	decimal inch
10.	⊢———	102.5	metric 1:1
11.	⊢———	14 13/16	1/4 size
12	⊢———	0'-11 1/2"	1/4" = 1'-0"
13.	⊢———	47.0 mm	metric 1:2
14.	⊢———	6 1/2"	half size
15.	⊢———	205.5 ft.	1" = 50'-0"

Ch2 Technical Drawing with Instruments			Problem
Section-Chapter-Problem 1 2 5	Scales: Architect's, Engineer's, and Metric.	Course______ Section____ Student________ Date____	1-2-5

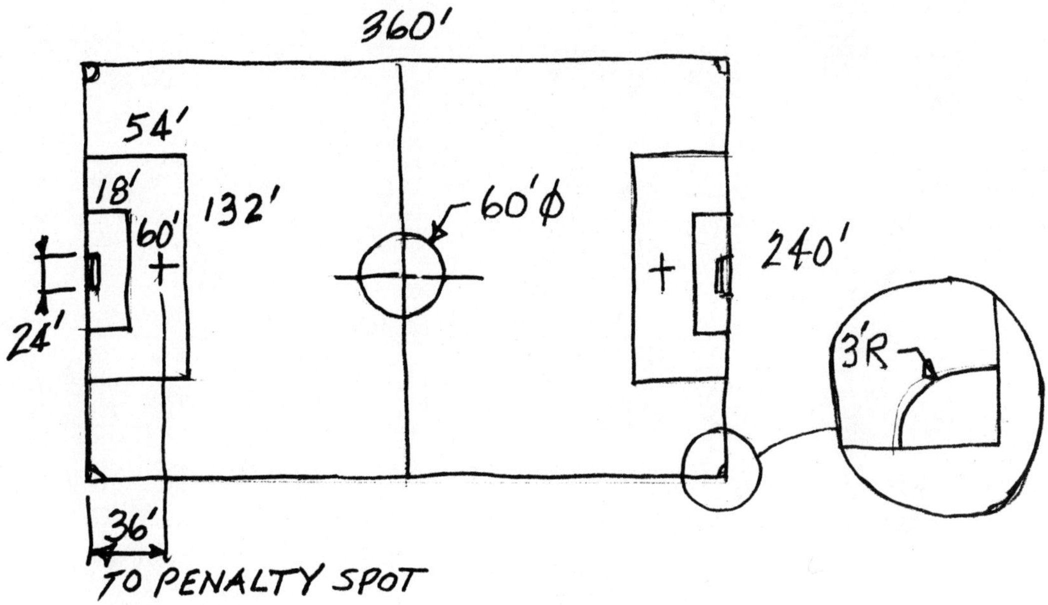

The freehand sketch shown describes the layout for a typical soccer field. Use drawing instruments to draw the layout to a scale of 1"=60' in the space below. Start your layout in the lower right corner as indicated.

Ch2 Technical Drawing with Instruments			Problem
Section-Chapter-Problem 1 2 6	Layout to scale information from a freehand sketch. Ch17 EX1 & EX2	Course__________ Section_____ Student________________ Date_____	1-2-6

Redraw the freehand sketches to scale using drawing instruments.
Also, answer the questions for the first sketching problem.

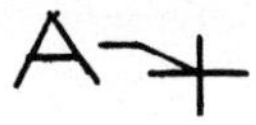

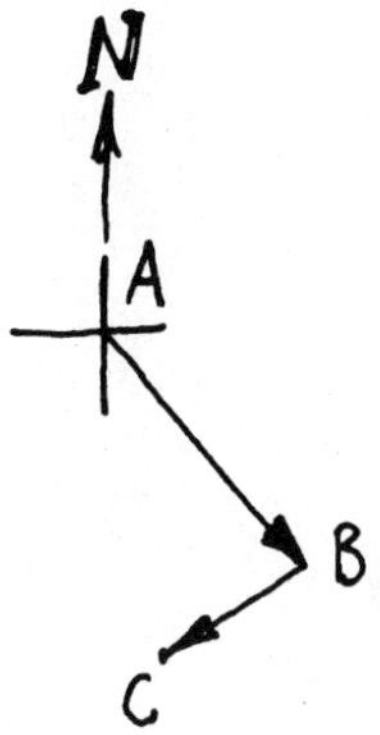

Scale 1:100
AB=S48°E, 7meters
BC=S63°W, 4meters

AC=_____, _____ ?
CA=_____, _____

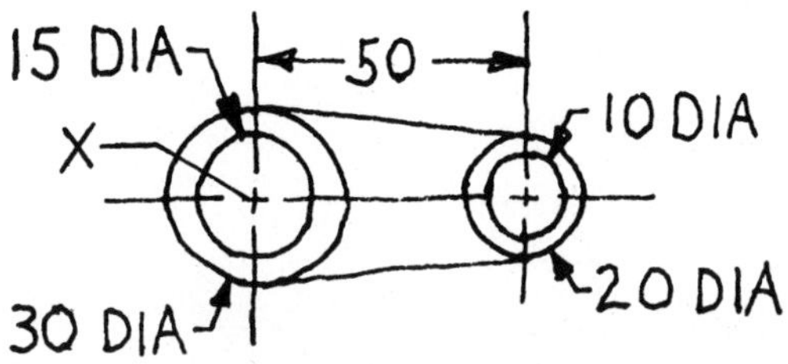

Full Scale

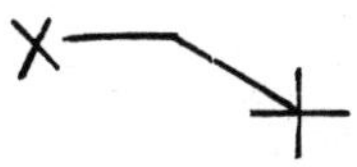

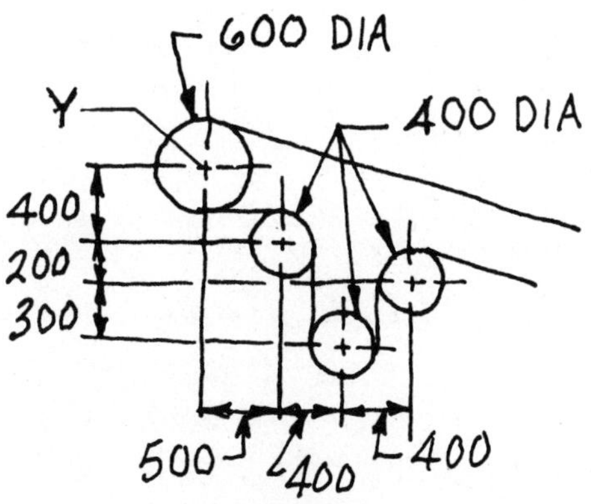

Scale 1:20

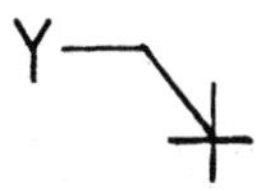

All dimensions in millimeters unless noted.

Ch2 Technical Drawing with Instruments			Problem 1-2-7
Section-Chapter-Problem 1 2 7	Layouts to scale of information in freehand sketches. Ch17 EX1	Course________ Section_____ **Student**__________ Date_____	

Redo, to twice the size, the AS DRAWN figures, angularity, perpendicularity, and parallelism, but revise the datum symbols as shown in the DATUM FEATURE drawing. Also, change the tolerances to .003" in the FEATURE CONTROL SYMBOLS. (Feature control symbols are 2X the lettering height.)

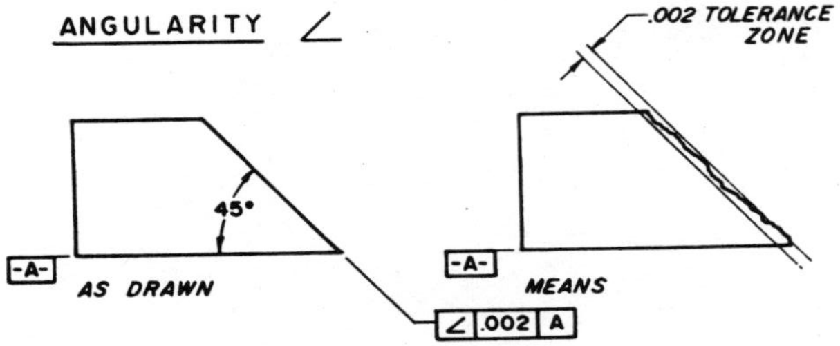

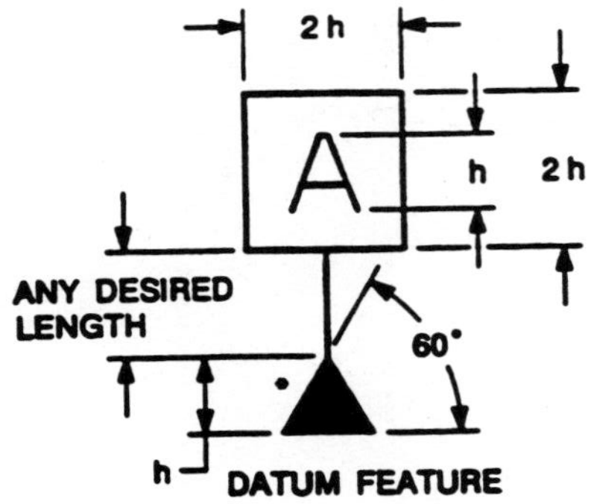

h=standard lettering height

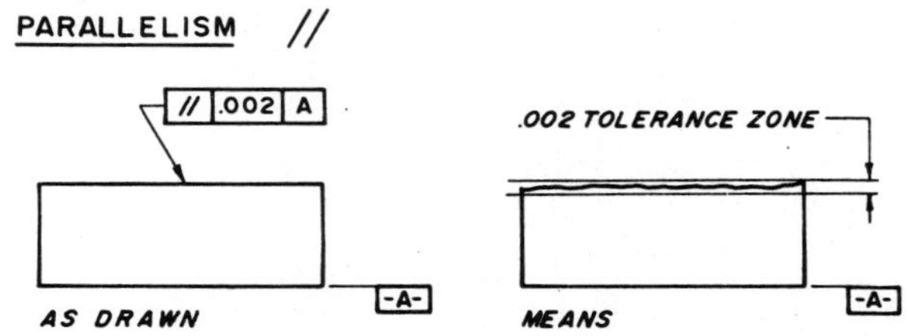

PARALLELISM

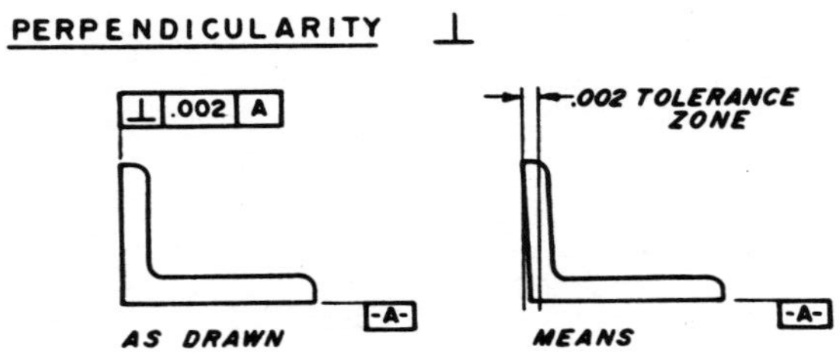

ANGULARITY

PERPENDICULARITY

Ch2 Technical Drawing with Instruments			Problem 1-2-8
Section-Chapter-Problem 1 2 8	ASME standard Y14.5M-1994 Angularity, Perpendicularity, & Parallelism Figure 12-20	Course______ Section____ Student______ Date____	

Use a French curve to draw a smooth curve through a series of points. Determine the general direction of the curve by connecting the points with light, short, straight lines. Then line-up the French curve along as many points as possible and draw a dark line connecting these points. Reposition the French curve along additional points and continue drawing the curved line in segments so that each segment is tangent to the next.

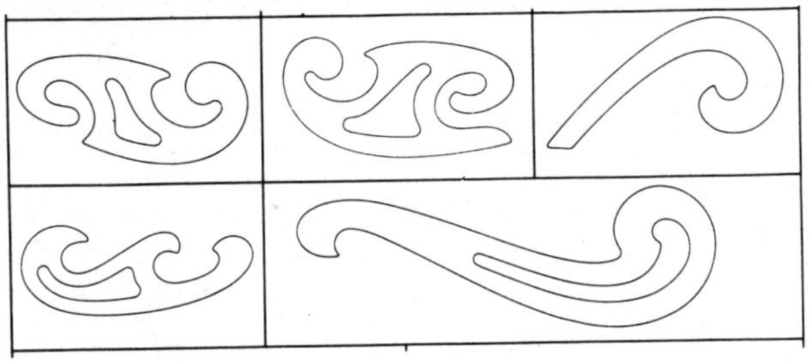

Assortment of French curves *(Courtesy Modern School Supplies Inc.)*

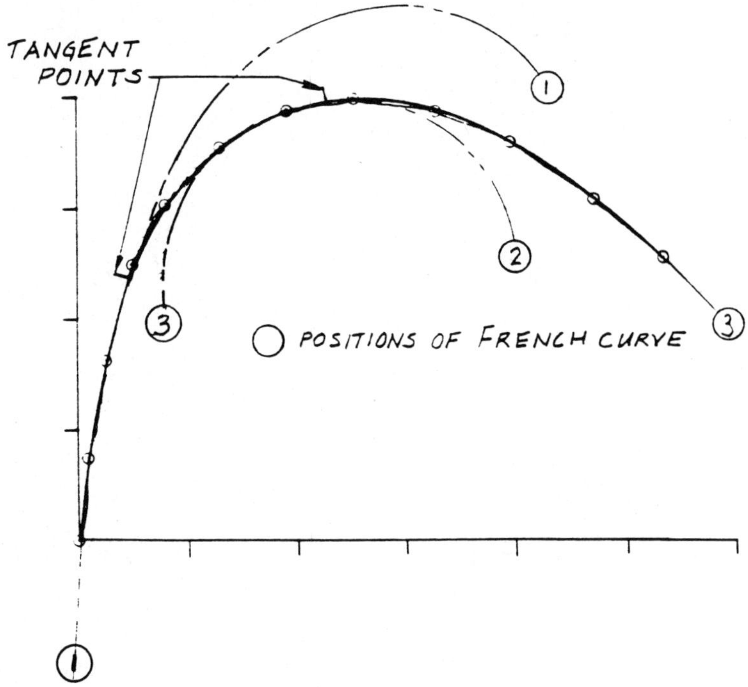

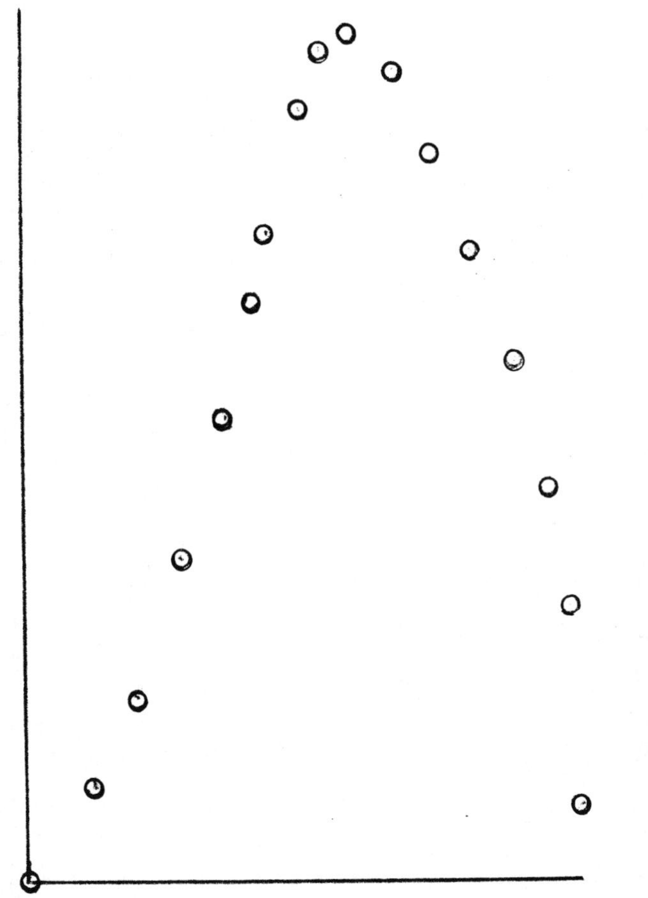

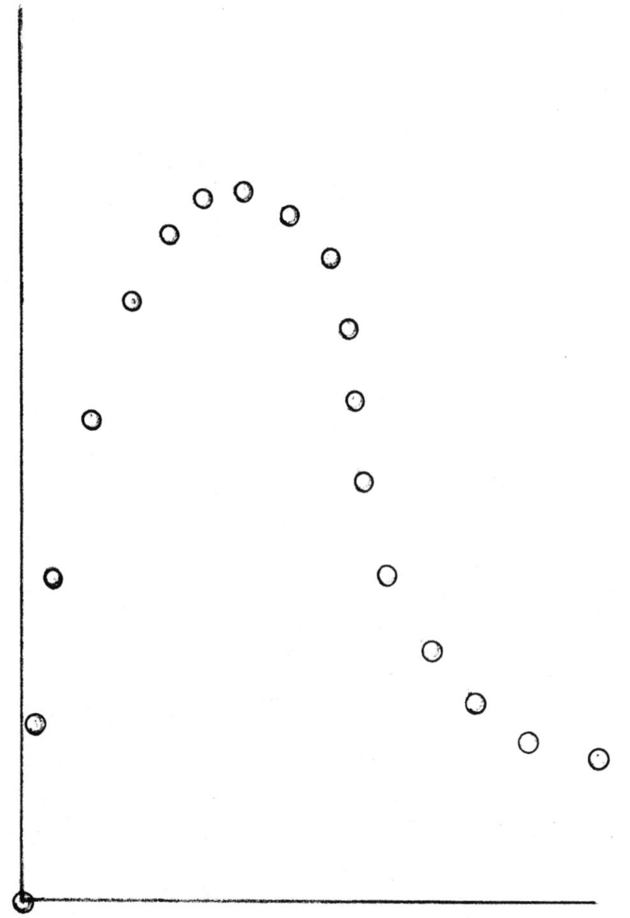

Ch2 Technical Drawing with Instruments

Section-Chapter-Problem	French curves Figures 2-35, 2-36	Course__________ Section_____	**Problem**
1 2 9		**Student**__________________ Date______	**1-2-9**

Use a protractor to measure the three angles shown. Next, calculate the same angles using trigonometry, "rise over run" equal to the tangent of the angle. Show your rise and run constructions.

Example:

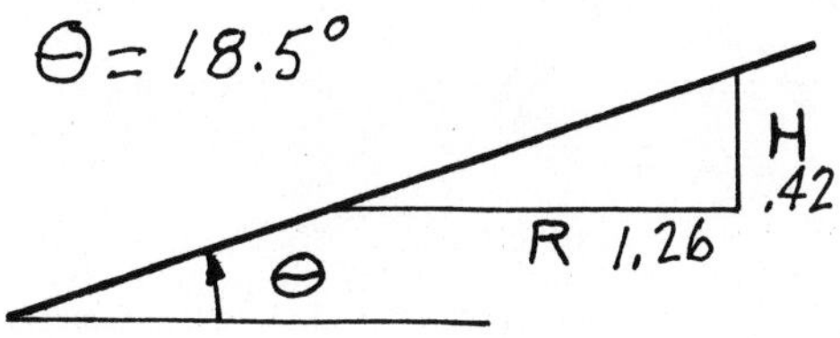

$\frac{H}{R} = \frac{.42}{1.26} = .333 = \tan\theta \qquad \theta = 18.43°$

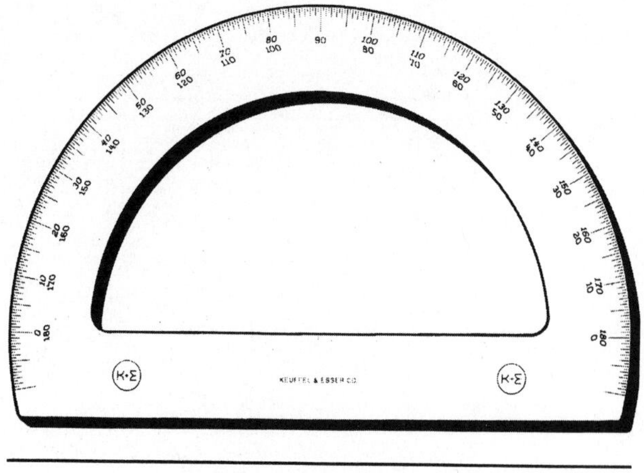

Protractor (Courtesy Keuffel & Esser Co.)

Use a protractor to layout the angles indicated.

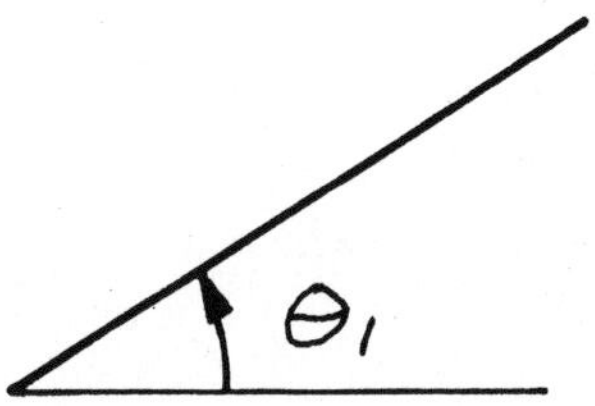

$\frac{H}{R} = \tan\theta_1 = \qquad \theta_1 =$

$\theta_4 = 54°$

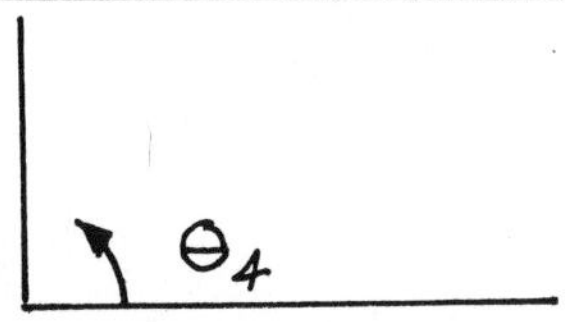

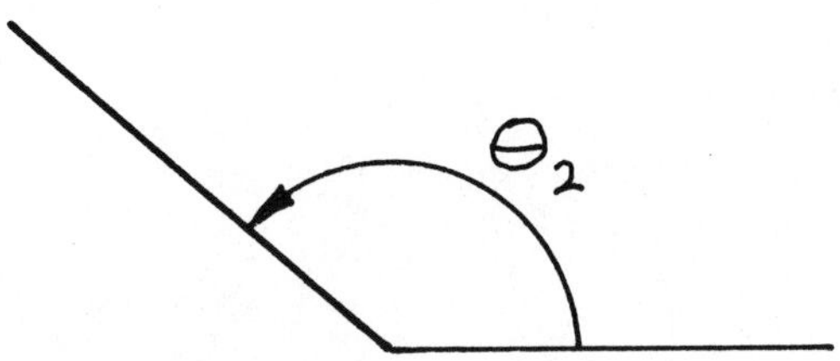

$\frac{H}{R} = \tan\theta_2 = \qquad \theta_2 =$

$\theta_5 = 232°$

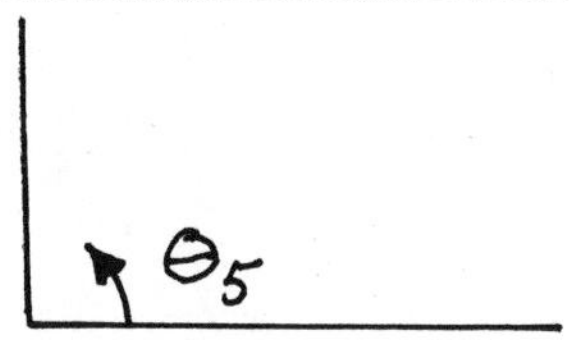

$\frac{H}{R} = \tan\theta_3 =$

$\theta_3 =$

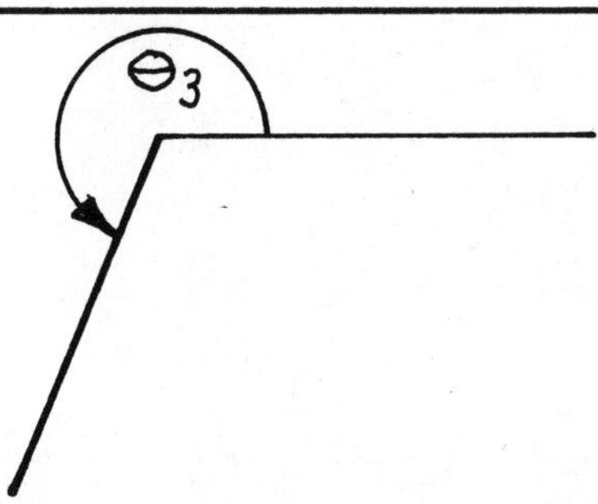

$\theta_6 = 311°$

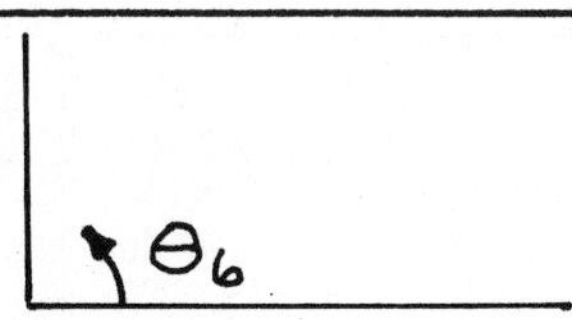

Ch2 Technical Drawing with Instruments			Problem
Section-Chapter-Problem 1 2 10	Protractor Figure 2-38	Course_________ Section_____ Student____________ Date_____	1-2-10

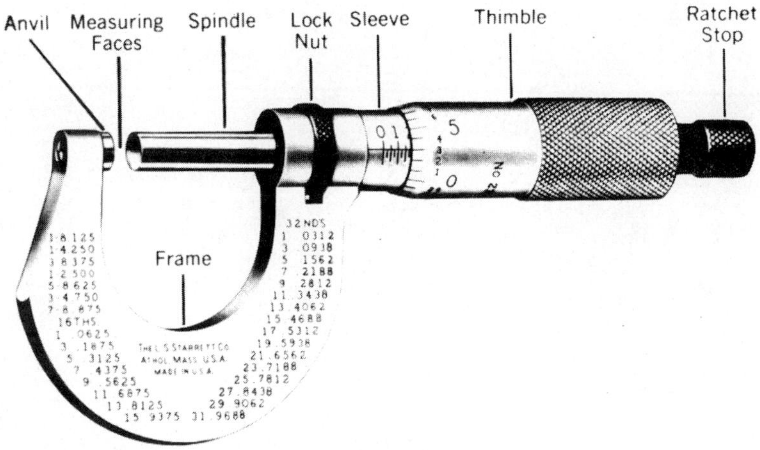

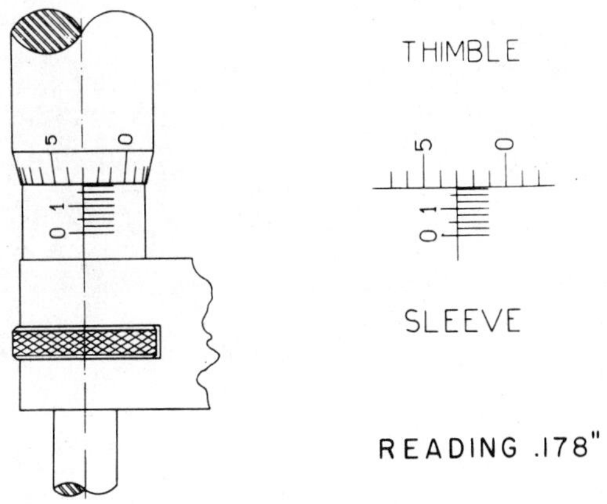

Example: READING: .178"=.100"+.075"+.003"

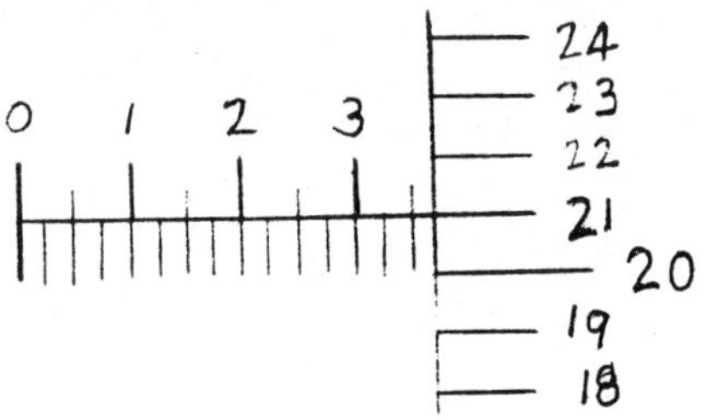

A. READING: ____=____+____+____

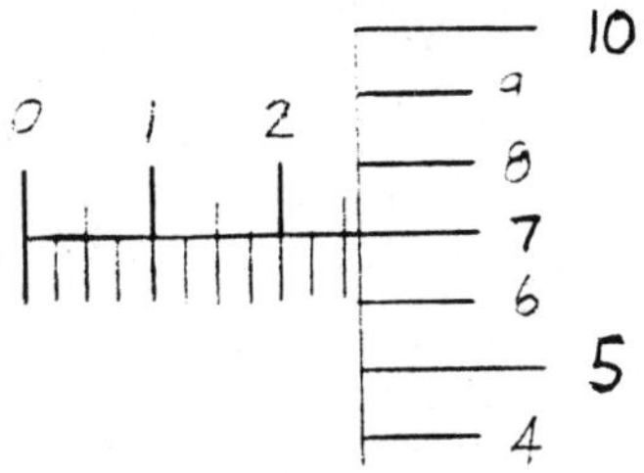

D. READING: ____=____+____+____

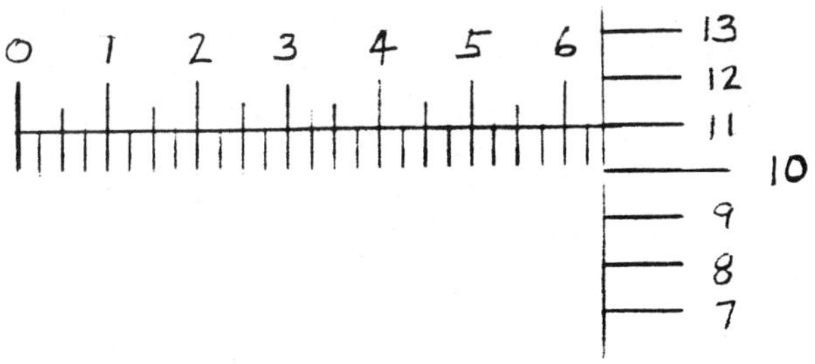

B. READING: ____=____+____+____

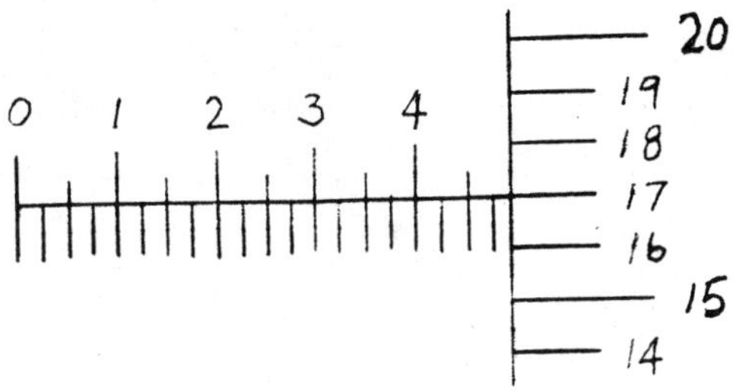

E. READING: ____=____+____+____

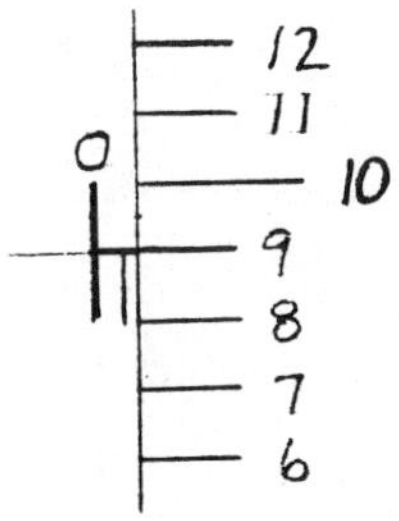

C. READING: ____=____+____+____

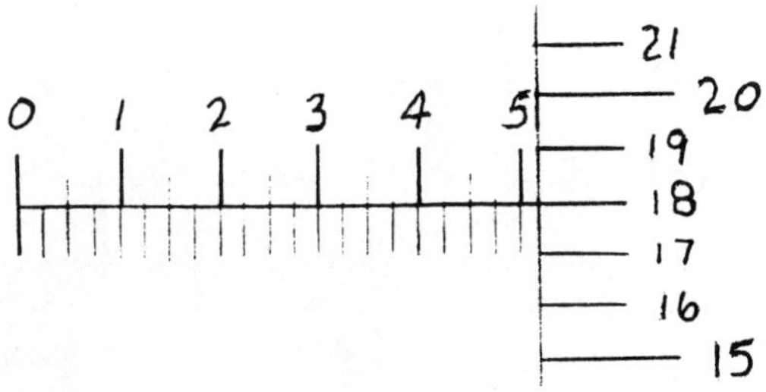

F. READING: ____=____+____+____

Ch2 Technical Drawing with Instruments			Problem 1-2-11
Section-Chapter-Problem 1 2 11	Micrometer measurements Figures 2-64 & 2-65	Course________ Section____ **Student**____________ Date____	

Construct each pattern to half size in the indicated spaces. Use black visible lines for the patterns, thin black lines for centerlines. Do not dimension.

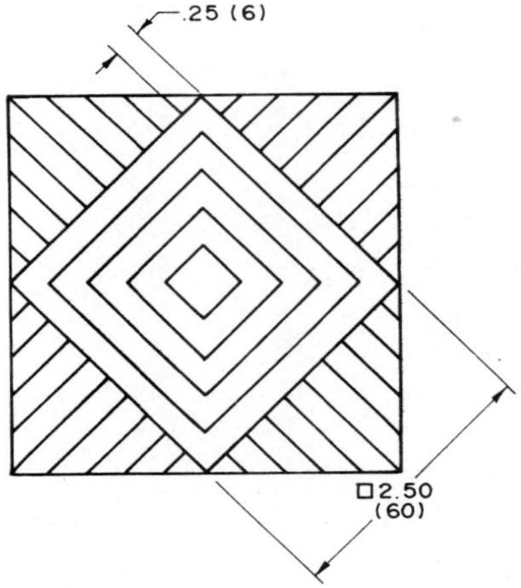

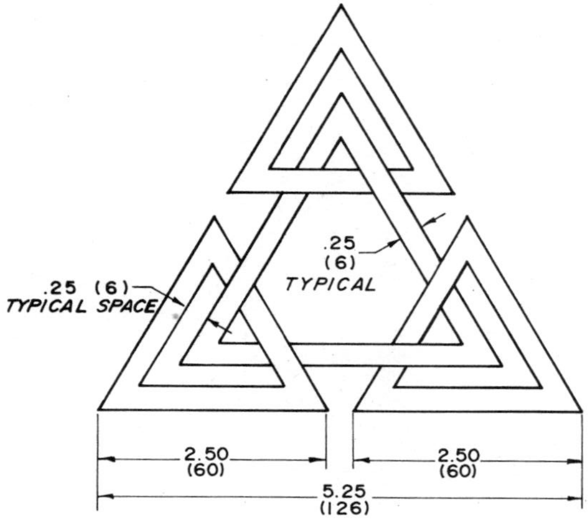

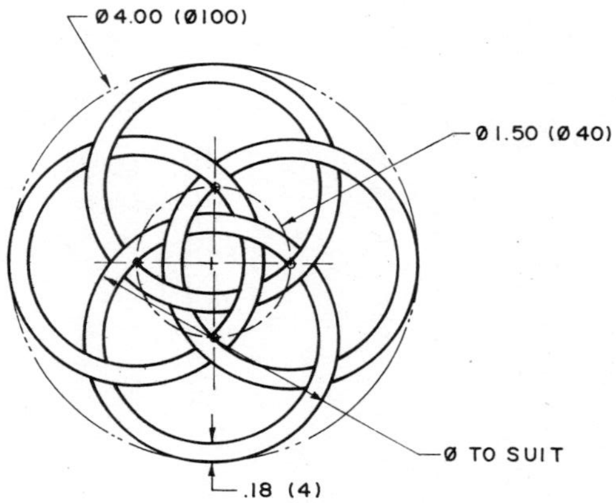

Ch2 Technical Drawing with Instruments			Problem
Section-Chapter-Problem 1 2 12	Patterns with instruments Figures 2-39 & 3-17	Course______ Section_____ Student______ Date_____	1-2-12

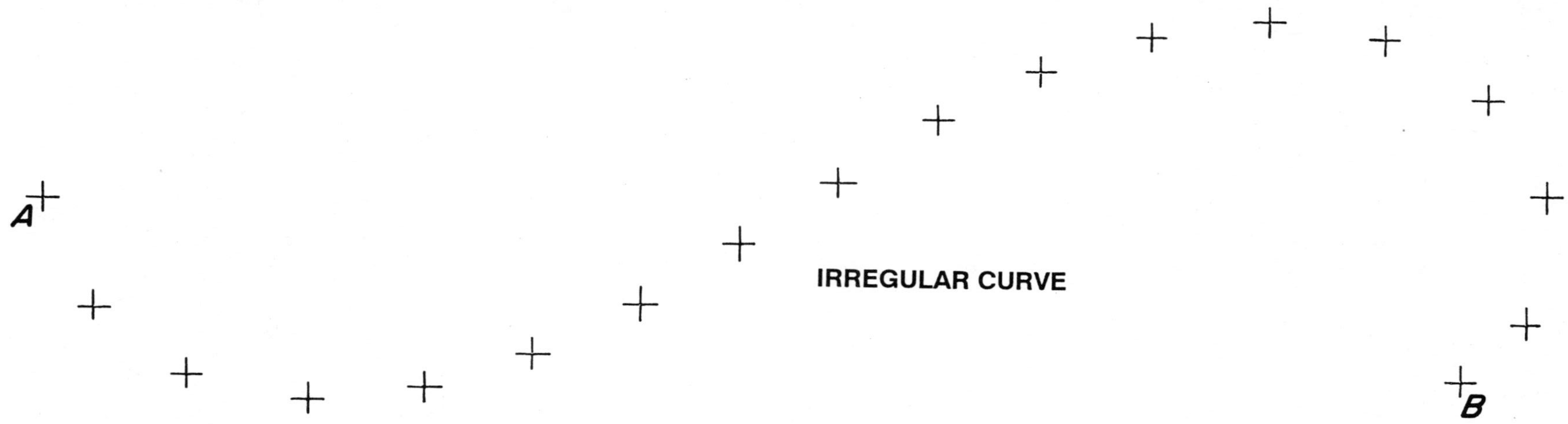

Use a French curve to draw a smooth curve through a series of points. Determine the general direction of the curve by connecting the points with light, short, straight lines. Then line-up the French curve along as many points as possible and draw a dark line connecting these points. Reposition the French curve along additional points and continue drawing the curved line in segments so that each segment is tangent to the next, for the irregular curve and the ellipse.

ELLIPSE

Ch2 Technical Drawing with Instruments			Problem
Section-Chapter-Problem 1 2 13	French curves Figures 2-35, 2-36	Course________ Section____ Student____________ Date____	1-2-13

With a sharp 2H or H lead practice each letter or number within the guidelines provided. Use the strokes illustrated along the left columns.

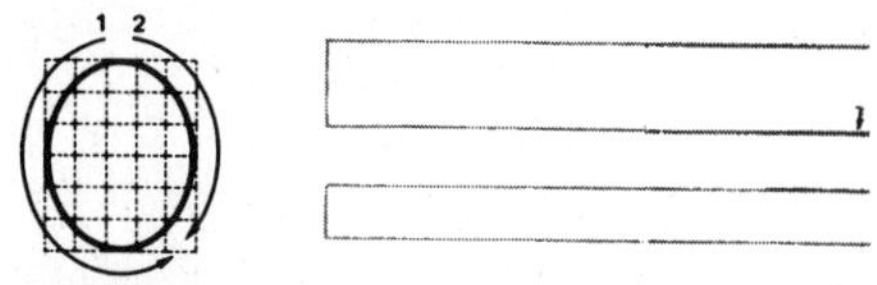

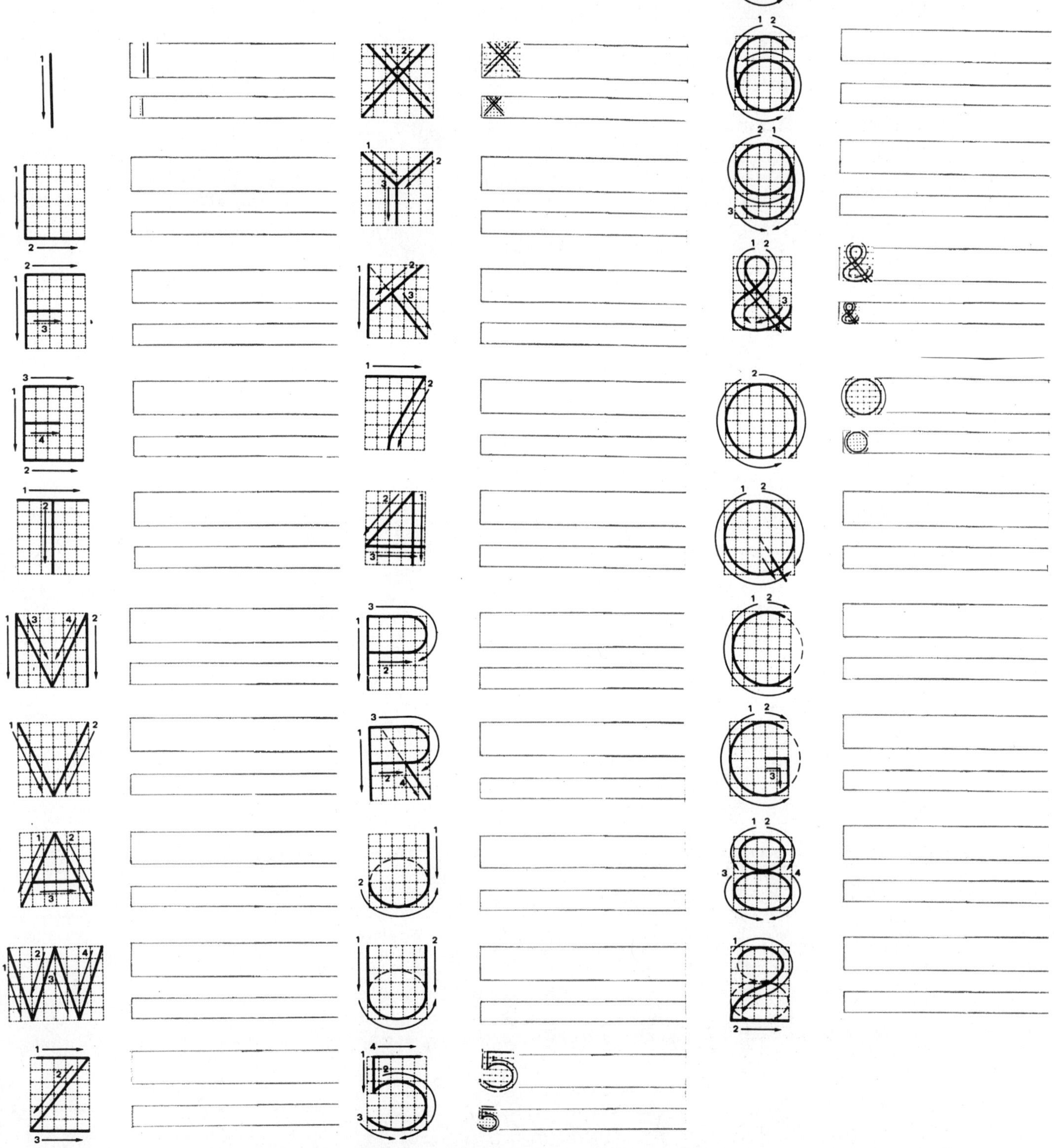

Ch3 Lettering, Sketching, and Line Techniques			Problem
Section-Chapter-Problem 1 3 1	Lettering upper case Gothic Figure 3-16	Course________ Section____ Student____________ Date____	1-3-1

Spur gears are mounted on parallel shafts to transmit rotary motion. The gear teeth are parallel to the axis of rotation and are cut with an involute profile. Spur gear terms are listed below and are referenced to the figure. Construct guide lines for lettering and print the terms in the spaces provided using a vertical Gothic style.

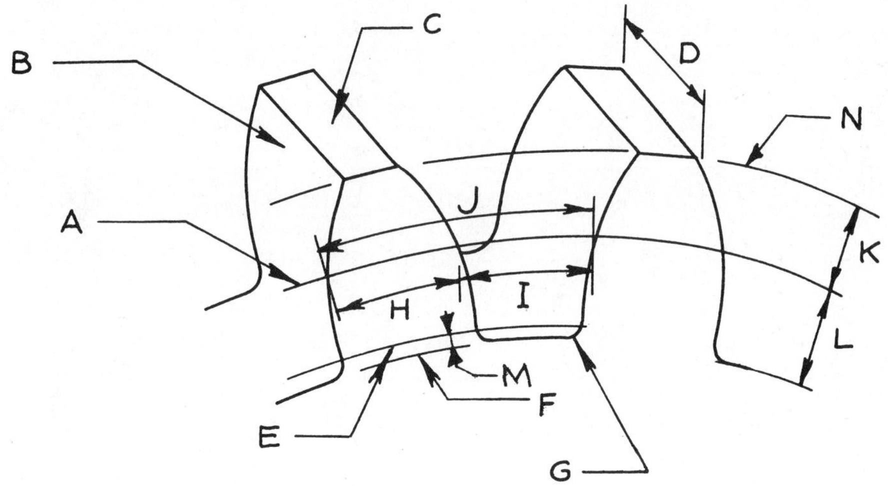

A. PITCH CIRCLE

B. FACE

C. TOP

D. FACE WIDTH

E. CLEARANCE CIRCLE

F. DEDENDUM CIRCLE

G. FILLET

H. TOOTH THICKNESS

I. SPACE

J. CIRCULAR PITCH

K. ADDENDUM

L. DEDENDUM

M. CLEARANCE

N. ADDENDUM CIRCLE

Ch3 Lettering, Sketching, and Line Techniques			Problem 1-3-2
Section-Chapter-Problem 1 3 2	Lettering upper case Gothic Spur Gears Figure 16-16	Course________ Section____ Student__________ Date____	

LINE TYPES FOR TECHNICAL DRAWINGS

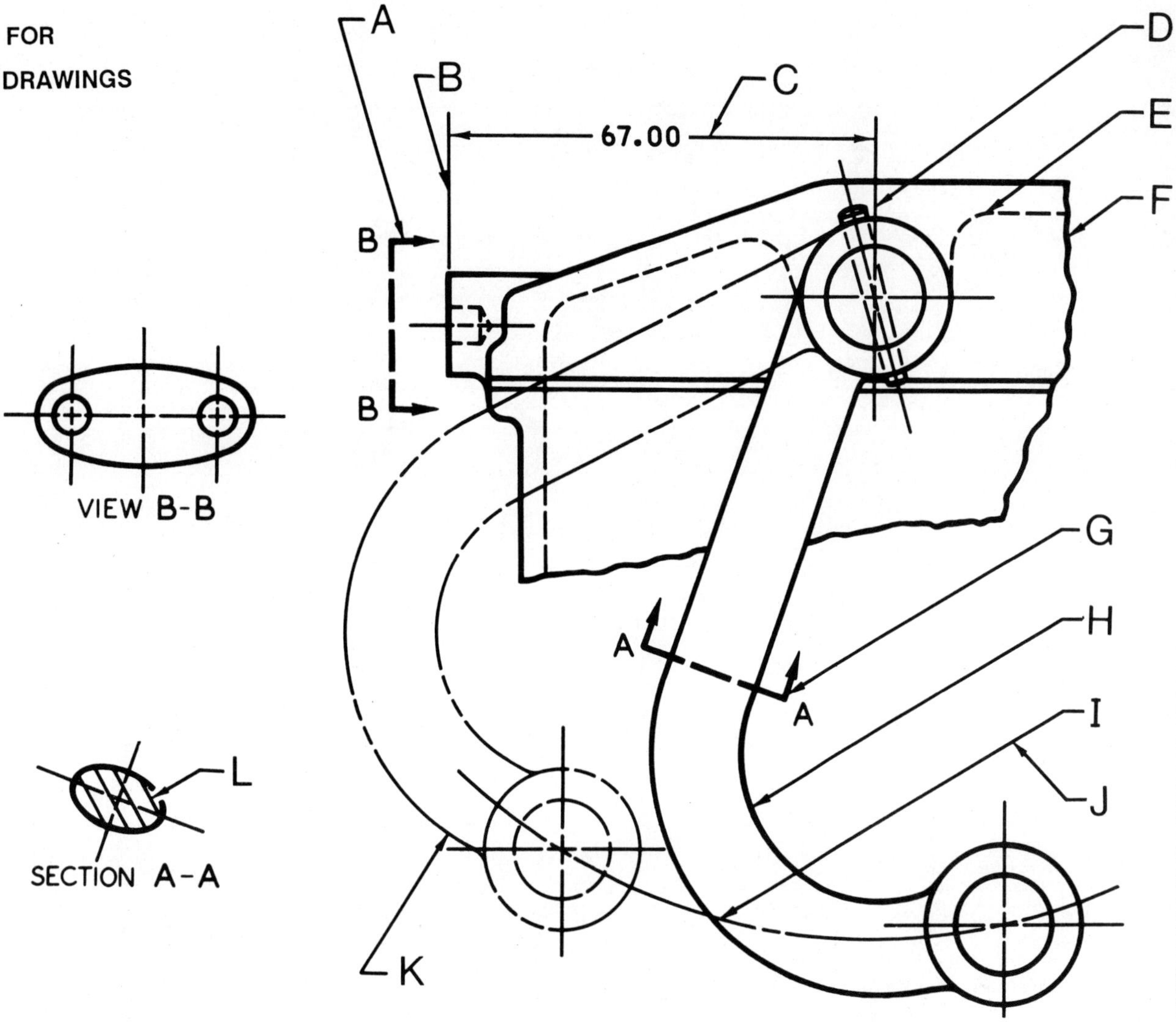

Letters identify each different type of line in the example drawing which shows line types used in technical drawing. From the list match the type of line to the appropiate letter and print the line type in the space provided. Use guidelines at .125" spacing, and vertical Gothic capitals. (Note: Two of the line types occur twice.)

A. __	G. __	VISIBLE
B. __	H. __	HIDDEN
C. __	I. __	CROSS HATCH
D. __	J. __	CENTER
E. __	K. __	DIMENSION
F. __	L. __	EXTENSION
		LEADER
		CUTTING PLANE
		BREAK
		PHANTOM

Ch3 Lettering, Sketching, and Line Techniques			Problem
Section-Chapter-Problem 1 3 3	Lettering upper case Gothic Line types Figure 3-17	Course__________Section_____ Student____________________Date_____	1-3-3

For the title block shown do the following:

1. Use guide lines for lettering and an H lead.
2. Use vertical Gothic lettering, .125” (3 mm) unless otherwise noted
3. Insert in the title block:

Your school’s name (in the largest space, .187”, 5mm letters)
Part name: CONNECTOR immediately above school name
Drawing number: GNC 311.1
Scale: Half
Sheet 1 of 2
Drawn: your initials
Checked: Instructor’s initials
Issued: current date
Contract: 00404

Tolerances ±1/16” .xx ±.02” .xxx ±.003”
Angles ±0.5°
Material AISI 1020 Steel
Finish: as rolled

Parts List
QTY Reqd 1
∠2x2x1/4
AISI 1020

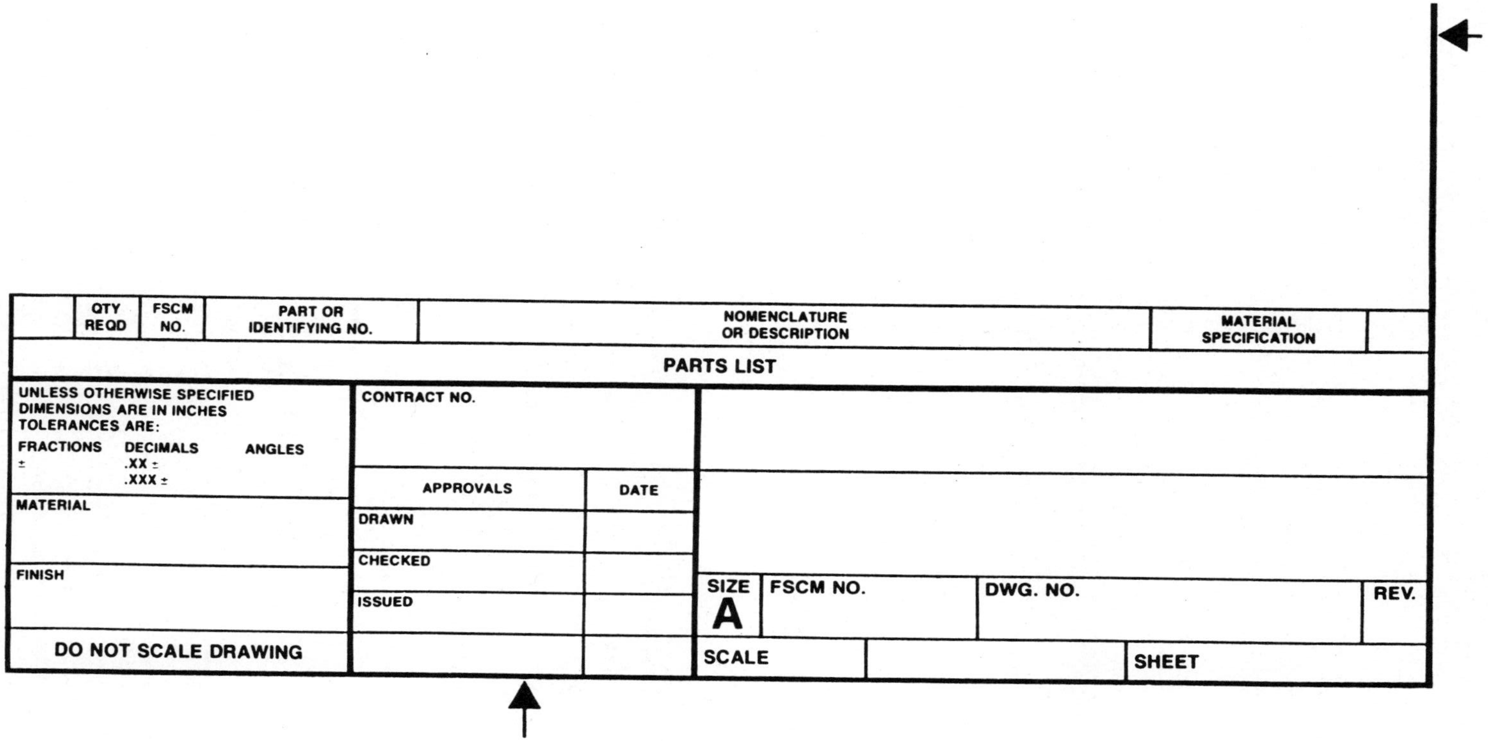

Ch3 Lettering, Sketching, and Line Techniques

Section-Chapter-Problem	Lettering upper case Gothic Title block Figure 17-10	Course________ Section_____ Student________ Date_____	Problem
1 3 4			1-3-4

If you systematically describe the attributes of an object you will have seen more of the object than if you casually observe it. The paper cup illustrates this concept which usually involves making a freehand sketch with several dimensions, and then listing the materials, the manufacturing processes, the functions and any special features.

Complete the guidelines and copy the attributes listed, using vertical Gothic lettering.

GEOMETRY AND SIZE

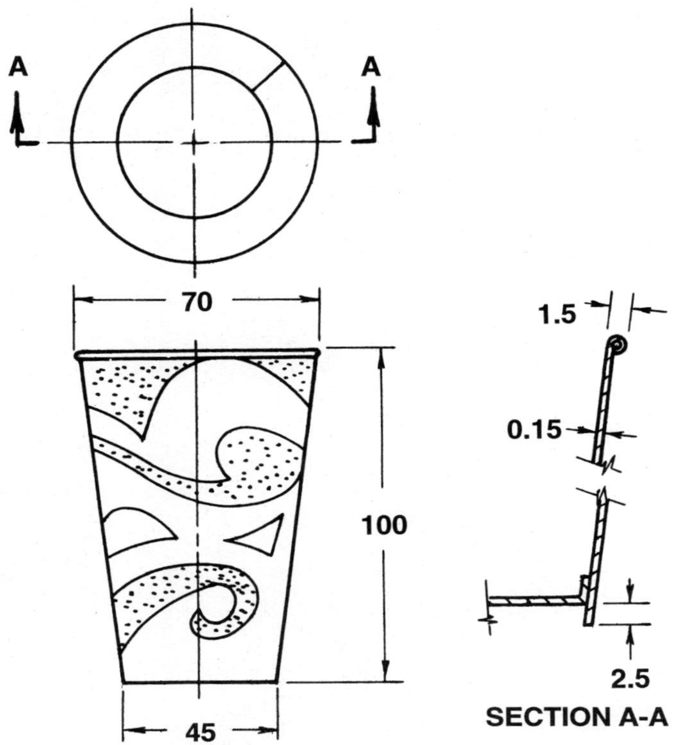

WAXED PAPER CUP
All Dimensions in mm

MATERIALS	MANUFACTURING PROCESSES
PAPER	PRINT DESIGN
WAX	CUT SIDE & BOTTOM
PIGMENT	ROLL SIDE INTO CONE
	GLUE

FUNCTIONS & FEATURES

HOLD COOL LIQUIDS

ROLLED EDGE TO STIFFEN CONE AND PROTECT LIPS

CONE TO STACK EASILY

"COOL" COLOR

Suggestion:
Use an attributes analysis to check detail and assembly drawings.

Ch3 Lettering, Sketching, and Line Techniques			Problem
Section-Chapter-Problem 1 3 5	Lettering upper case Gothic Attributes Figure 24-7	Course_____ Section_____ Student_____ Date_____	1-3-5

SPECIFICATIONS FOR DESIGN AND DEVELOPMENT

Purpose: Provide a positive, absolute identification of the desired item or system.

Desirable Characteristics: Statements should be:

BRIEF _

CONCISE _

SIMPLE _

COMPLETE _

Construct guidelines at .125" (3mm) and print the four words by the dash lines, using a vertical Gothic style..

Construct guidelines at .125" (3mm) and print the following example specification statement, using a vertical Gothic style.

Example: Unless otherwise specified the conductor shall be copper per MIL-F-55561 Type 1 and shall weigh not less than 1/2 ounce per square foot at .0007" thick.

Unless _

_

_

_

_

_

_

_

Note that an item produced from a well written "spec" can usually be verified by testing; for example, the 1/2 ounce can be checked on a scale and the .0007" thickness, by a micrometer.

Avoid vague statements such as "easily assembled", "safe", aesthetically pleasing" "inexpensive", etc.

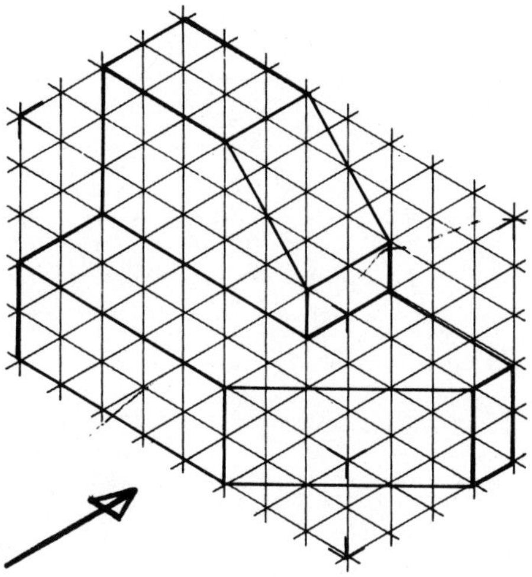

Sketch freehand three views of the objects shown. Use the isometric grid for dimensions, and use the suggested starting corner for the front view.

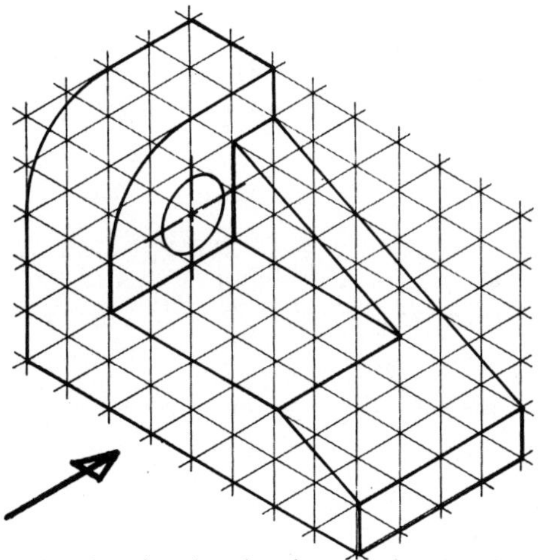

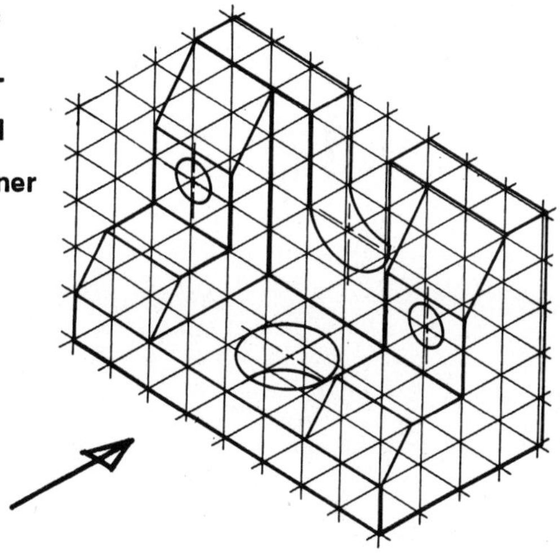

Ch3 Lettering, Sketching, and Line Techniques

Section-Chapter-Problem				
1 3 7	Freehand sketching Figures 3-49 through 3-52	Course__________ Section_____	Student________________ Date_____	Problem 1-3-7

The flat pattern shown is .018" thick cardboard with typical coloring and printing (on the side not seen) for a commercial product---logo, product name, instructions for brewing (tea bags), bar code, company name, etc. All dash lines indicate a fold line up so that a rectangular box is formed. Prepare a freehand, isometric sketch of the assembled box with the "notch" as a front view. Use the flat pattern for dimensions, and start your sketch on the point indicated.

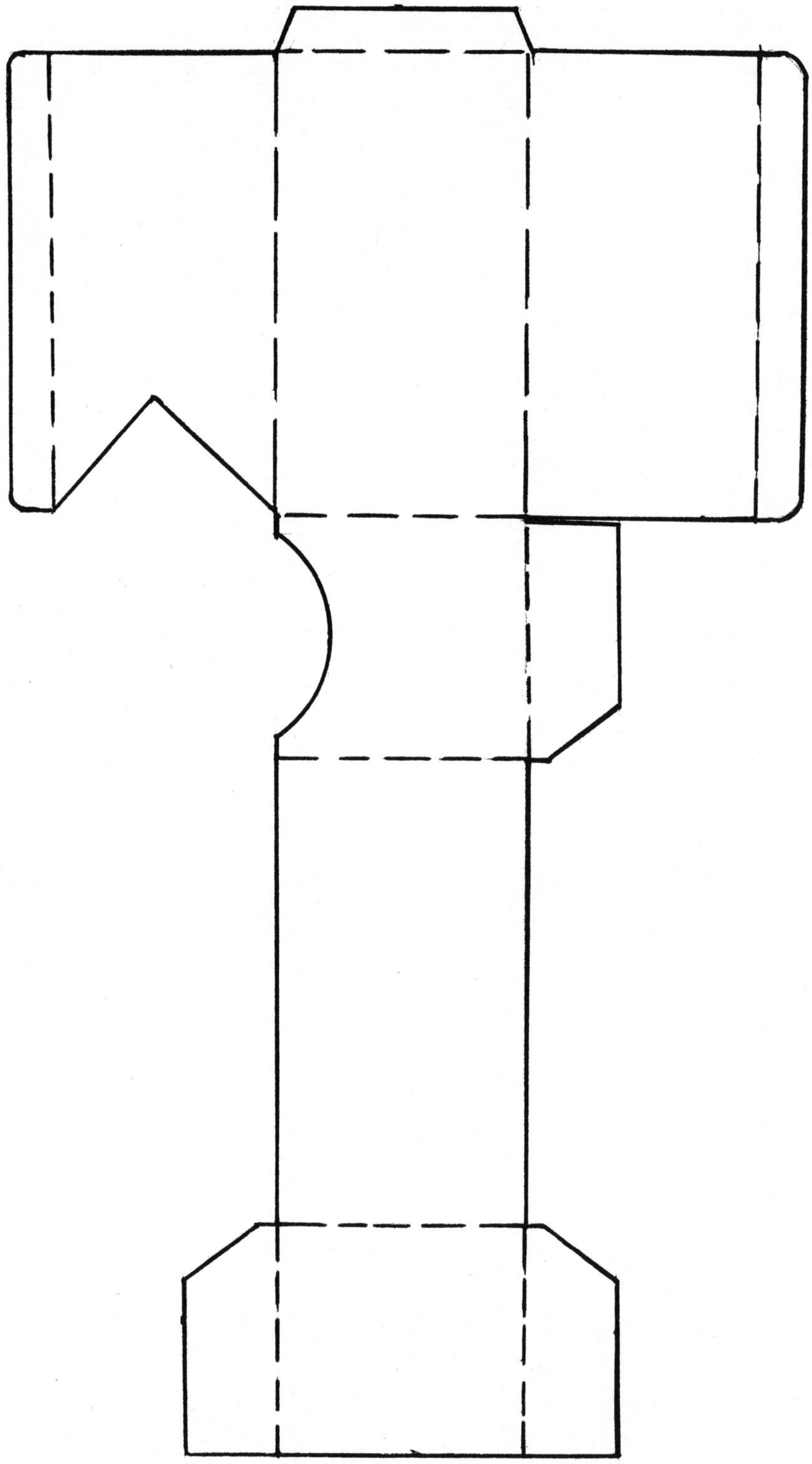

Ch3 Lettering, Sketching, and Line Techniques			Problem
Section-Chapter-Problem 1 3 8	Freehand isometric sketching Figure 3-36 & 9-7	Course__________ Section_____ **Student**________________ Date_____	**1-3-8**

From the dots to the borders draw the suggested horizontal, visible, thick lines with an H lead.

From the dots to the borders draw the suggested 45°, visible, thick lines with an H lead.

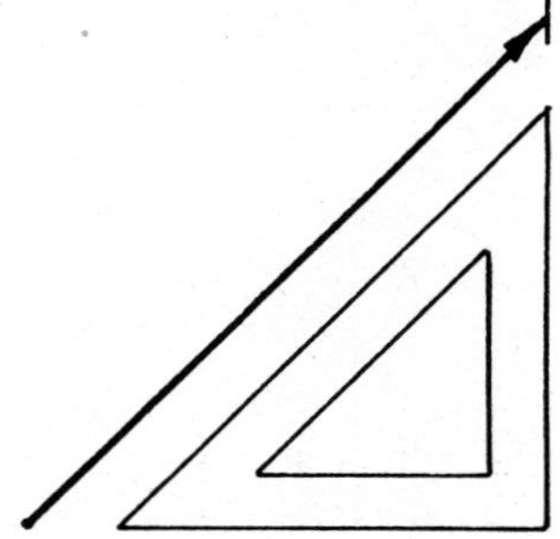

From the dots to the borders draw the suggested vertical, visible, thick lines with an H lead.

From the dots to the borders draw the suggested 60°, visible, thick lines with an H lead.

Ch3 Lettering, Sketching, and Line Techniques			Problem
Section-Chapter-Problem 1 3 9	Drawing straight lines with instruments Figures 3-23,25 & 26	Course__________ Section____ Student______________ Date____	1-3-9

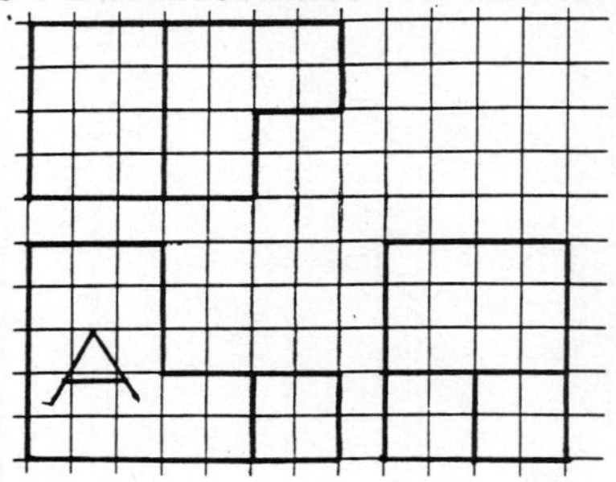

Visualization of an object drawn in the traditonal, three orthographic views can be enhanced by sketching the object in anyone of several three-dimensional views. Isometric, oblique, and one-point perspective sketches are illustrated here.

Start each type with an outline "box" (blocking-in) and then lightly sketch in the object. Finish the sketch by darkening appropriate lines. Omit hidden lines. Refer to the three illustrations at left, and then prepare three similar sketches for the given object.

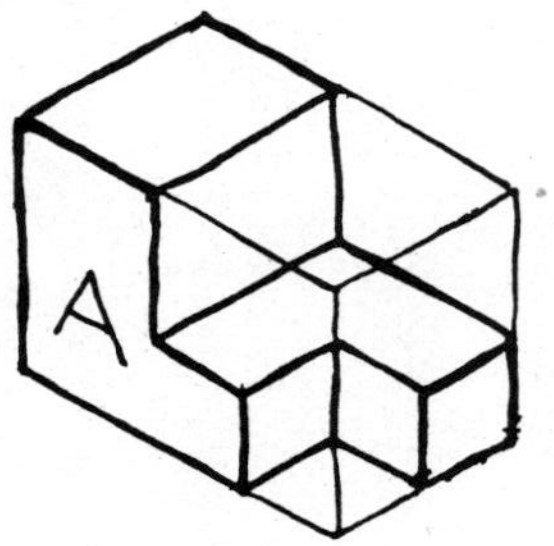

ISOMETRIC

OBLIQUE

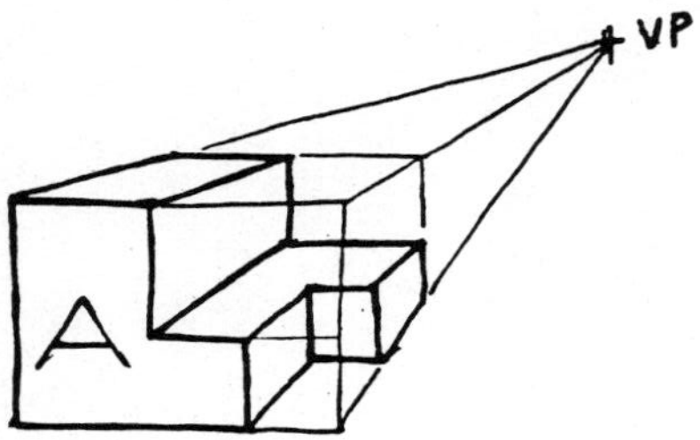

ONE POINT PERSPECTIVE

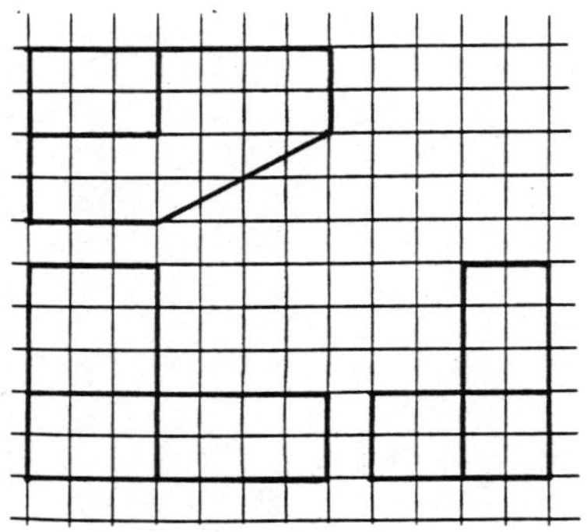

+VP

ONE POINT PERSPECTIVE

ISOMETRIC

OBLIQUE

Note that any one of the types could be used as a "talking sketch", but an oblique type with an approximate perspective seems to be the most popular.

Ch3 Lettering, Sketching, and Line Techniques			Problem
Section-Chapter-Problem 1 3 10	Freehand sketching: isometric, oblique & perspective Figures 3-53 through 3-67	Course________ Section____ Student__________ Date____	1-3-10

Visualization of an object drawn in the traditional, three orthographic views can be enhanced by sketching the object in any one of several three-dimensional views. Isometric, oblique, and one-point perspective sketches are common ones.

Start each type with an outline "box" (blocking-in) and then lightly sketch in the object. Finish the sketch by darkening appropriate llines. Omit hidden lines. Refer to the three examples on the preceding page of this workbook.

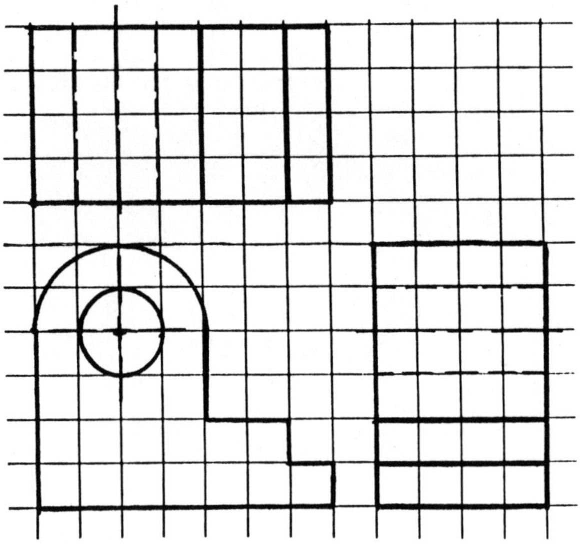

ISOMETRIC

+VP

OBLIQUE

ONE POINT PERSPECTIVE

Ch3 Lettering, Sketching, and Line Techniques			Problem
Section-Chapter-Problem 1 3 11	Freehand isometric, oblique & perspective Figures 3-53-67	Course________ Section____ Student__________ Date_____	1-3-11

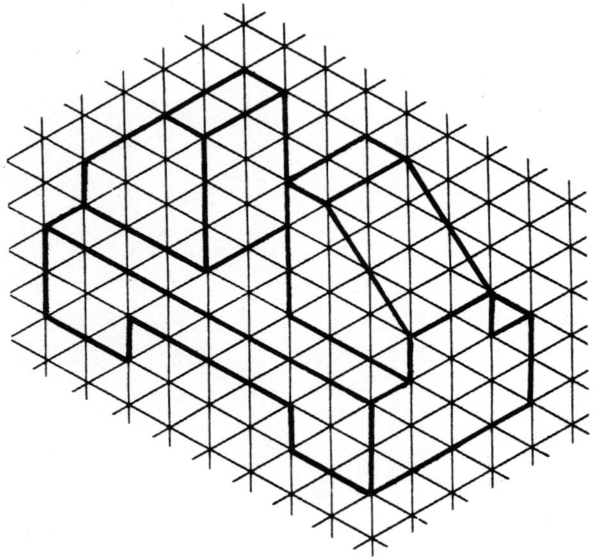

Given: An isometric drawing of an object. In the space provided sketch freehand a two point perspective drawing of the object.

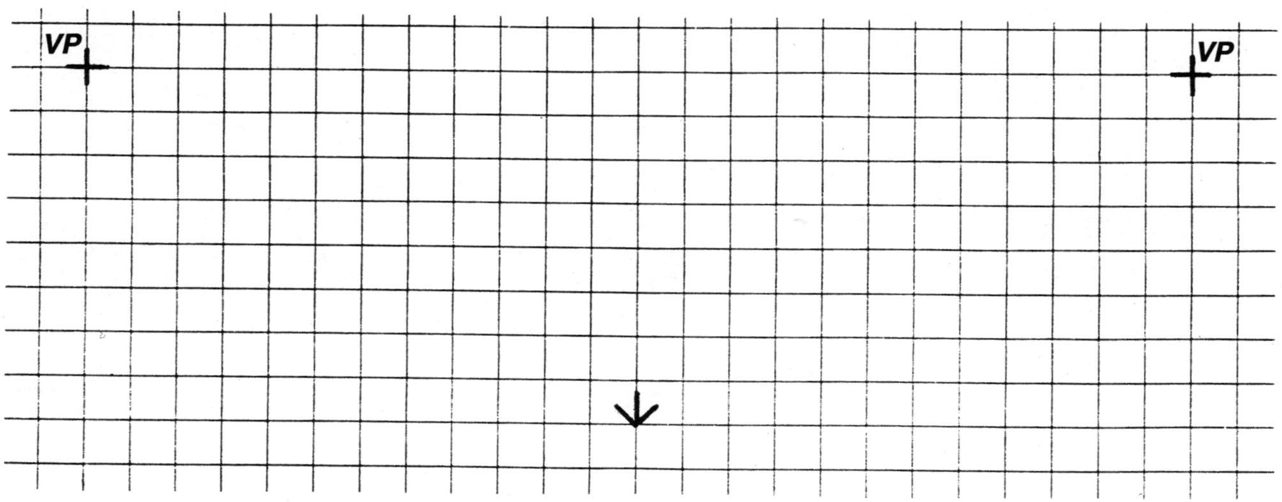

Ch3 Lettering, Sketching, and Line Techniques			Problem 1-3-12
Section-Chapter-Problem 1 3 12	Freehand two point perspective Figure 3-68	Course______ Section______ Student______ Date______	

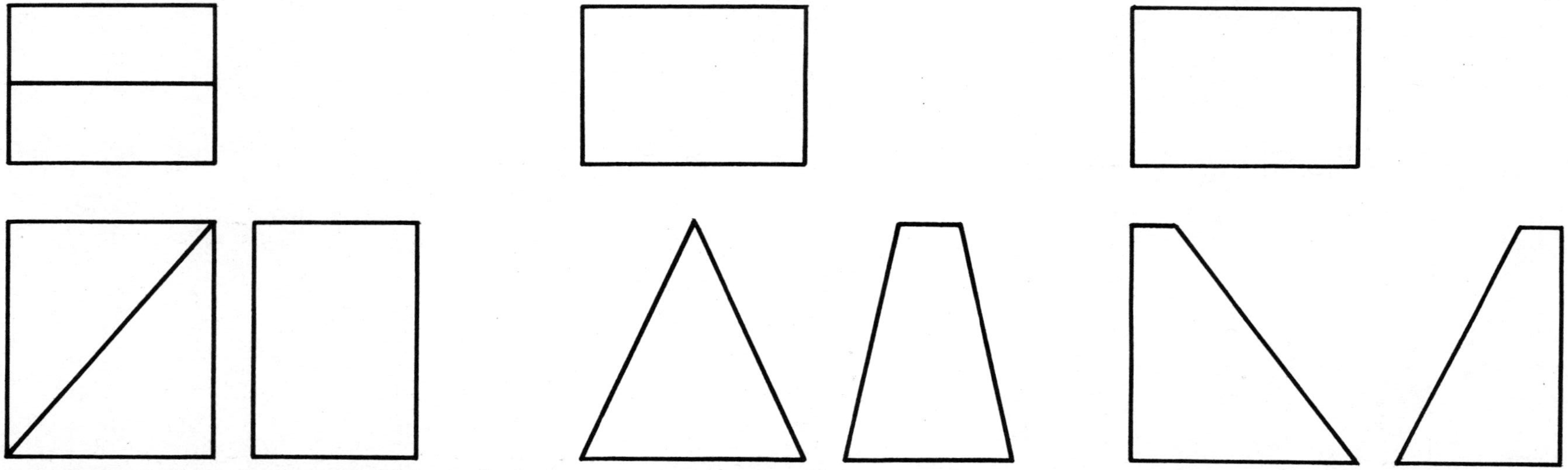

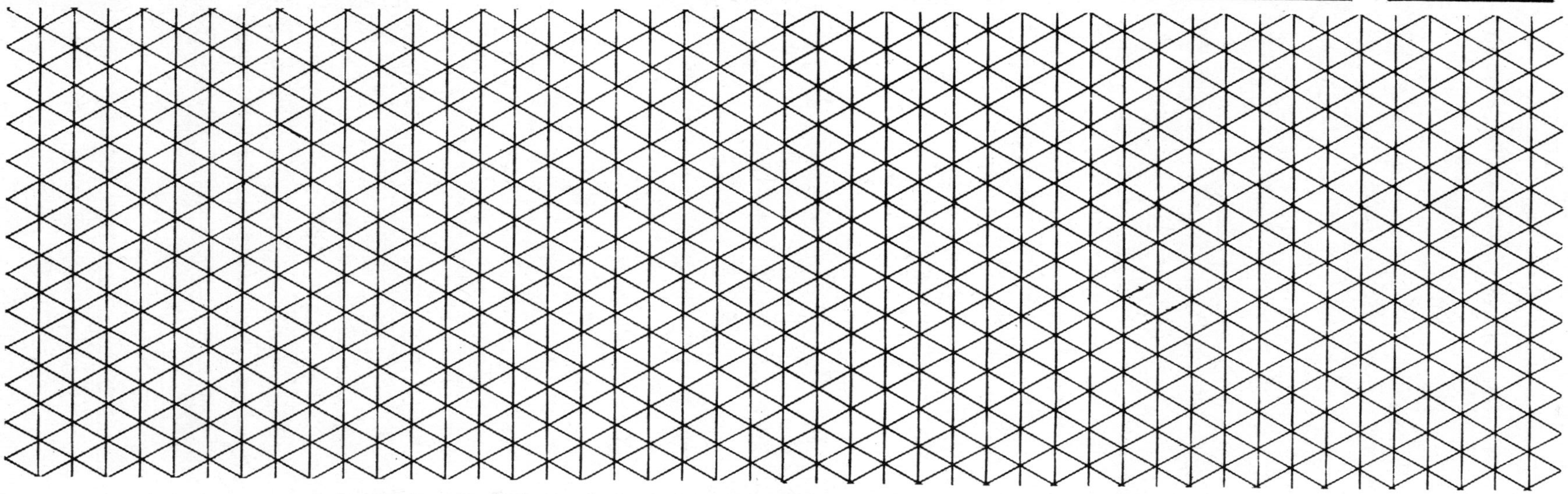

1. In each set of the orthographic drawings add the missing lines to the incomplete views.
2. Make a freehand isometric drawing of the object in the space provided.

Suggestion: Do 1. and 2. concurrently. Assume full size.

Ch3 Lettering, Sketching, and Line Techniques			Problem 1-3-13
Section-Chapter-Problem 1 3 13	Freehand three view & isometric Figures 3-50 & 3-59	Course______ Section____ Student______ Date____	

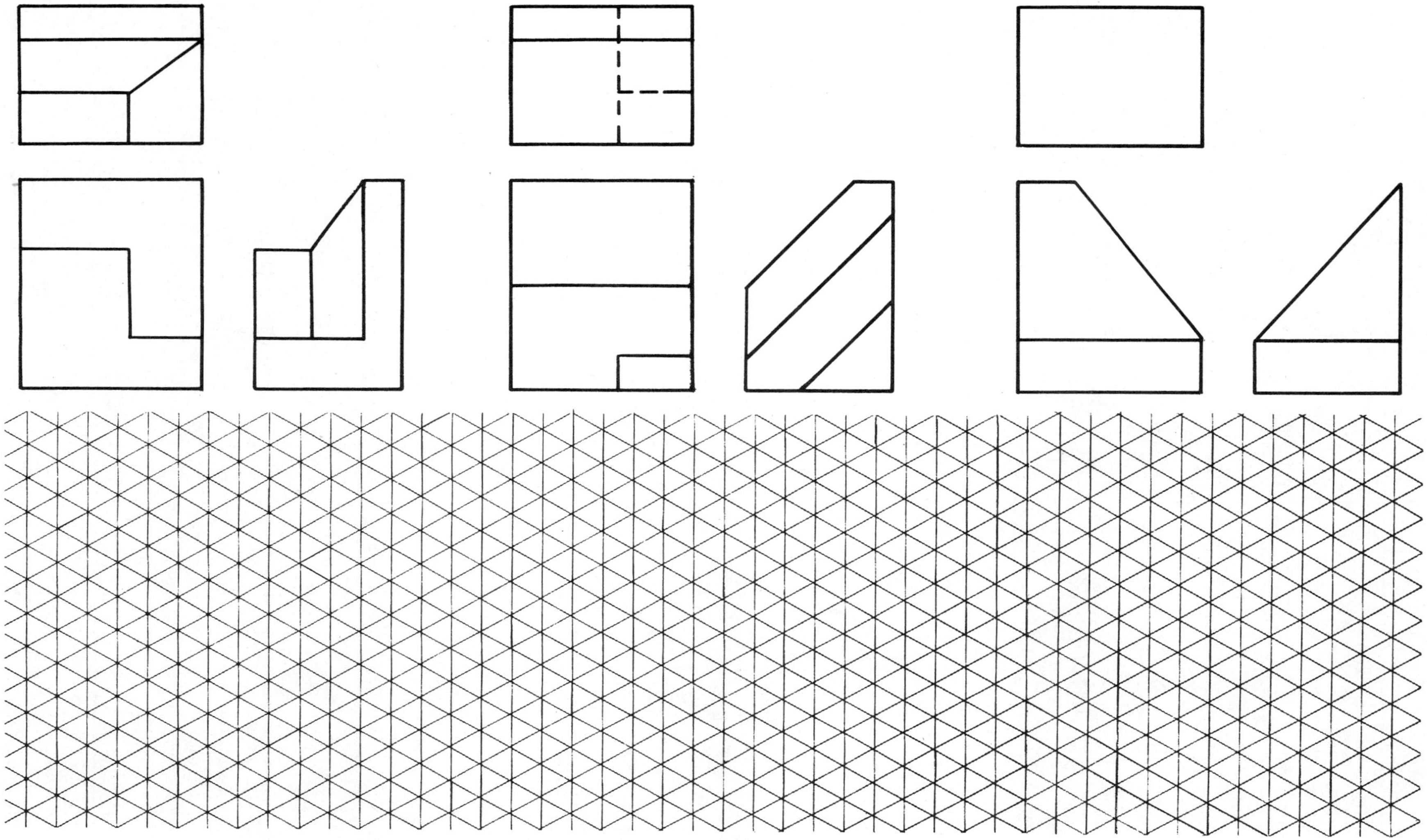

1. In each set of the orthographic drawings add the missing lines to the incomplete views.
2. Make a freehand isometric drawing of the object in the space provided.

Suggestion: Do 1. and 2. concurrently. Assume full size.

Ch3 Lettering, Sketching, and Line Techniques			Problem
Section-Chapter-Problem 1 3 14	Freehand three view & isometric Figures 3-50 & 3-59	Course______ Section______ Student______ Date______	1-3-14

A Use file "03-1.dxf" to enter the text into the "A" size title block. Follow the instructions on the page to determine what information to include.

B The object shown in file "03-2.dxf" has been sketched using a CAD program. On the drawing, there are three simple changes that need to be made. All changes should be made using the sketching capabilities of the CAD program.

Ch3 Lettering, Sketching, and Line Techniques			Problem
Section-Chapter-Problem 1 3 15	Title Block Drawing Changes	Course__________Section_____ **Student**____________Date______	**1-3-15**

In the space indicated prepare the top and front views of the piano shown in the freehand sketch. Use drawing instruments and a 1/2"=1'-0 scale. Leave a 1" space between the views. Omit dimensions.

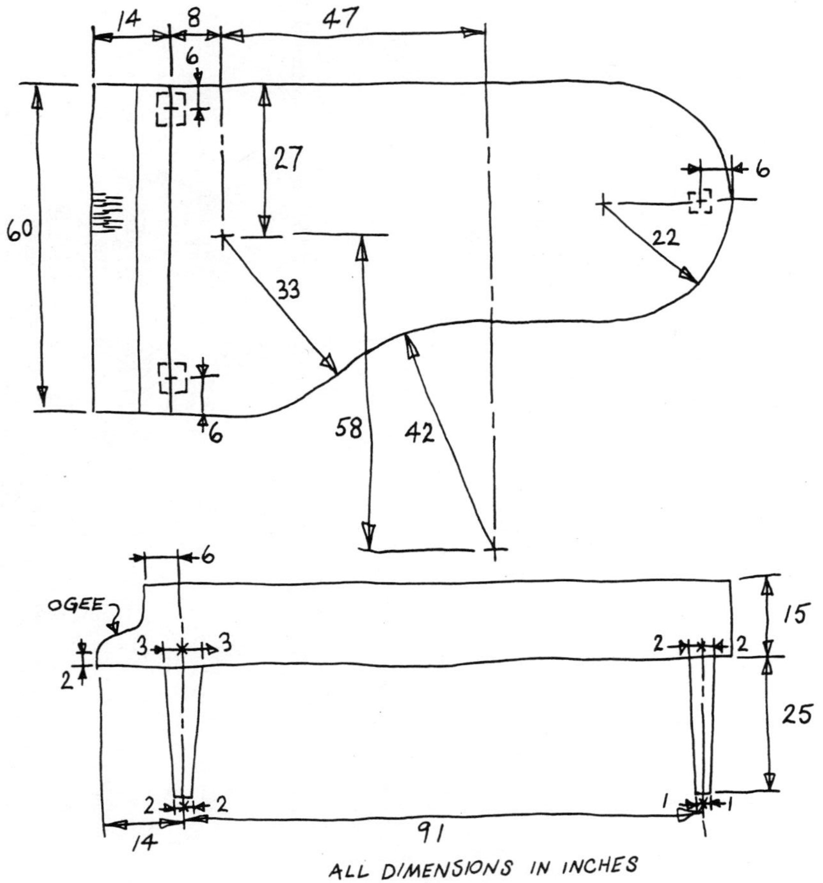

Ch4 Geometric Construction			Problem
Section-Chapter-Problem 1 4 1	Line types Figure 3-17, Ogee curves Figure 4-46	Course________ Section____ Student________ Date____	1-4-1

Freehand sketch of a proposed stairway. All dimensions are in millimeters.

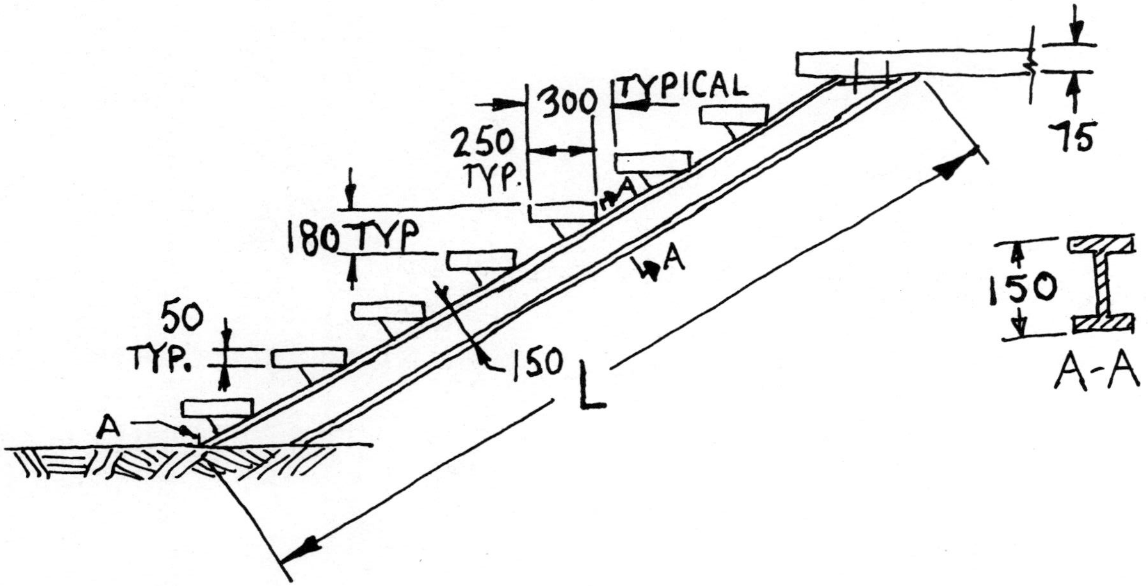

Layout to 1:20 scale the proposed stairway and determine the length L of the 150mm beam. Do not erase construction lines. Show the dimension for the length of the beam.

Ch4 Geometric Construction			Problem
Section-Chapter-Problem 1 4 2	Line types Figure 3-17, Design development layout Figure 24-9B	Course__________ Section_____ Student______________ Date______	1-4-2

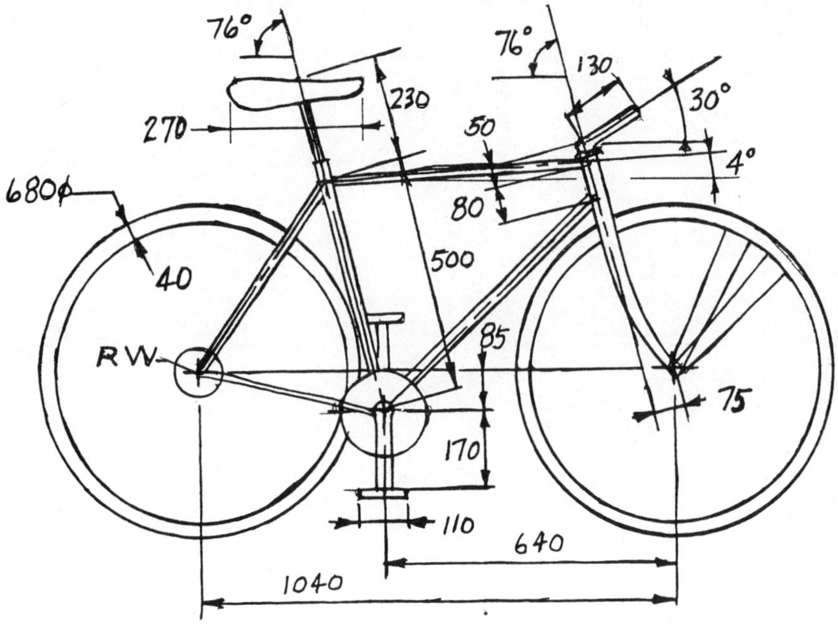

In the space indicated prepare the front view of the bicycle shown in the freehand sketch. Dimensions are to centerlines. Assume that the top of the seat is flat and that the handle bars are 20mm in diameter. Use drawing instruments and a scale of 1mm=10mm.
Omit dimensions,but show the dimensions from the ground to the top of the seat, and from the ground to the top of the handle bars.

Ch4 Geometric Construction			Problem
Section-Chapter-Problem 1 4 3	Design development layout, Figure 24-9B	Course____________Section______ Student____________________Date______	1-4-3

In the spaces provided below each symbol (from ASME Y14.5M-1994) construct the given symbol assuming that h=.125" in height. Use drawing instruments.

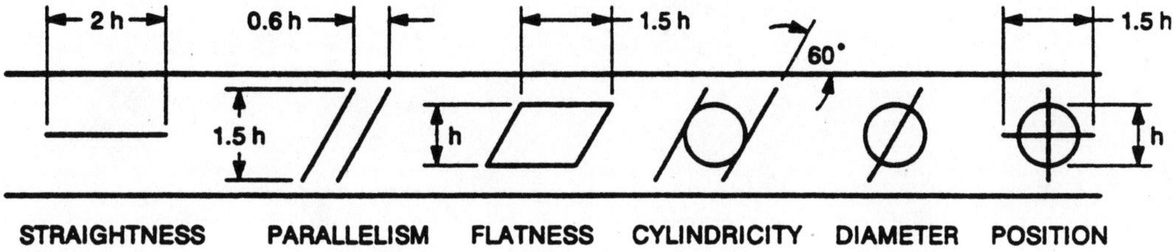

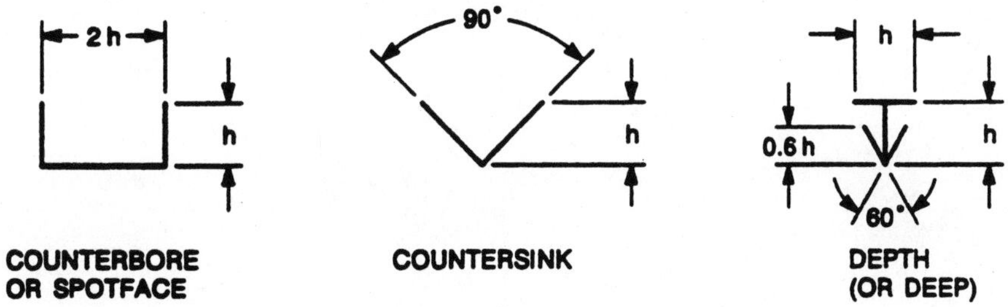

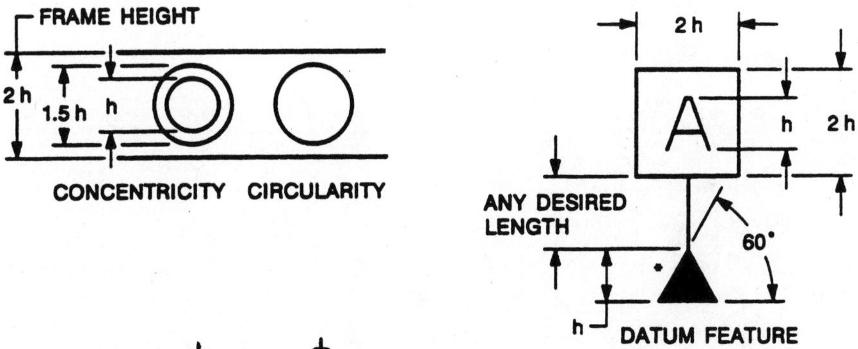

Ch4 Geometric Construction			Problem 1-4-4
Section-Chapter-Problem 1 4 4	Figure 12-7 B & C	Course__________ Section_____ Student________________ Date______	

Solve these three problems graphically: (1) Find the center of the circle. (2) Draw an arc to pass through C, D, and E. (3) Construct the largest circle within triangle F-G-H. (Hint: Use bisectors to locate the center of the triangle.) Do not erase construction lines.

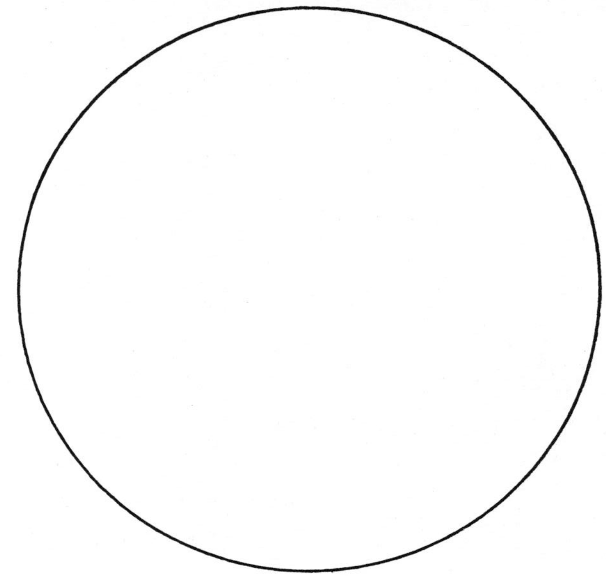

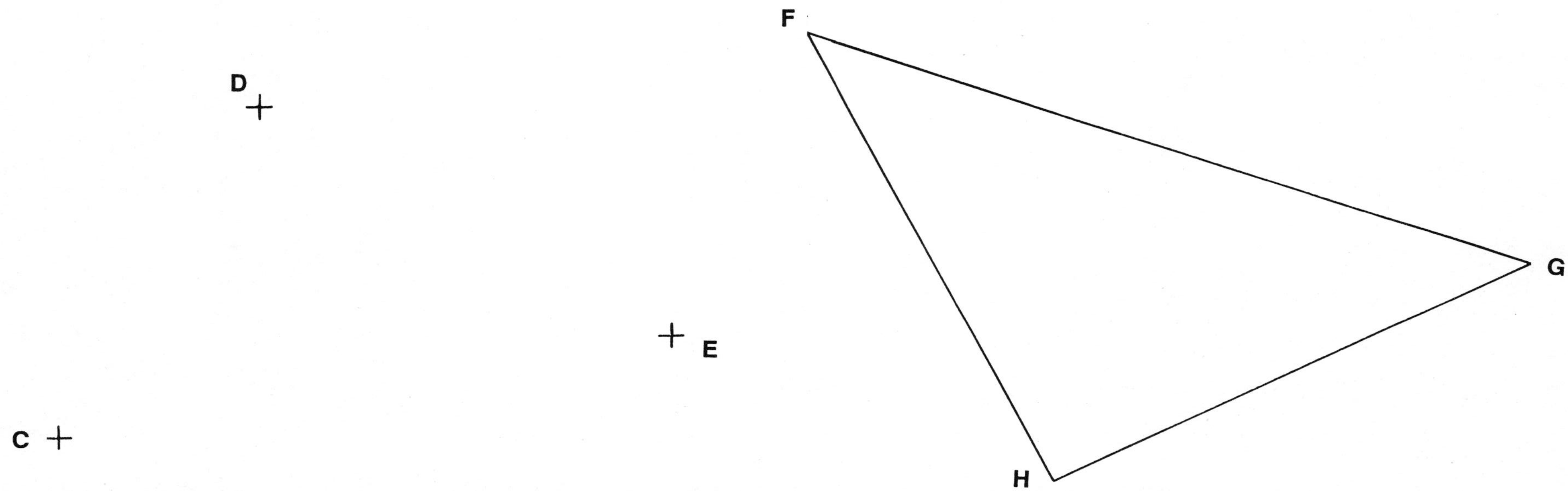

Ch4 Geometric Construction			Problem
Section-Chapter-Problem 1 4 5	Figures 4-19 (bisecting angles), 4-20 (circle through three points), & 4-38 (finding center of a circle)	Course__________Section_____ **Student**________________Date______	1-4-5

Solve these three problems graphically: (1) Construct a perpendicular to line M-N at 1.5” from point N. (2) Divide line C-D into seven equal segments. and (3) Locate a point X at three quarters of the distance from point F to point G. Do not erase construction lines.

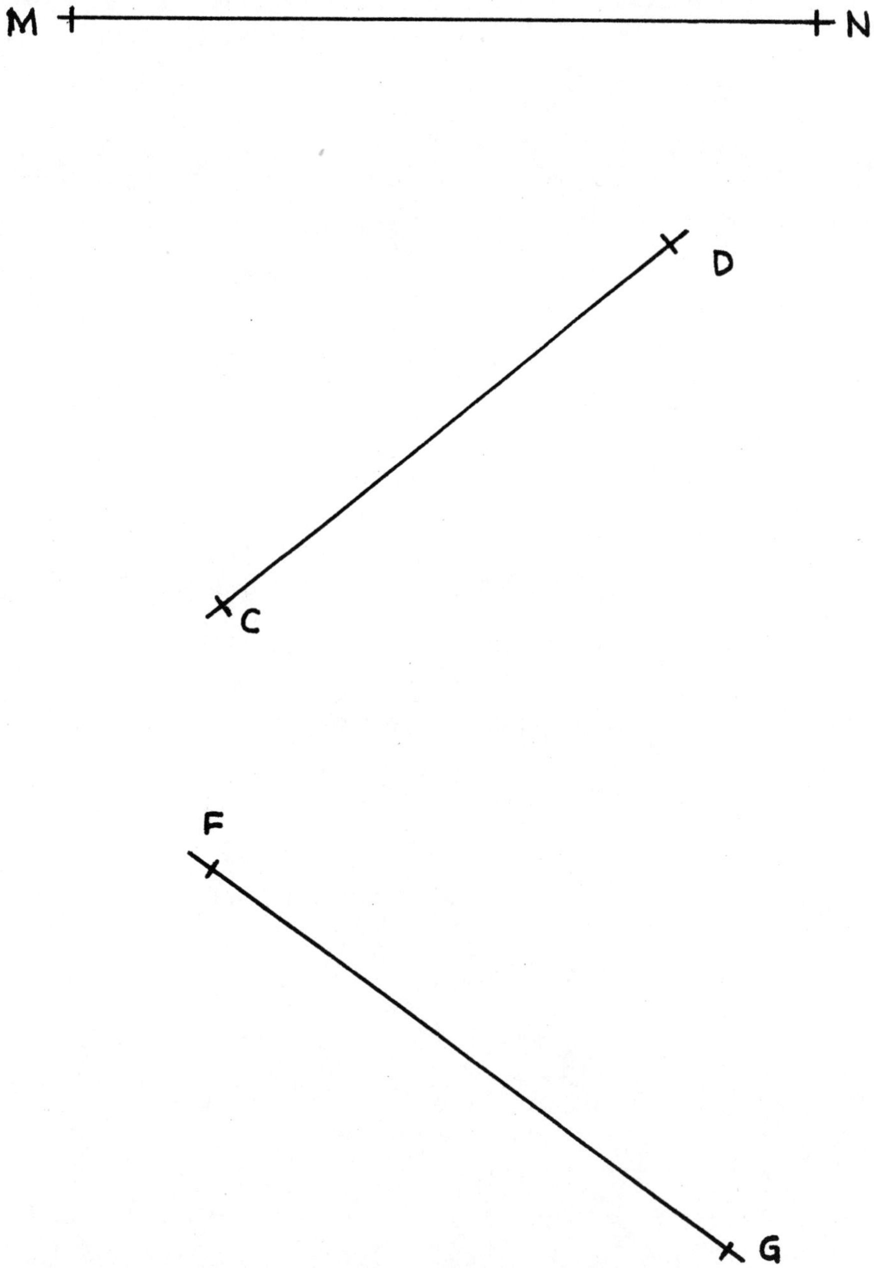

Ch4 Geometric Construction			Problem
Section-Chapter-Problem 1 4 6	Figures 4-23 or 4-24, 4-26 & 4-27	Course____________ Section_____ **Student**____________________ Date______	1-4-6

Solve these three problems graphically: (1) Transfer the shape with corners X and Y to the new position X-Y. (2) Double the size of the two figures shown at right. Do not erase construction lines.

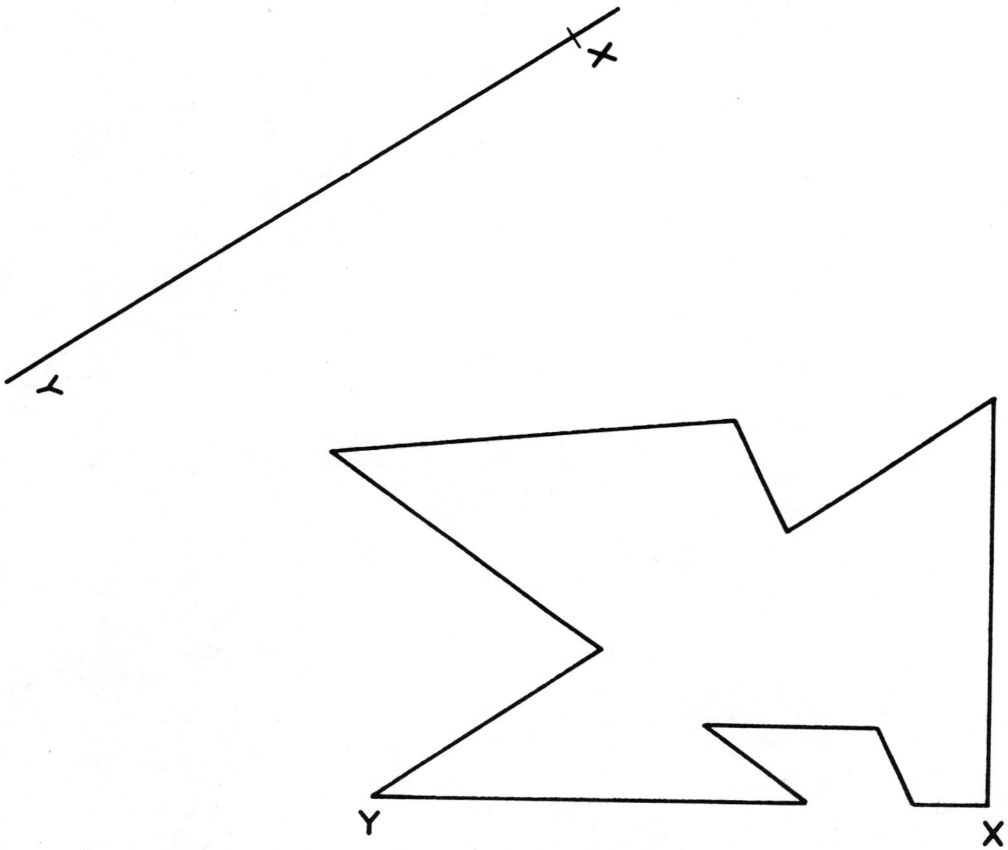

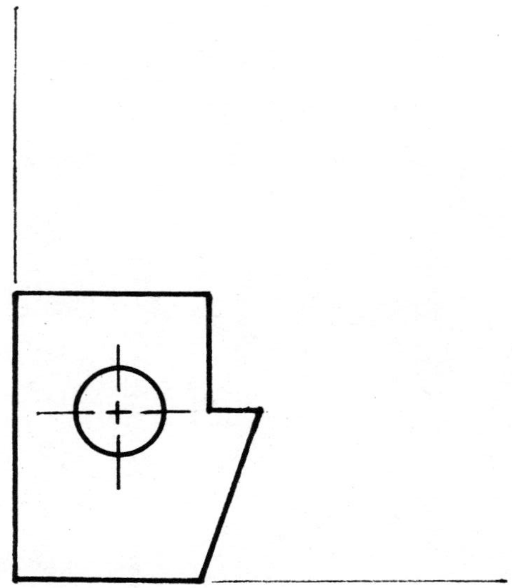

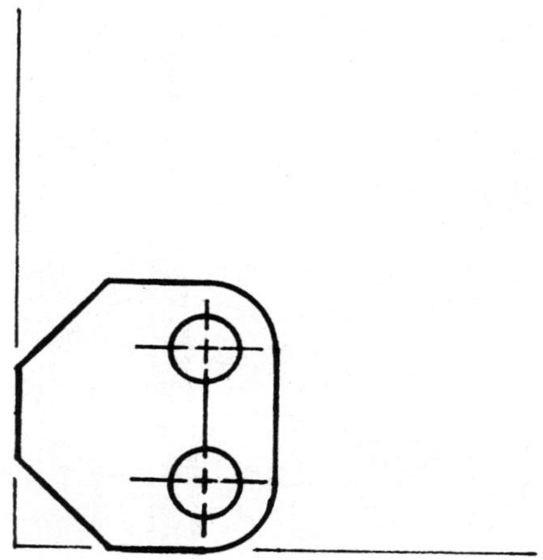

Ch4 Geometric Construction			Problem
Section-Chapter-Problem 1 4 7	Transfer complicated shapes Figure 4-30 Enlarge a shape Figure 4-31	Course______ Section____ Student______ Date____	1-4-7

Solve these six problems graphically as indicated in the spaces. Do not erase construction lines.

A. Construct a pentagon within a 1.5" circle.

+

B Construct one hexagon with 36mm across the flats.

+

C. Construct an octgon with 1.8" across the flats.

+

D. Construct an arc tangent to the angle at a radius of 14mm.

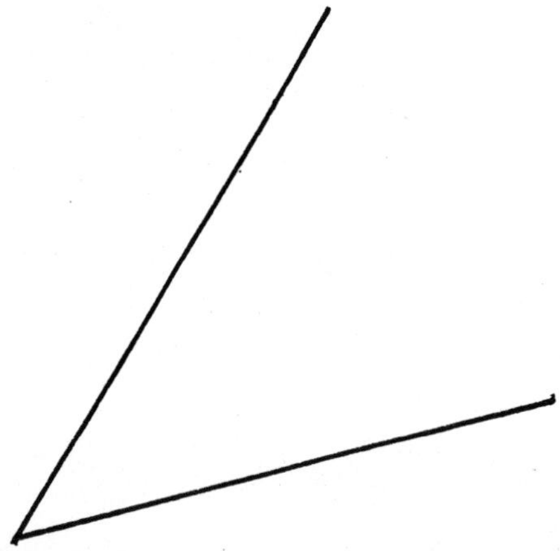

E. Construct an arc tangent to the circle and line at a radius of 18mm.

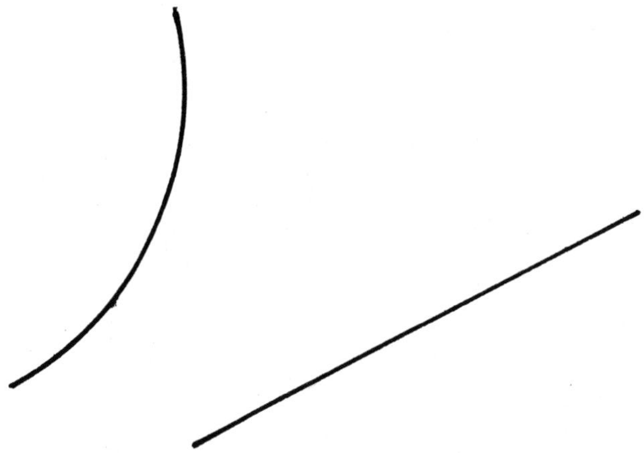

F. Construct an arc tangent to the circles at a radius of 1 inch.

Ch4 Geometric Construction			Problem
Section-Chapter-Problem 1 4 8	Figures 4-35,36,37,42,44,45	Course__________ Section_____ Student______________ Date______	1-4-8

Construct the pattern shown below, centered in the space provided, using consistent black visible lines. Identify tangent points. Do not erase construction lines. Do not dimension.

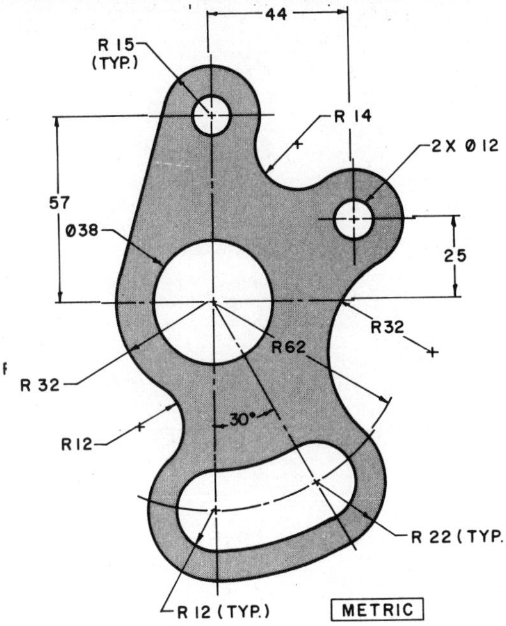

Construct the pattern shown below, centered in the space provided, using consistent black visible lines. Identify tangent points. Do not erase construction lines. Do not dimension.

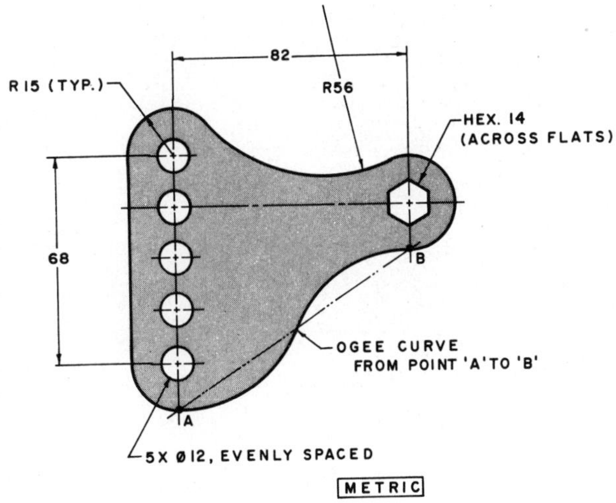

Ch4 Geometric Construction

Section-Chapter-Problem		Course____________ Section_____	Problem
1 4 9	Arc tangents Figure 4-45 Ogee curves Figure 4-46	Student____________________ Date________	1-4-9

Complete the figure shown below with a 1.125" radius at the top between arcs A and B, and a 4.50" radius at the bottom between arcs A and B. Locate swing points above the object. Add the five holes on the centerline as indicated. Do not erase construction lines.

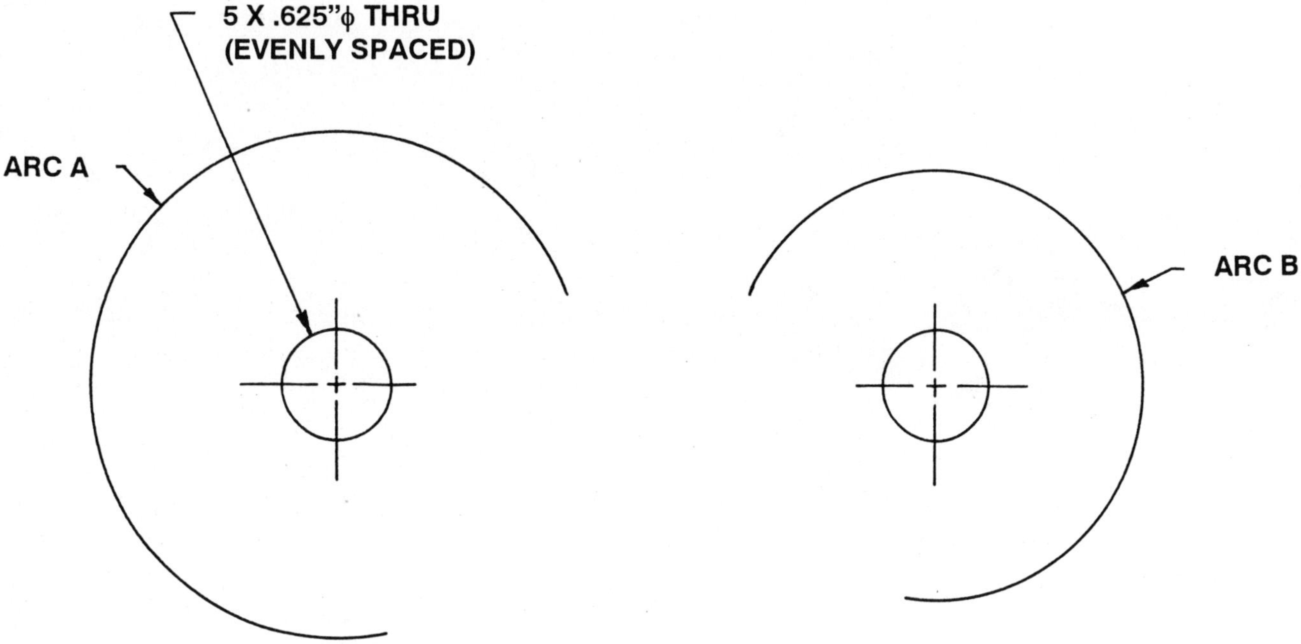

Redraw the gasket shown for four evenly spaced tabs and four evenly spaced holes .38"ϕ (instead of the five shown). Use Half Scale. Do not erase construction lines.

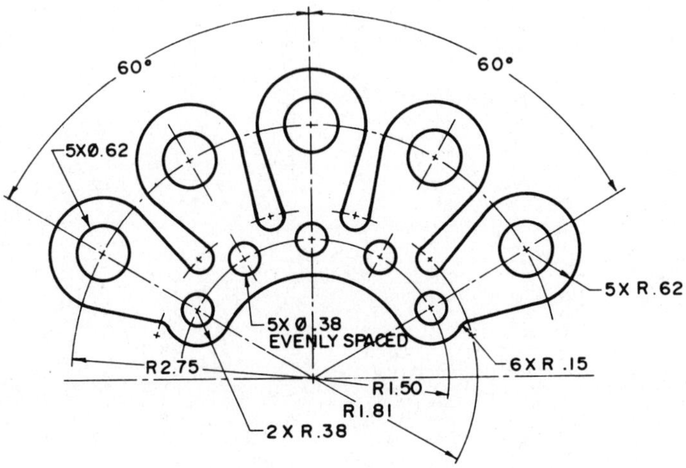

Ch4 Geometric Construction			Problem
Section-Chapter-Problem 1 4 10	Arc tangents Figure 4-40 Arc tangents Figure 4-45	Course________ Section____ Student__________ Date____	1-4-10

Complete this wrench with an ogee curve between points A and B. Do not erase construction lines.

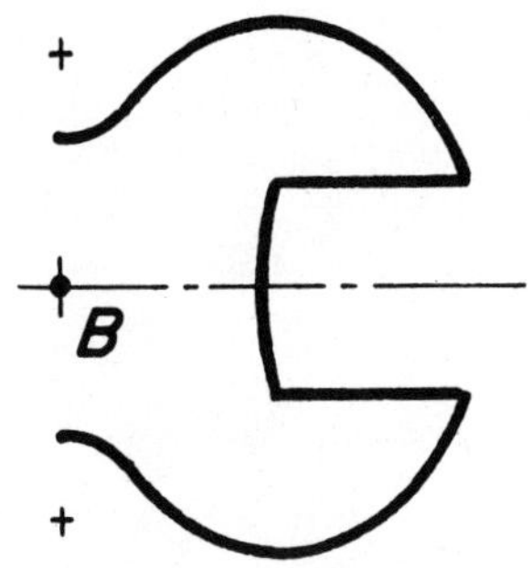

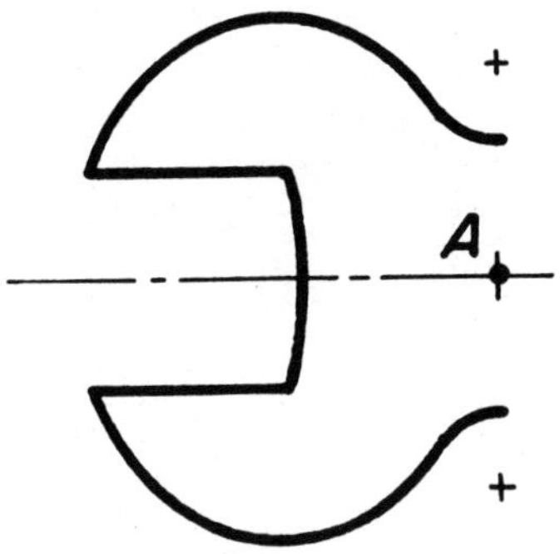

Safe chairlift operation requires that chairs plus passenger weights be centered under the cable. Ogee curves are employed as shown in the sketch. Complete the support partially shown at left with an ogee curve between points A and B. Do not erase construction lines.

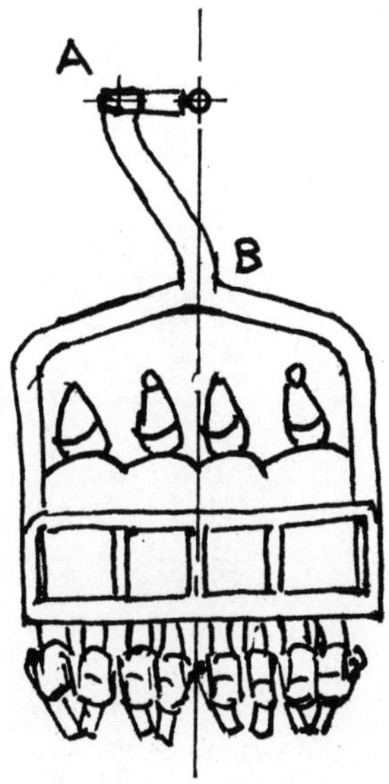

Ch4 Geometric Construction			Problem
Section-Chapter-Problem 1 4 11	Ogee curves Figure 4-46	Course__________ Section_____ Student______________ Date______	1-4-11

Construct parabolic curves for each of the three given pairs of lines, A-B. Use the ends of the lines, at A and B, as the tangent points. Do not erase construction lines.

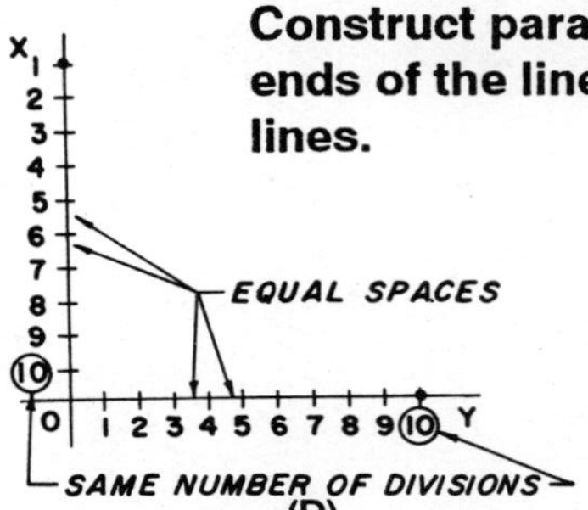

(D)

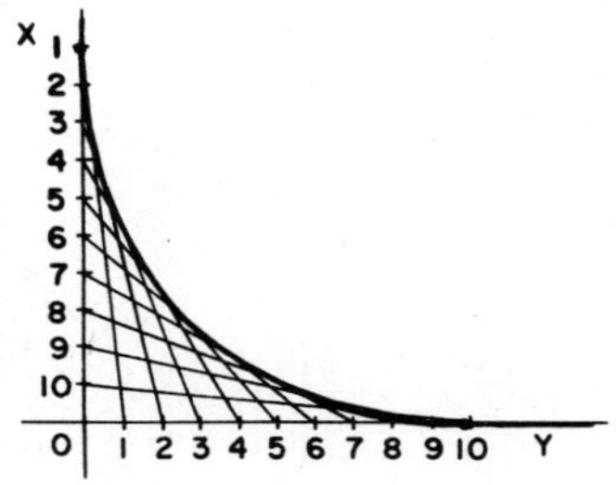

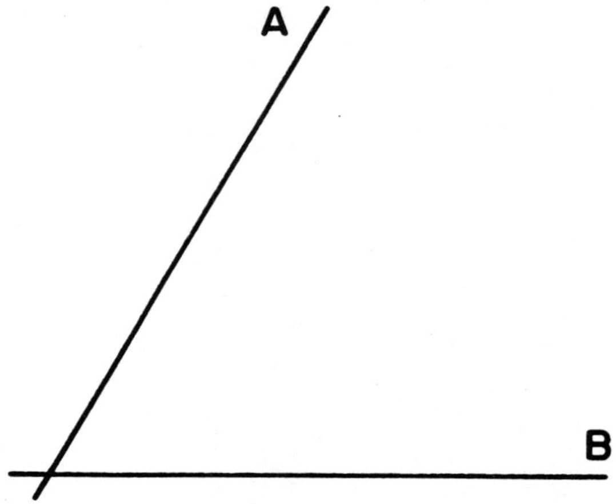

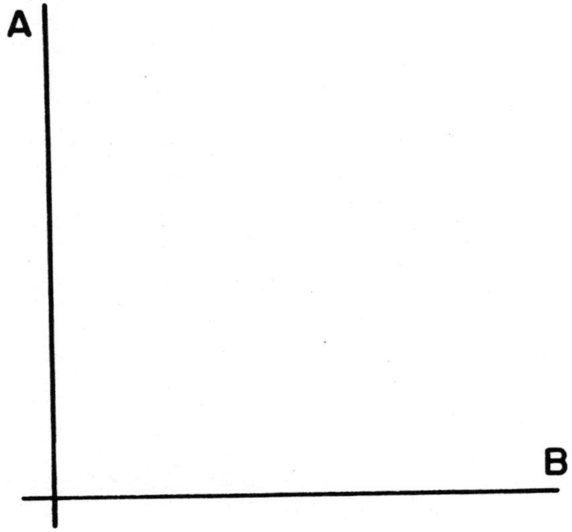

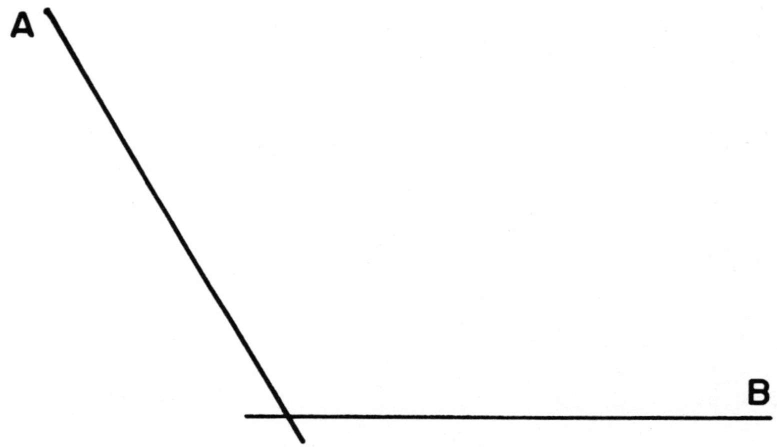

Ch4 Geometric Construction			Problem
Section-Chapter-Problem 1 4 12	Parabolic Curves between two points Figure 4-59	Course________ Section_____ Student__________ Date_____	1-4-12

First angle projection is used for the views of the cones shown below to facilitate construction of the parabola and hyperbola. Construct the parabola and the hyperbola using the guidelines shown. Do not erase construction lines.

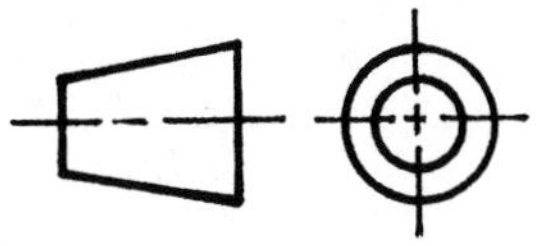

FRONT VIEW

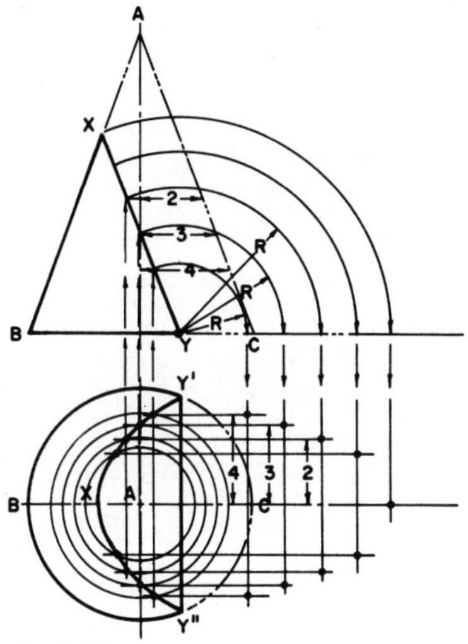

PLAN VIEW

FRONT VIEW

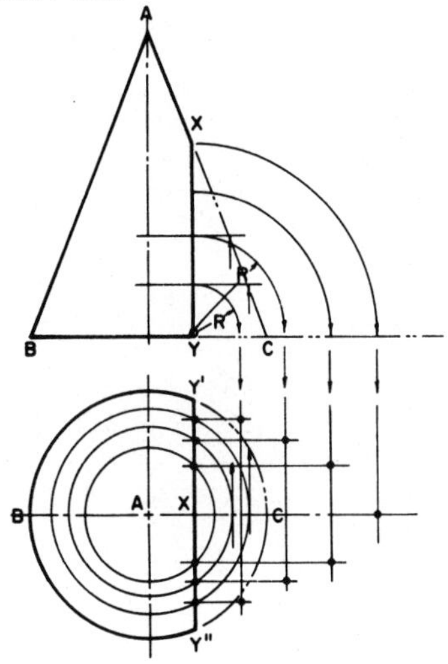

PLAN VIEW

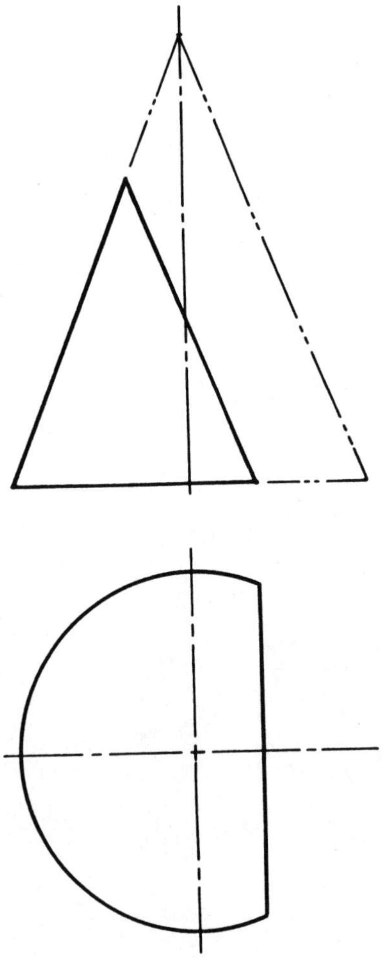

PARABOLA

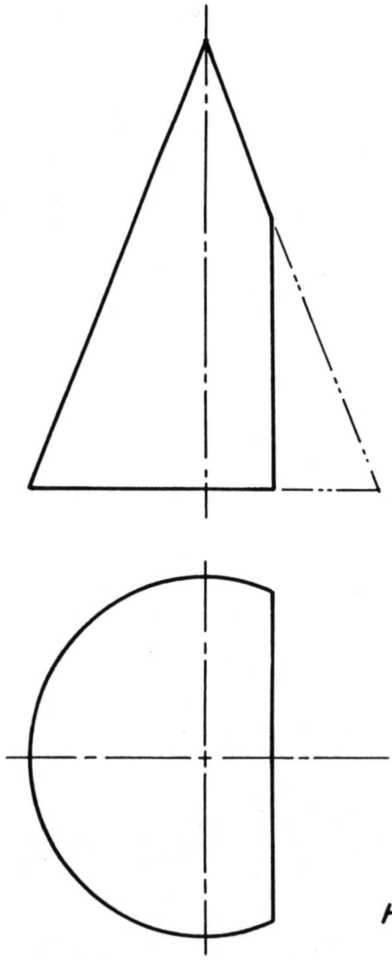

HYPERBOLA

Ch4 Geometric Construction			Problem
Section-Chapter-Problem 1 4 13	Parabola Figure 4-54 Hyperbola Figure 4-57	Course__________ Section____ **Student**____________ Date_____	1-4-13

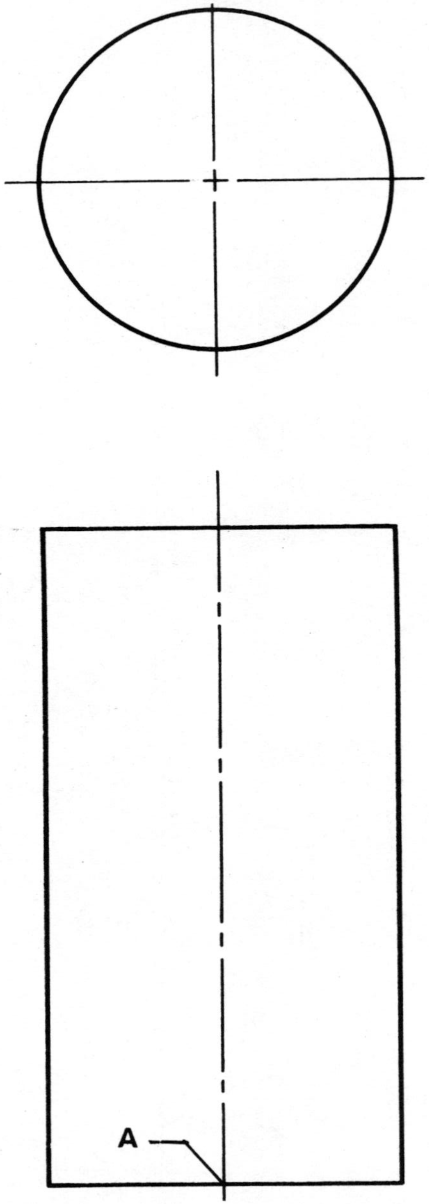

Construct a right-hand helix with two complete revolutions starting at Point A. Do not erase construction lines.

Ch4 Geometric Construction

Section-Chapter-Problem		Course________ Section____	**Problem**
1 4 14	How to draw a helix Figure 4-61	**Student**__________ Date____	**1-4-14**

Finish the involutes started in the figures below for the (a) straight line, (b) triangle, (c) circle, and (d) square. Do not erase construction lines.

(a)

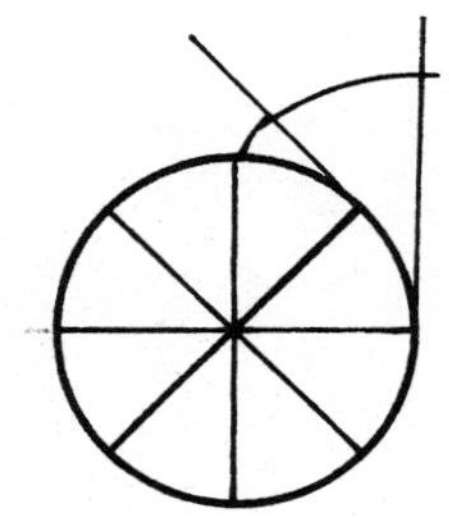

(c)

(b)

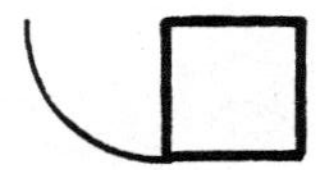

(d)

Ch4 Geometric Construction

Section-Chapter-Problem	How to draw involutes Figures 4-62,63,64,65 (line, triangle, circle, and square respectively)	Course____________ Section_____	Problem
1 4 15		Student____________________ Date_______	1-4-15

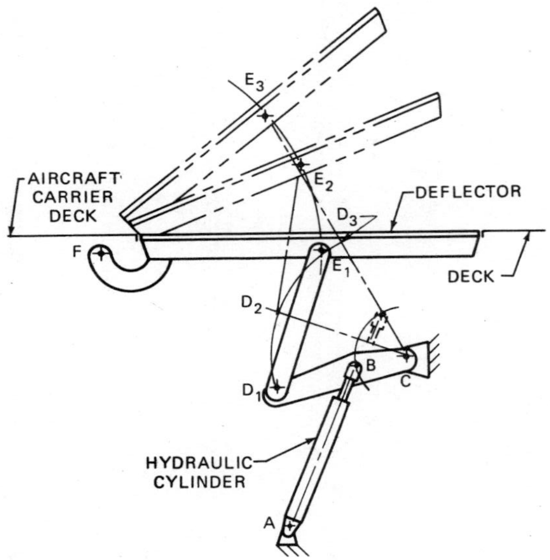

Cylinder A-B on the front loader extends 585mm (that is, has a stroke of 585mm) from the position shown in the drawing. Use layout techniques to determine:

a. The maximum dump truck height T that the loader can accomodate. See sketch A.

b. The change in length J-H needed to cause the scoop to rotate 105° CCW

(Suggestions to get started: Construct an arc C-B about C. Add 585mm to A-B and construct an arc A-B about A. Where C-B and A-B intersect determines the new positon of arm C-B-D.)

Do not erase construction lines.

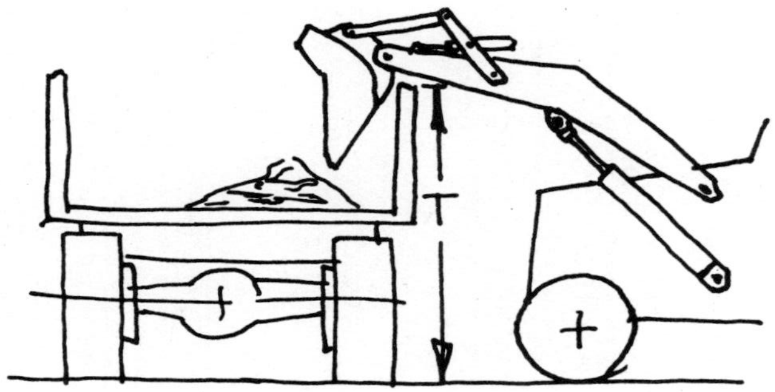

Sketch A

Scale 1mm=30mm

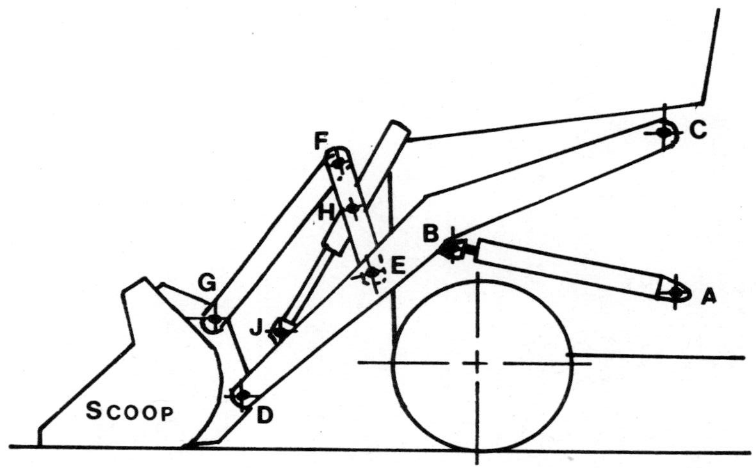

Ch4 Geometric Construction

Section-Chapter-Problem	How to draw kinemtic layouts (See upper left figure.)	Course____________ Section_____	Problem
1 4 16		Student__________________ Date______	1-4-16

A Shown in file "04-1.dxf" are a series of lines, circles, arcs, and angles. Follow the directions on the page to modify the items.

B In file "04-2.dxf" there are two fictitious companies and their logos. Use the principles of geometric construction to design new logos for each of the companies. For instance, incorporate spirals, octagons, text, various line styles, and colors as appropriate.

Ch4 Geometric Construction			Problem 1-4-17
Section-Chapter-Problem 1 4 17	Geometric Shapes Logo Design	Course________ Section____ **Student**____________ Date_____	

Section 2
Objectives:

After studying and doing the exercises and problems in Section 2 of the workbook you should be able to:

- prepare one, two, three or more views of an object in order to completely communicate its geometry
- construct sectional views as needed to show additional details
- construct auxiliary views to further describe an object
- project points, lines, and planes in the necessary views to find true lengths, true distances, true angles, and true sizes
- handle basic vector constructions and operations
- develop and layout patterns of common shapes
- completely dimension objects, add realistic tolerances, and make proper notes

Supplemental Comments: Section 2

Chapter 5 Multiview Drawings

Problem 2-5-1 The ubiquitous "glass box" showing projected images appears in practically all graphics and technical drawing texts, but students rarely are asked to prepare a drawing of one. This exercise has the student complete a pictorial of a "glass box".

Problems 2-5-2 and **2-5-3** require visualization skills to complete **F** or **A** views from a given **T** view, and to match an isometric drawing of an object to the three views of the object.

Problems 2-5-4 and **2-5-5** are freehand sketching exercises to prepare three views of an object from an isometric.

Problem 2-5-6 includes an isometric drawing to assist in the analysis of locations of surfaces of an object. The numbered planes also illustrate the projection fundamental that says that if a plane is parallel to a fold line in one view, the plane will appear in true size in the adjacent view. For example, 10-11 represents a plane on edge in the top view, and appears in true size 1-2-3-4 in the front view. Similarly, 6-7 in the front view is in true size in the side view as 21-24-27-26.

Problem 2-5-7: While completing the missing views, keep in mind that lines that are truly parallel will be parallel in all views.

Problems 2-5-8, 2-5-9, 2-5-10, and **2-5-12** give practice in centering views in a drawing space and leaving 1" between views for facilitating dimensioning.

Problem 2-5-11: Third-angle projection was used to prepare the assigned object shown. Use the space at right to redraw and complete the drawing in third-angle projection. Note the truncated cone on the right used to illustrate the third-angle projection, and the one on the left, for first angle. Drawings from Europe are often done in first angle.

Problems 2-5-13, 2-5-14, and **2-5-15** require dimensioning of radii and holes, and drawing just portions of an object to clarify details.

Problems 2-5-16, 2-5-17, 2-5-18B, and **2-5-19** require views to be added or finished. Isometric drawings are shown to enhance visualization.

Problems 2-5-20 and **2-5-21:** Note that a cast object usually needs machined surfaces to be used as datums (reference planes). The datums may be flat or cylindrical.

Problem 2-5-22: Long objects with few or no changes along most of their lengths can be shortened with "Z" breaks and enlarged for clarity.

Problem 2-5-23: Cut out the die pattern, fold it, and tape the edges to form a cube; check your answers to the three questions.

Problem 2-5-24 has you use a CAD program to draw three views of a house, and to create a door hinge.

Chapter 6 Sectional Views

Problems 2-6-1, 2-6-2, 2-6-3, 2-6-4, and **2-6-5** are examples of sectioning used in industry. Check for aligning considerations. Note the isometric drawings for assisting visualization.

Problem 2-6-6 introduces you to a useful drawing technique for communicating information that is a **detail** of a portion of an object, or a **detail** of a partial assembly.

Problems 2-6-7 and **2-6-8** involve fasteners, shafts, and keys in sections.

Problem 2-6-9 requires several calculations to ensure equal cross-sectional areas along a nozzle.

Problem 2-6-10 shows additional section conventions.

Problem 2-6-11 requires the use of a section drawing to produce a top and front view of an object.

Problem 2-6-12 uses a CAD program to draw a full section and an offset section.

Chapter 7 Auxiliary Views

Problem 2-7-1 reviews basic orthographic projection principles used to produce auxiliary views. Discussions of these fundamentals are also in Chapter 8 (Descriptive Geometry).

Problem 2-7-2 introduces two methods of constructing auxiliary views—the fold line and reference line methods.

Problem 2-7-3 requires the reference line method.

Problem 2-7-4 requires the fold line method.

Problem 2-7-5 requires using an auxiliary view to produce orthographic views.

Problems 2-7-6, 2-7-7, and **2-7-8** are exercises using the reference line and fold line methods.

Problem 2-7-9 requires the following two descriptive geometry principles for solving the problem:

1. If a line is parallel to a fold line in one view the line will appear in true length in the adjacent view.
2. If a line appears in true length in one view, it will appear as a point in an adjacent view if the fold line is perpendicular to the direction of the true length line. (Note: If the line lies in a plane then the plane will be seen as an edge.)

Problem 2-7-10 shows an auxiliary view to be used to produce a front view. Also, the area of the auxiliary view is to be calculated.

Problem 2-7-11 introduces an O-Ring application by requiring a groove for the O-Ring.

Problem 2-7-12 has a CAD program to draw a paper airplane and do auxiliary views.

Chapter 8 Descriptive Geometry

Problem 2-8-1 emphasizes that descriptive geometry constructions almost always begin with, or involve, finding the true length of a line.

Problem 2-8-2 Four steps in basic descriptive geometry constructions lead to the true shape of a plane surface.

Problems 2-8-3 and **2-8-4** focus on projections to discover several fundamentals of descriptive geometry.

Problem 2-8-5 involves finding the shortest distance between two lines, and from a point to a plane. Note that the shortest distance is always perpendicular to the line, or the plane.

Problem 2-8-6 contains two practice problems and one practical application involving the sparking distance from a power transmission line.

Problems 2-8-7 and **2-8-8** require the "Four Steps" to finding the true size of a plane.

Problems 2-8-9 and **2-8-10:** To find the dihedral angle (true angle between two plane surfaces), the line of intersection between the two surfaces is needed. Then find the line of intersection as a point.

Problem 2-8-11: Descriptive geometry also involves finding intersections of lines, planes, prisms, and cylinders.

Problem 2-8-12: True slopes of lines and planes, with respect to the earth, can only be seen in views adjacent to T views (i.e., elevation views).

Problem 2-8-13: Finding the true length of a line using revolution provides an efficient use of drawing space.

Problem 2-8-14 illustrates descriptive geometry applied to topography problems: a profile cut (similar to a full section), and a "cuts and fills" problem related to highway excavations.

Problem 2-8-15: Graphical force analyses in frame members require a scaled drawing of the frame and scaled force diagram.

Problem 2-8-16 requires CAD to do two descriptive geometry problems: an intersection between two objects, and a topography consideration for a house.

Chapter 9 Patterns and Developments

Problems 2-9-1 and **2-9-2:** As a challenge to visualization skills, you are asked to study a flat development and deduce the original three-dimensional shape of (1) a hollow prism (box) and (2) a truncated cylinder.

Problems 2-9-3 and **2-9-4** are straightforward applications of the parallel line development method.

Problem 2-9-5 The backstop shown is symmetrical, so only half needs to be developed to determine the total screen area, and total tubing length.

Problems 2-9-6 and **2-9-7** require the parallel line development method to develop these intersecting cylinders and prisms.

Problems 2-9-8, 2-9-9, and **2-9-10A** require the radial line development method. **Problem 2-9-10B** requires triangulation to do the development.

Problem 2-9-11 includes an application of revolution to aid in the development that also uses triangulation.

Problem 2-9-12 requires visualization skills. You're asked to study a flatcardboard development of a commercial package and deduce the original three-dimensional shape.

Problem 2-9-13: Bend allowance charts are needed in the calculations of the total lengths of the bent metal bars.

Problem 2-9-14: You are asked to use a CAD program to do two flat patterns, one of a candy box and one of a food container.

Chapter 10 Dimensioning and Notation

Problem 2-9-1 calls for a variety of standard dimensioning practices.

Problems 2-10-2, 2-10-3, 2-10-4, and **2-10-5** focus on geometric breakdowns of objects, and present a checklist for dimensioning.

Problem 2-10-6: When you are dimensioning irregular curves, use relatively short intervals for rapidly changing curves and longer intervals for flatter curves.

Problems 2-10-7 and **2-10-8:** Standard practices: Rectangular coordinates, bolt center dimensioning, and dimensioning internal slots are shown in examples to aid in doing the problems.

Problem 2-10-9: Size and location dimensioning emphasizes geometric breakdown of objects.

Problem 2-10-10 uses typical callout notation for technical drawings.

Problem 2-10-11 provides more practice in dimensioning holes.

Problem 2-10-12 gives typical callout notation directions, and contains examples of dimensioning grooves and threads.

Problems 2-10-13, 2-10-14, 2-10-15, and **2-10-16** emphasize tolerancing for mating parts, both for relative motion (RC) and for interferences (FN). Note that all of the tolerances for these problems (.0003" to .0017") are less than the thickness of this page, which is approximately .004" thick! Close tolerances are expensive; therefore, keep tolerances as loose as practical without compromising the functions of the parts.

Problem 2-10-17: ALL dimensions must have a tolerance, either included in the dimension or specified in the title block. The MOUNTING GUIDE drawing is an excellent illustration.

Problem 2-10-18: Using a CAD program, add notes and dimensions to a drawing and complete an engineering change order (ECO).

For the object shown complete the Front, Top and Side views on the projecton surfaces. Lable the corners as needed. Show all the necessary projection lines to complete the Front View. Show a few of the projection lines for the Top and Side views.

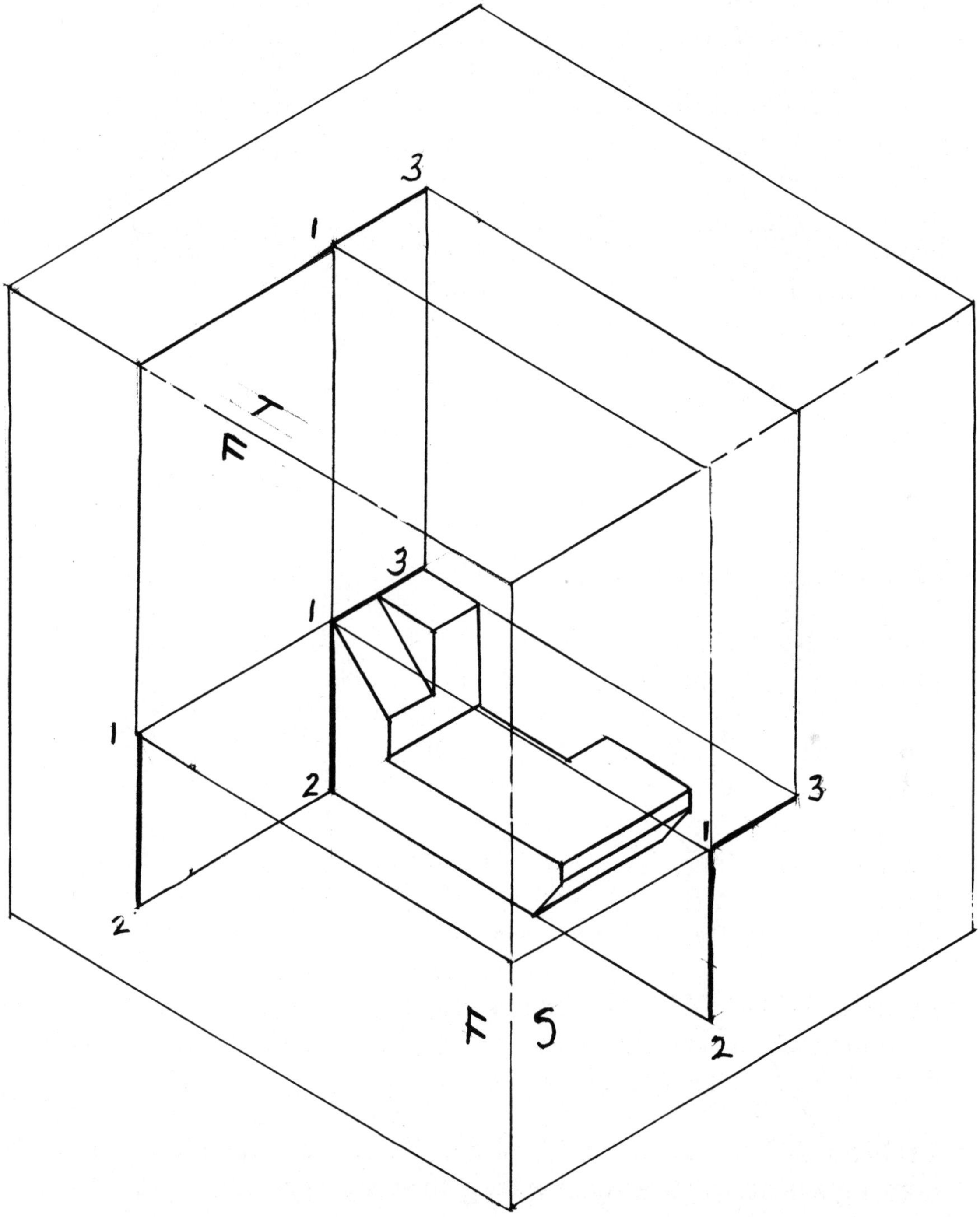

Ch5 Multiview Drawings			Problem
Section-Chapter-Problem	Principal viewing planes: Transparent	Course____________ Section_____	**2-5-1**
2 5 1	"box" Figures 5-12, 5-18, 5-19	**Student**__________________ Date______	

In each set of three views, the T view is correct as given. Complete the A or F view as needed. Show all hidden lines. There are no inclined or oblique surfaces, and no lines need to be erased.

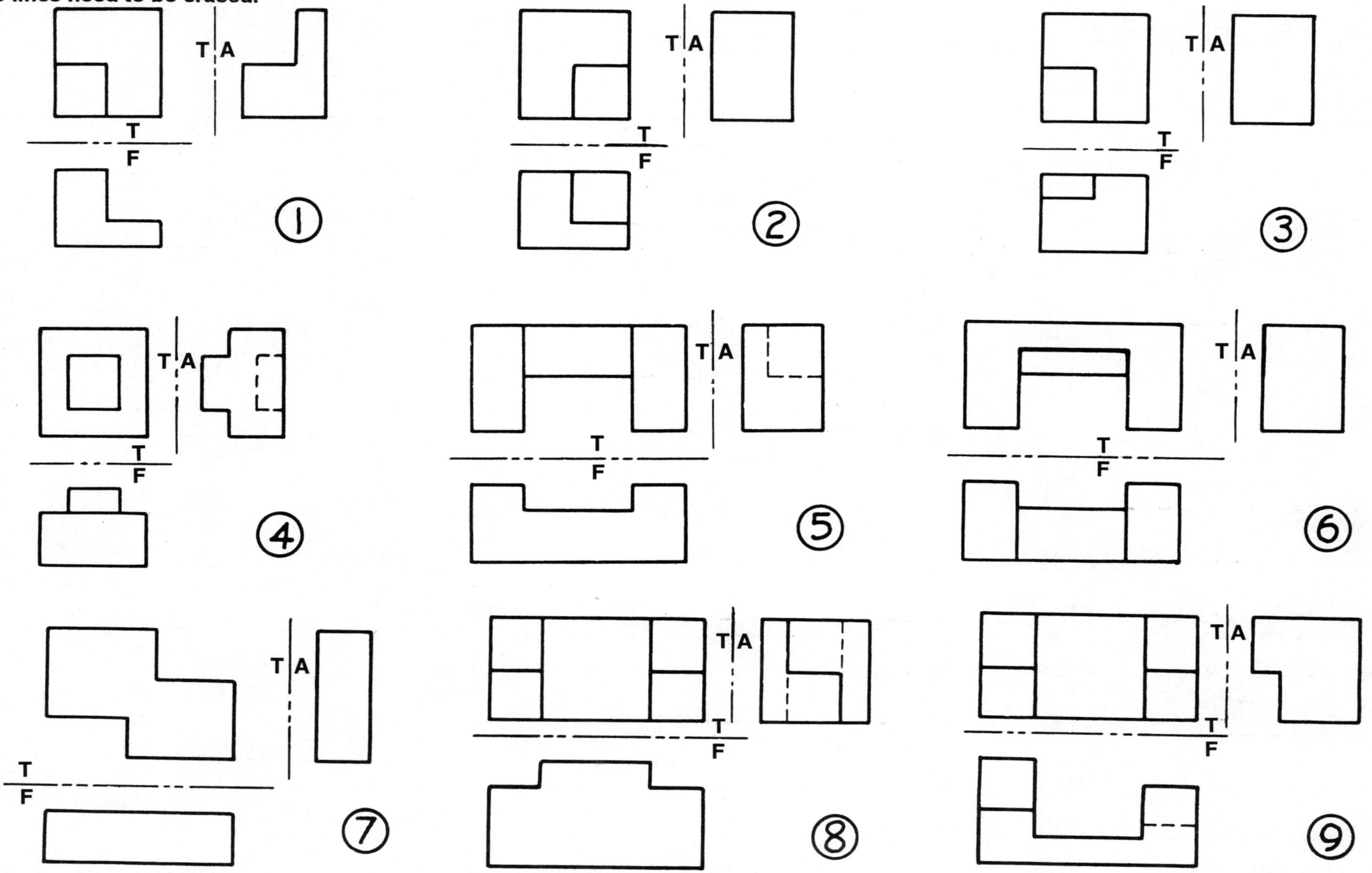

Ch5 Multiview Drawings

Section-Chapter-Problem		Course_____ Section_____	Problem
2 5 2	Multiview Drawing Problems Figures 5-19, 5-22	Student_____ Date_____	2-5-2

Match the isometric drawings of objects to the appropriate three view drawings of the objects. Note: arrows point toward the front views.

Lable the grid below as follows:

	A	B	C	D
1				
2				
3				
4				

Record your answers:

EXAMPLE: A-1 matches B-3

A-2 matches ____

B-1 matches ____

B-2 matches ____

C-1 matches ____

C-2 matches ____

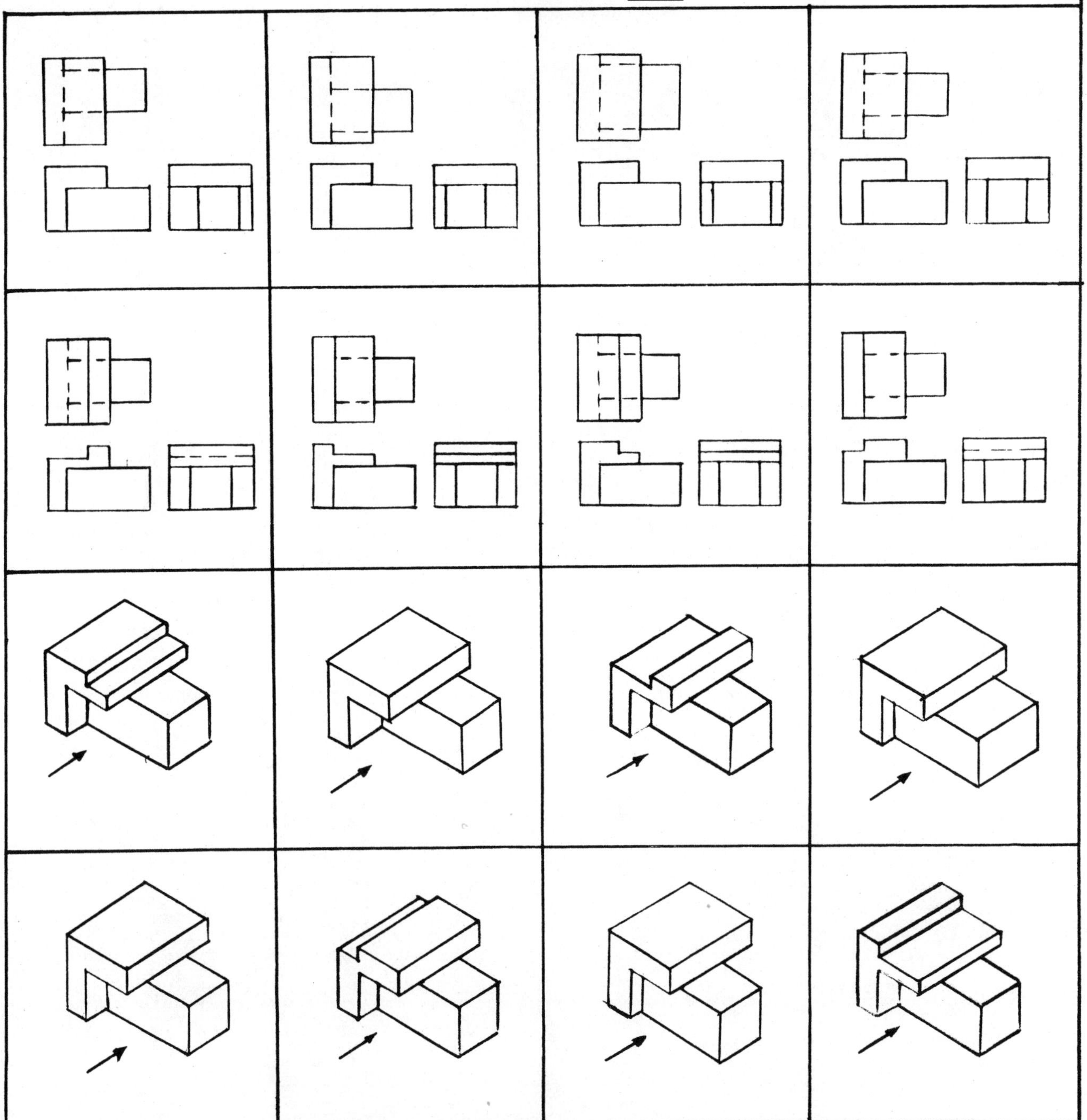

Ch5 Multiview Drawings

Section-Chapter-Problem	Multiview Drawing Problems	Course________ Section____	Problem
2 5 3	Figures 5-18, 5-19	Student__________ Date_____	2-5-3

A

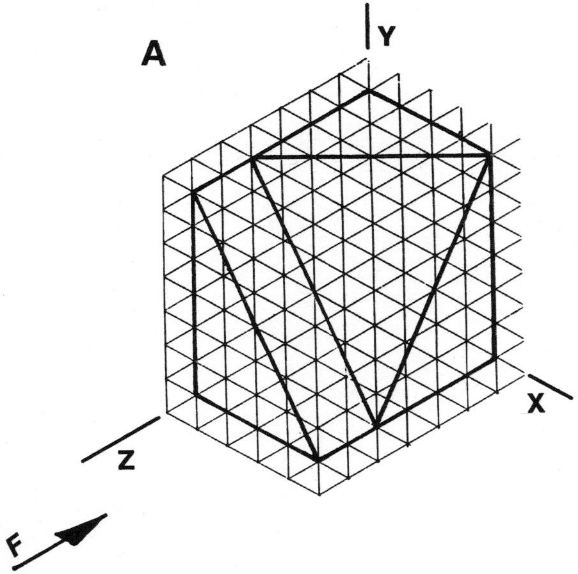

Use standard lines and in the spaces indicated, sketch freehand, three views of the objects given. Leave three spaces between views.

B

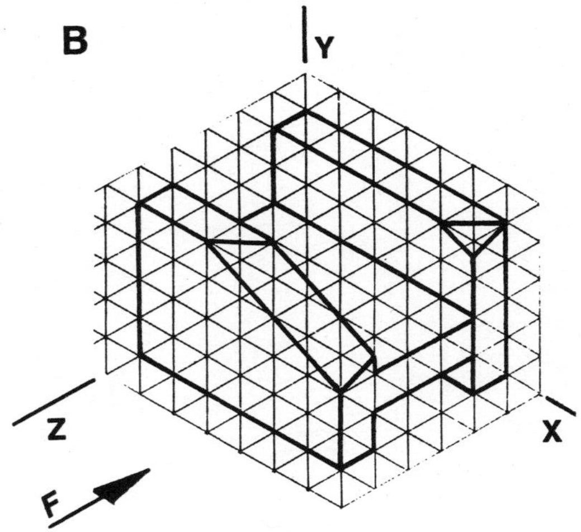

Ch5 Multiview Drawings			Problem 2-5-4
Section-Chapter-Problem 2 5 4	Multiview Drawing Problems: Freehand sketching Figures 3-17, 5-19	Course________ Section____ Student__________ Date____	

Use standard lines and in the spaces indicated, sketch freehand, three views of the objects given. Leave three spaces between views.

A

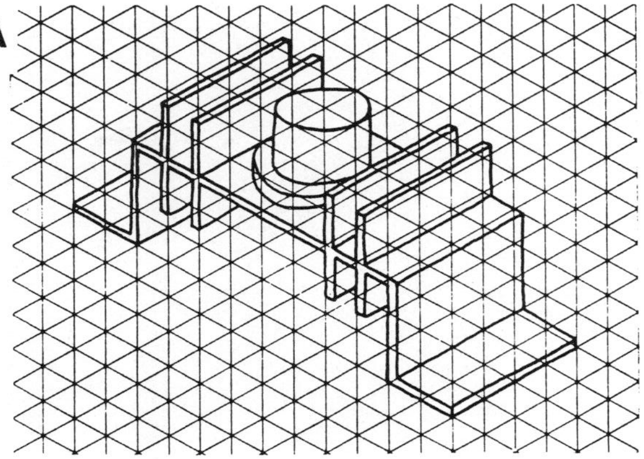

B

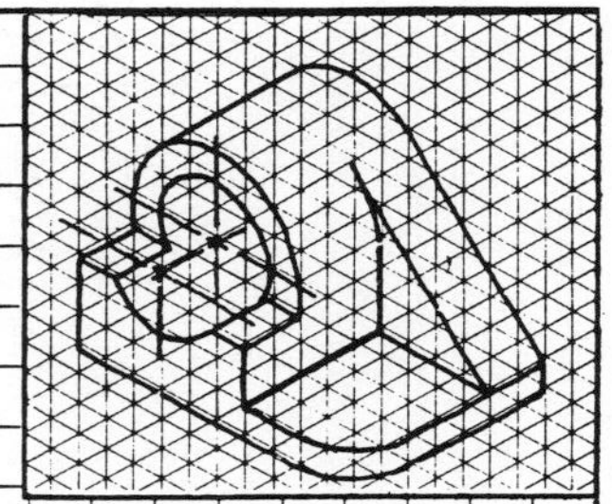

C

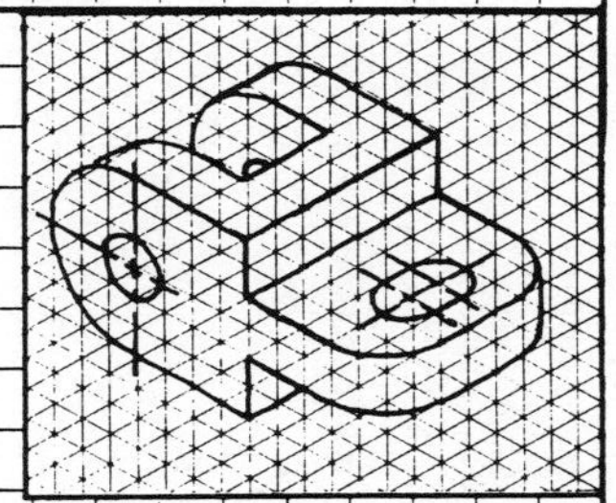

Ch5 Multiview Drawings

Section-Chapter-Problem	Multiview Drawing Problems: Freehand sketching	Course__________ Section_____	Problem
2 5 5	Figures 3-17, 5-19	Student________________ Date______	2-5-5

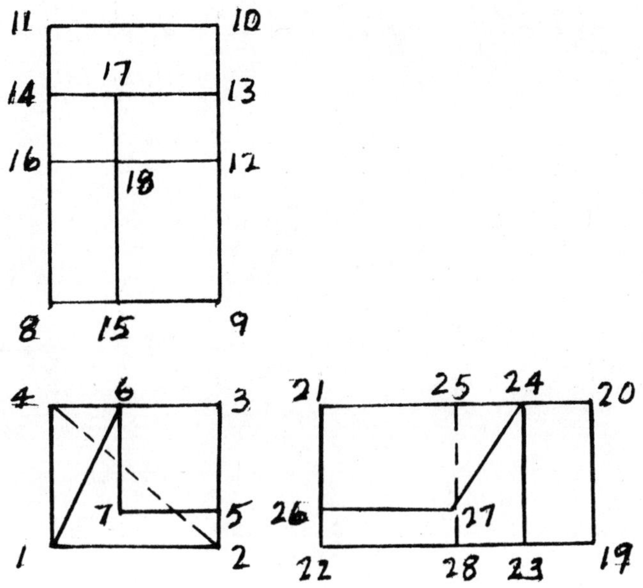

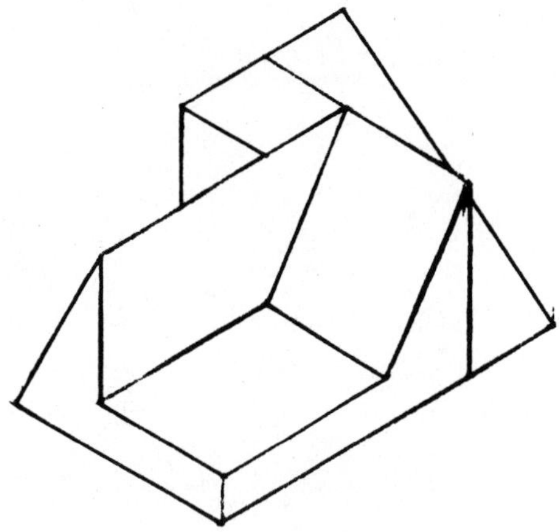

Use numbers to describe surfaces and answer the following:

(1. an 2. are examples:)

1.	**Surface**	**1-2-5-7-6**	**in the F view is**	**8-15-9**	**in the T view**
2.	**"**	**15-9-12-18**	**in the T view is**	**26-27**	**in the S view**
3.	**"**	**18-12-13-17**	**T**	**is__________**	**F**
4.	**"**	**18-12-13-17**	**T**	**is__________**	**S**
5.	**"**	**4-2**	**F**	**is__________**	**T**
6.	**"**	**1-6**	**F**	**is__________**	**S**
7.	**"**	**1-6**	**F**	**is__________**	**T**
8.	**"**	**16-18-17-14**	**T**	**is__________**	**S**
9.	**"**	**25-28**	**S**	**is__________**	**F**
10.	**"**	**25-28**	**S**	**is__________**	**T**
11.	**"**	**6-7**	**F**	**is__________**	**T**
12.	**"**	**6-7**	**F**	**is__________**	**S**
13.	**"**	**23-24**	**S**	**is__________**	**T**
14.	**"**	**23-24**	**S**	**is__________**	**F**
15.	**"**	**4-2**	**F**	**is__________**	**S**

Ch5 Multiview Drawings			**Problem**
Section-Chapter-Problem 2 5 6	Inclined surfaces: Reading Drawings Figures 5-23, 5-36	Course____________Section_____ **Student**____________________Date______	**2-5-6**

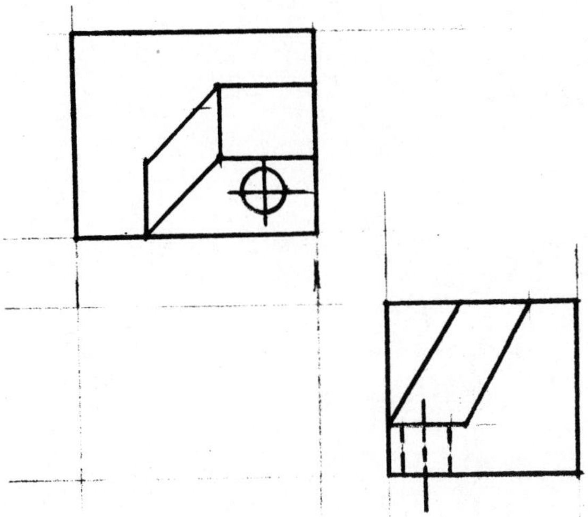

A

B

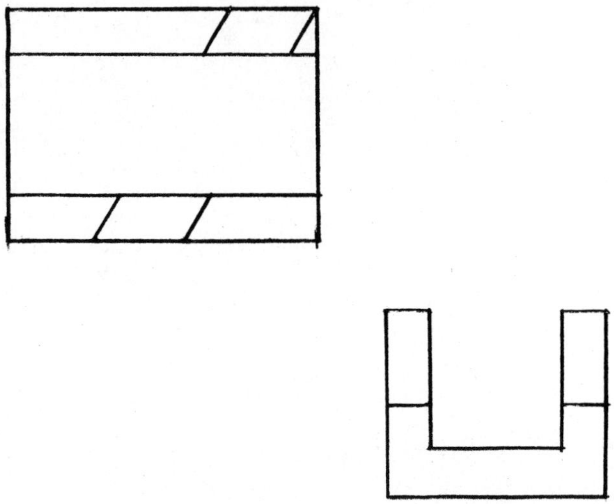

C

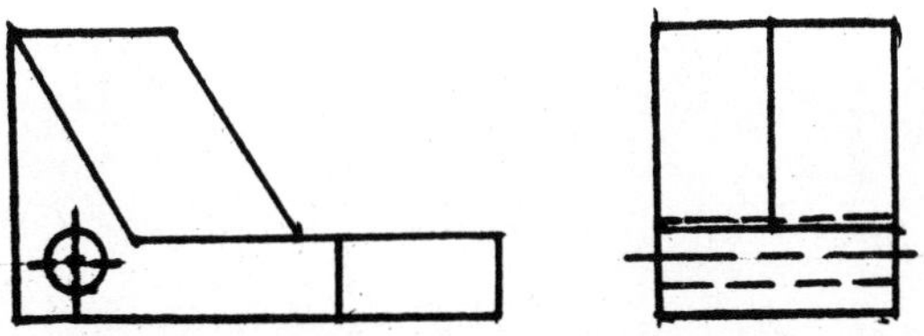

D

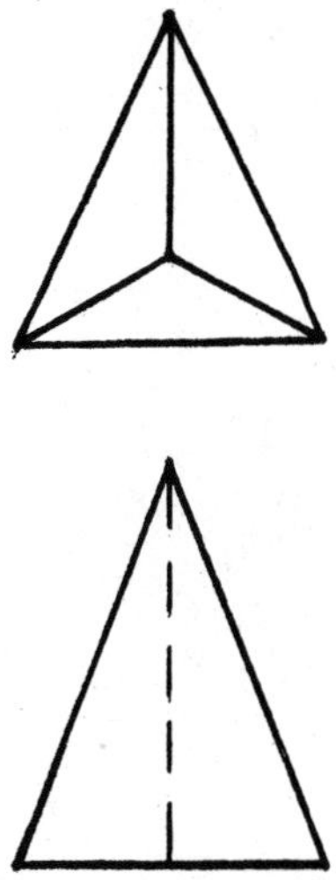

Complete the missing F, T, or S views.
Show all hidden lines.
Holes are only curved surfaces.

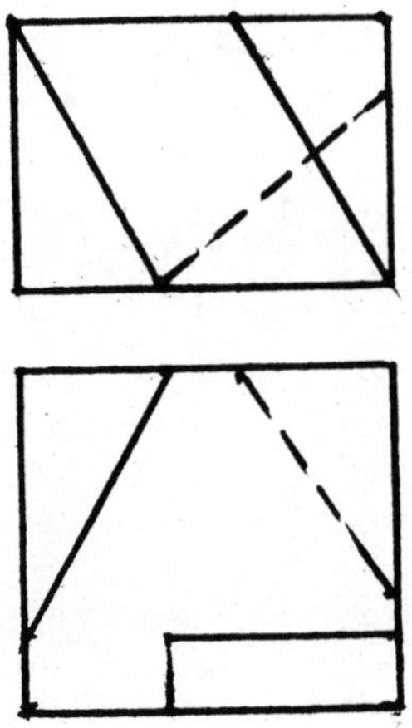

E

F

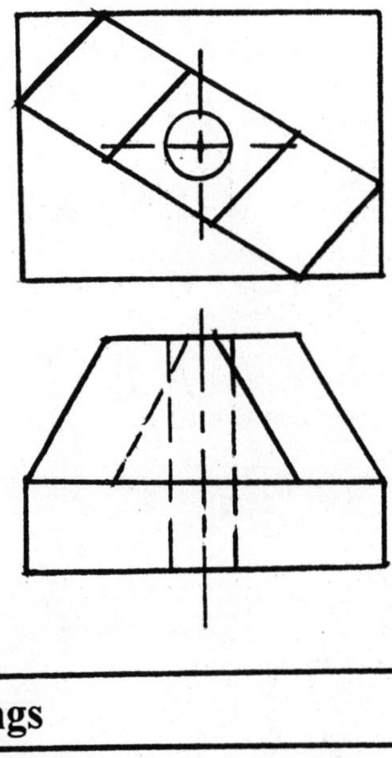

Ch5 Multiview Drawings			Problem
Section-Chapter-Problem 2 5 7	Inclined & Oblique surfaces Figures 5-16, 5-17 & 5-23	Course____________Section_____ **Student**____________________Date______	2-5-7

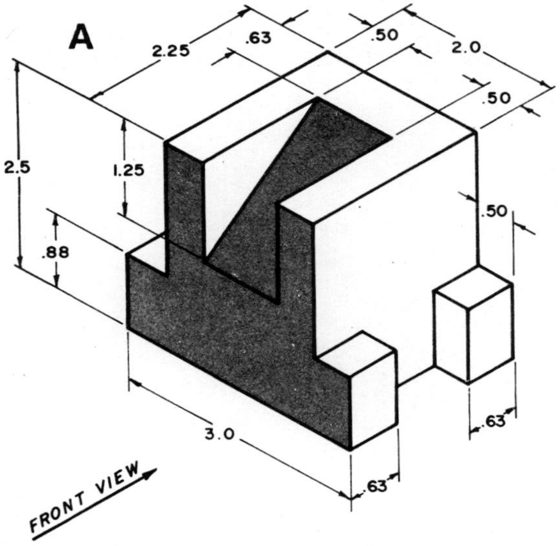

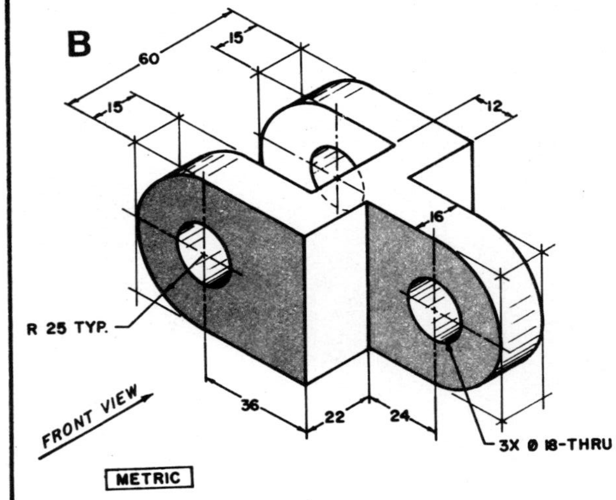

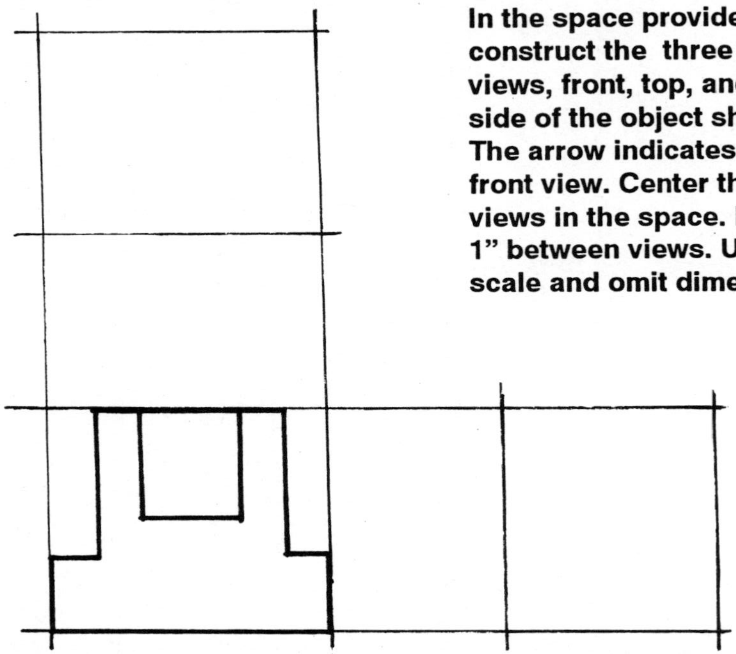

In the space provided construct the three main views, front, top, and right side of the object shown. The arrow indicates the front view. Center the views in the space. Leave 1" between views. Use half scale and omit dimensions.

In the space provided construct the three main views, front, top, and right side of the object shown. The arrow indicates the front view. Center the views in the space. Leave 1" between views. Use half scale and omit dimensions.

Ch5 Multiview Drawings

Section-Chapter-Problem	Centering orthographic views	Course________ Section____	Problem
2 5 8	Figures 5-33 & 5-34	Student________ Date____	2-5-8

A

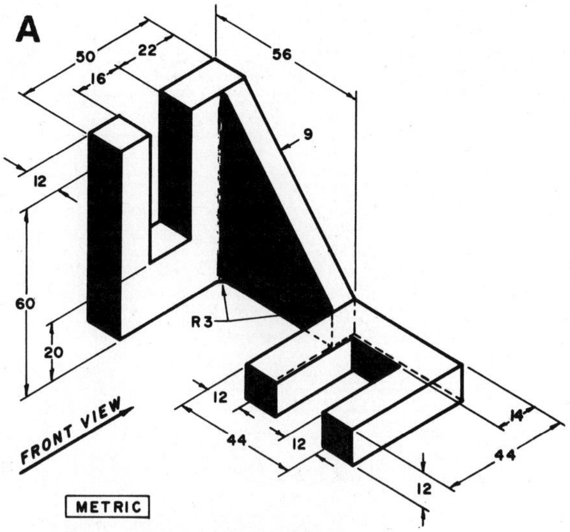

In the space provided construct the three main views, front, top, and right side of the object shown. The arrow indicates the front view. Center the views in the space. Leave 1" between views. Use half scale and omit dimensions.

B

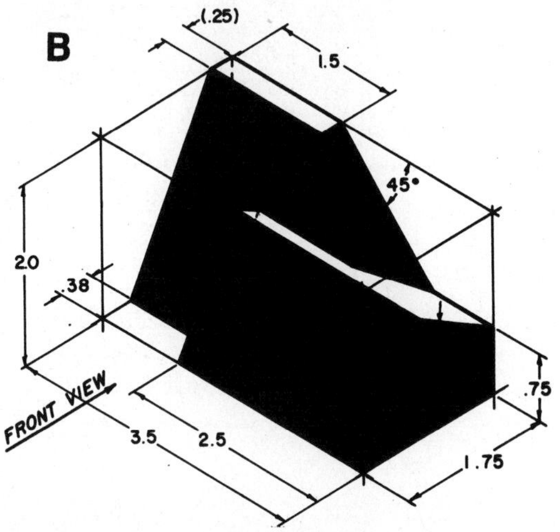

In the space provided construct the three main views, front, top, and right side of the object shown. The arrow indicates the front view. Center the views in the space. Leave 1" between views. Use half scale and omit dimensions.

Ch5 Multiview Drawings			Problem
Section-Chapter-Problem 2 5 9	Centering orthographic views Figures 5-33 & 5-34	Course__________ Section_____ Student______________ Date_____	2-5-9

A

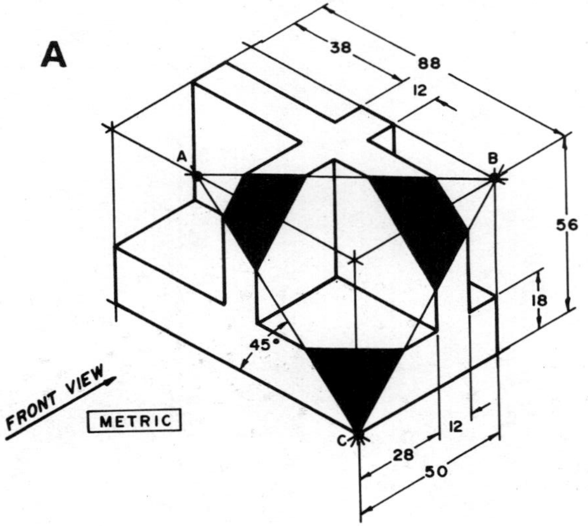

In the space provided construct the three main views, front, top, and right side of the object shown. The arrow indicates the front view. Center the views in the space. Leave 1" between views. Use half scale and omit dimensions.

B

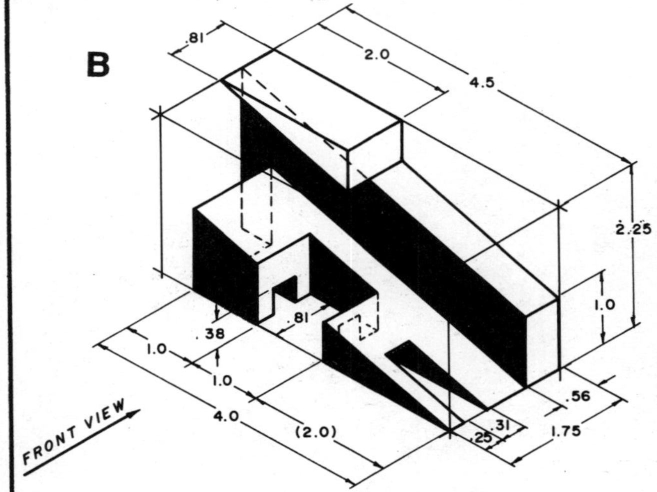

In the space provided construct the three main views, front, top, and right side of the object shown. The arrow indicates the front view. Center the views in the space. Leave 1" between views. Use half scale and omit dimensions.

Ch5 Multiview Drawings

Section-Chapter-Problem			Centering orthographic views Figures 5-33 & 5-34	Course________ Section____ Student__________ Date____	Problem
2	5	10			2-5-10

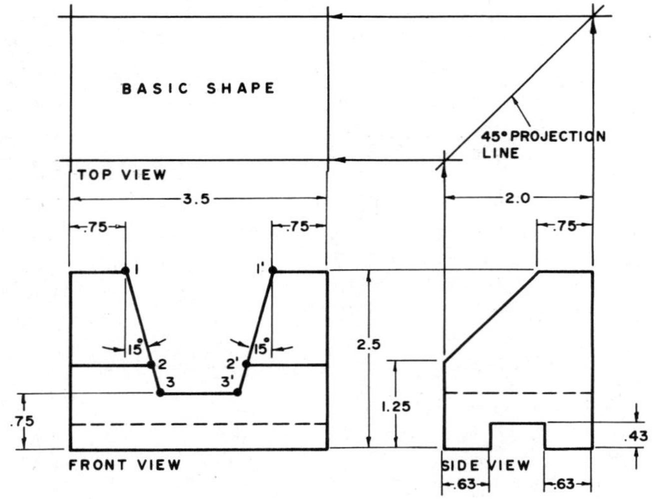

In the space provided construct the three main views, front, top, and right side of the object shown using the FIRST angle projection system.

In the space provided construct the three main views, front, top, and right side of the object shown using the THIRD angle projection system.

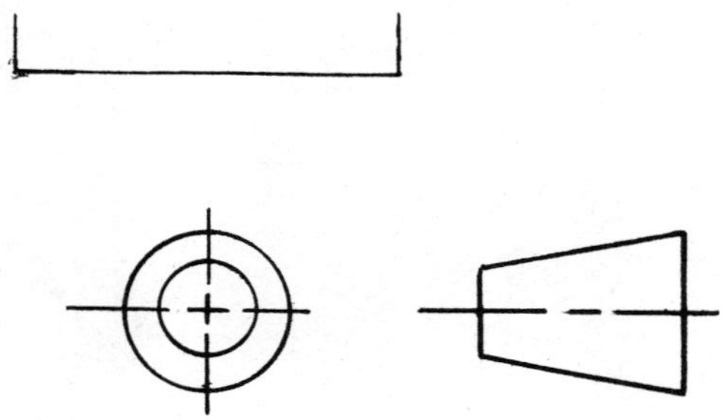

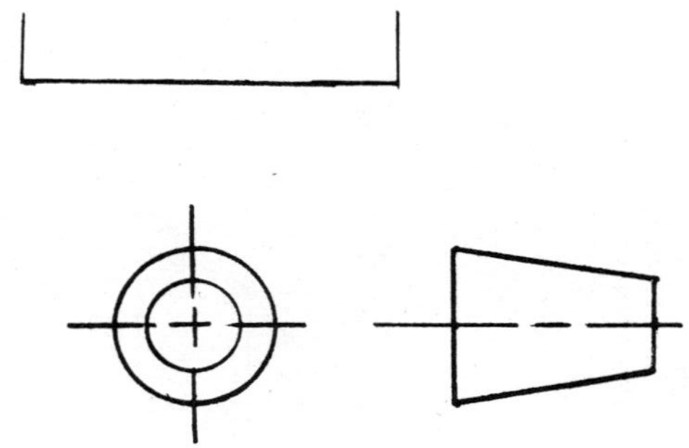

Ch5 Multiview Drawings

Section-Chapter-Problem	Orthographic projections:	Course______ Section______	Problem
2 5 11	Third Angle Figure 5-71 (A), First Angle Figure 5-71	Student______ Date______	2-5-11

A

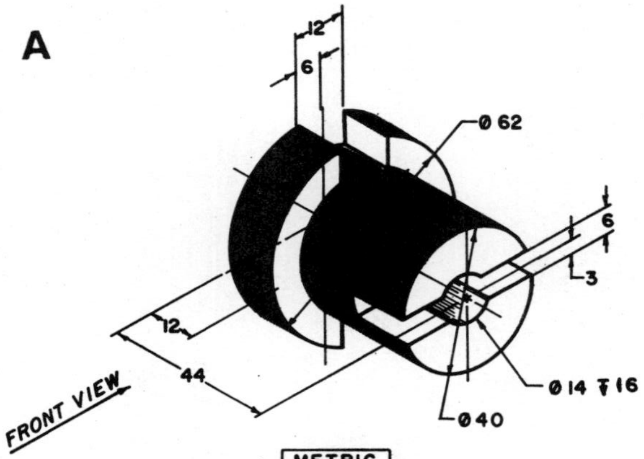

METRIC

In the space provided construct the three main views, front, top, and right side of the object shown.

The arrow indicates the front view. Center the views in the space. Leave 1" between views. Use half scale and omit dimensions.

B

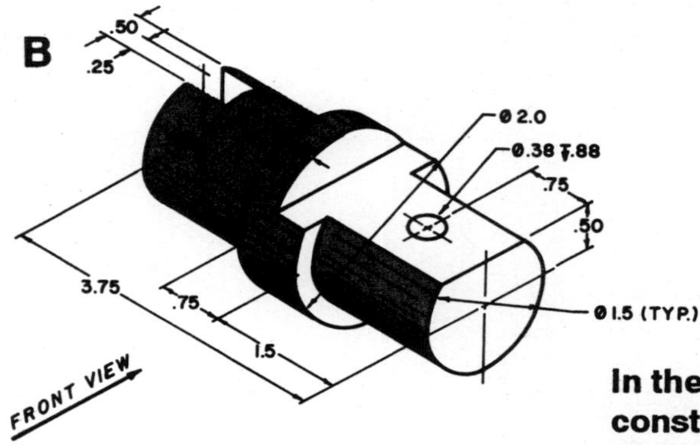

In the space provided construct the three main views, front, top, and right side of the object shown. The arrow indicates the front view. Center the views in the space. Leave 1" between views. Use half scale and omit dimensions.

Ch5 Multiview Drawings

Section-Chapter-Problem	Centering orthographic views	Course______ Section____	Problem
2 5 12	Figures 5-33 & 5-34	Student______ Date____	2-5-12

A

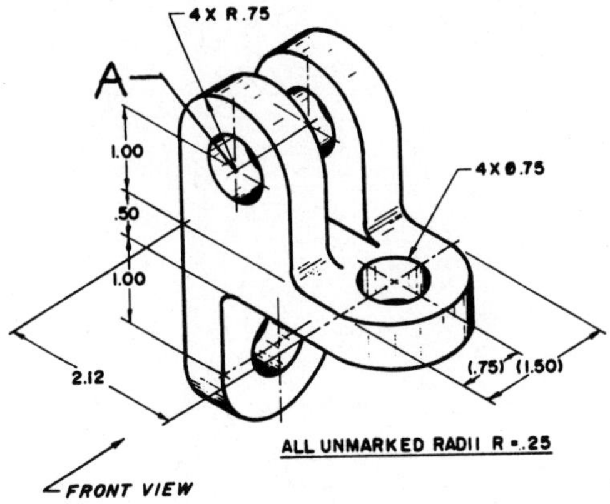

In the space provided construct the three main views, front, top, and right side of the object shown. Locate the front view as indicated. Leave 1" between views. Use half scale and omit dimensions.

B

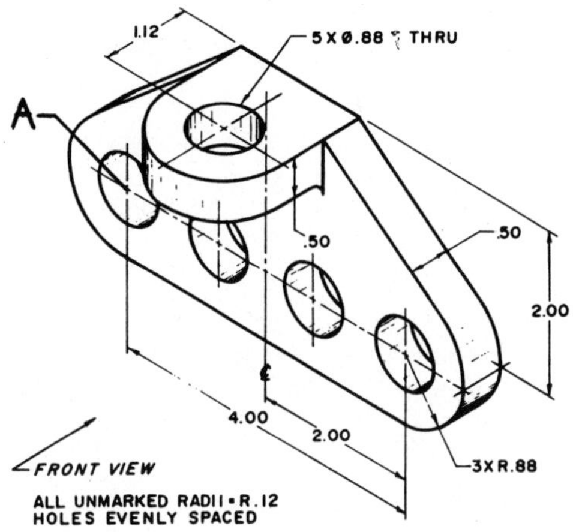

In the space provided construct the three main views, front, top, and right side of the object shown. Locate the front view as indicated. Leave 1" between views. Use half scale and omit dimensions.

Ch5 Multiview Drawings			Problem
Section-Chapter-Problem 2 5 13	Orthographic views Figures 5-15 & 5-17	Course__________ Section_____ Student_______________ Date______	2-5-13

A

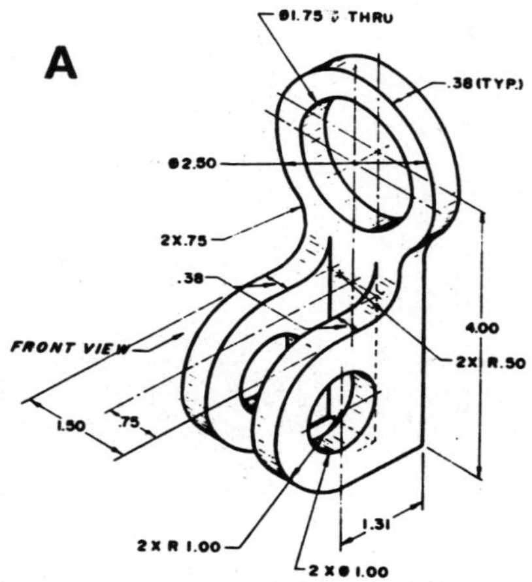

In the space indicated construct the three main views, front, top, and right side of the object shown. The arrow indicates the front view. Leave .6" between views. Use half scale and omit dimensions.

B

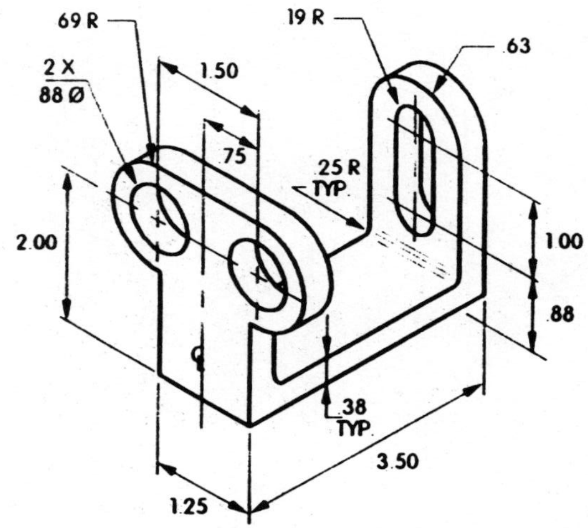

In the space indicated construct four views, front, top, right side, and left side of the object shown. Leave .6" between views. Use half scale and omit dimensions.

Ch5 Multiview Drawings				Problem
Section-Chapter-Problem 2 5 14	Orthographic views Figures 5-15 & 5-17	Orthographic views Figures 5-15, 5-17, & 5-58	Course_____ Section_____ Student_____ Date_____	2-5-14

In the space indicated construct four views, front, top, right side, and left side of the object shown. Leave 1" between views and center the views in the space provided. Use half scale and omit dimensions.

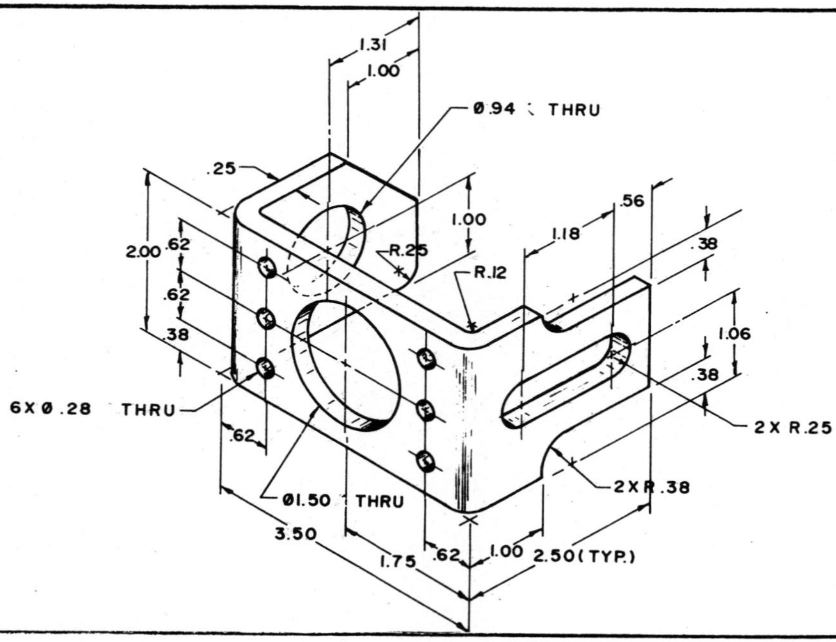

Ch5 Multiview Drawings

Section-Chapter-Problem	Orthographic views: better practices	Course______ Section______	Problem
2 5 15	Figures 5-15, 5-17, & 5-58	Student______ Date______	2-5-15

Complete the Top view using correct line types and widths. Do not erase constructon lines.

A

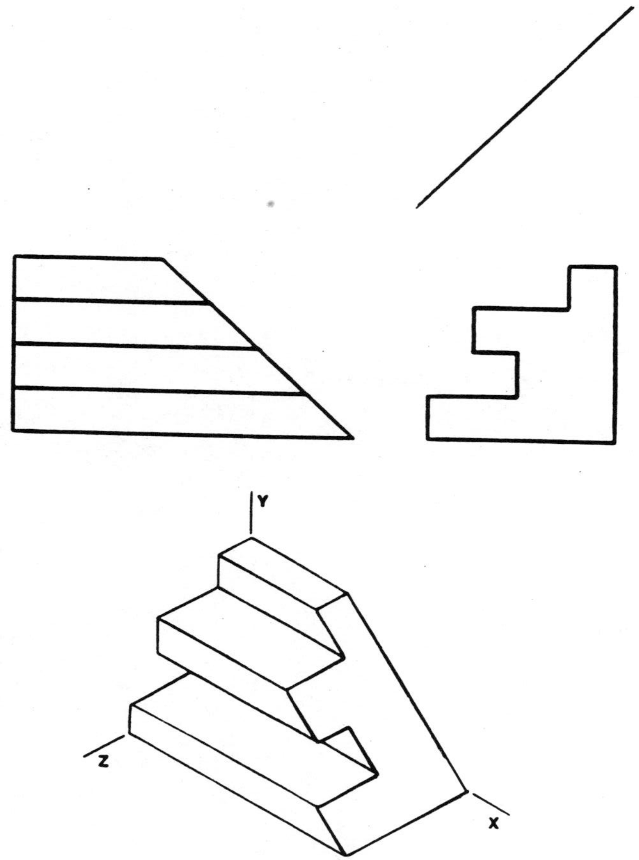

B

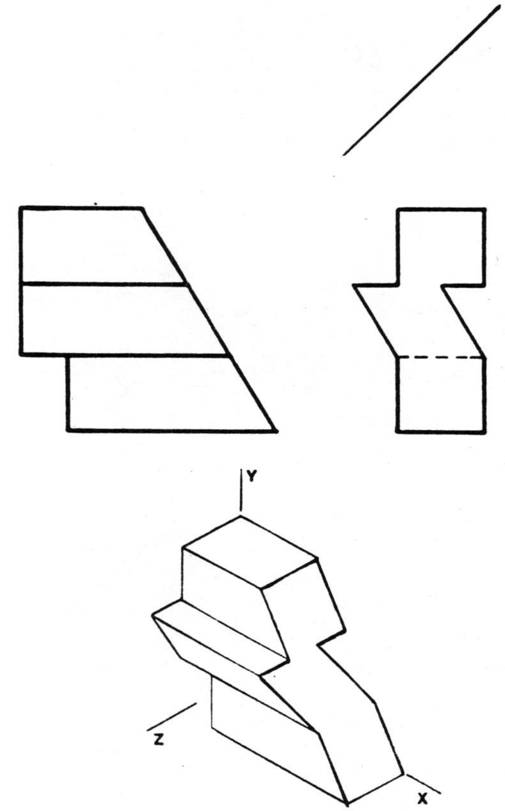

Ch5 Multiview Drawings

Section-Chapter-Problem	Orthographic views: Inclined surfaces	Course____________Section_____	Problem
2 5 16	Figures 5-15, 5-17 & 5-23	Student____________________Date______	2-5-16

Complete the Top view using correct line types and widths. Do not erase constructon lines.

A

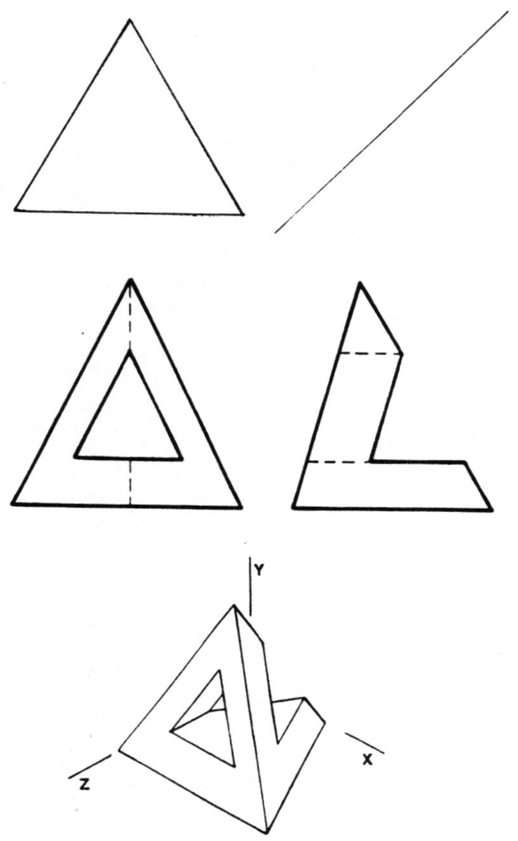

Complete the Front view using correct line types and widths. Do not erase constructon lines.

B

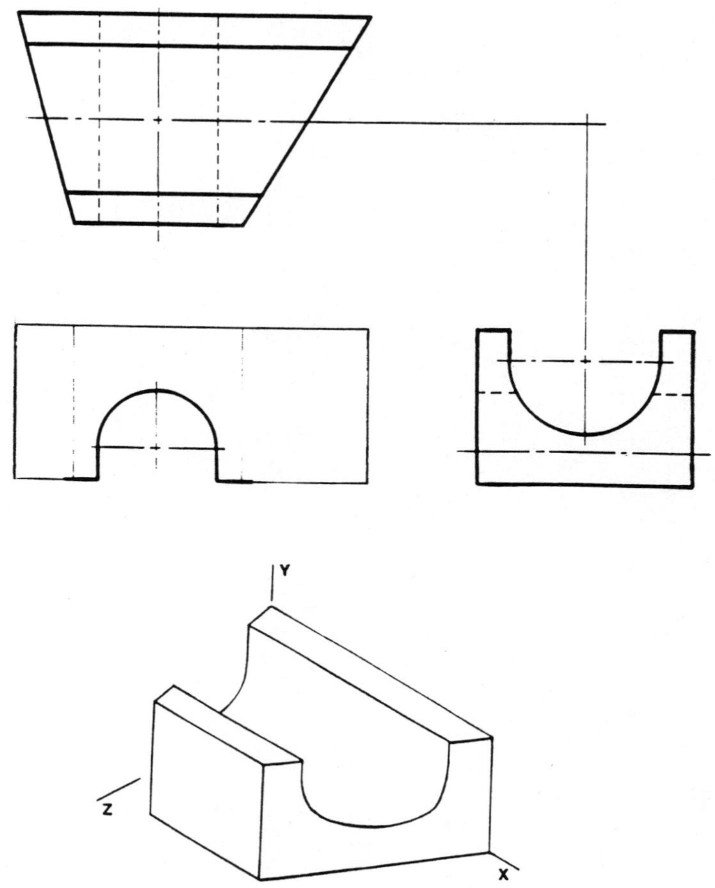

Ch5 Multiview Drawings

Section-Chapter-Problem	Orthographic views: Oblique, Inclined, and Curved surfaces	Course________ Section____	Problem
2 5 17	Figures 5-15, 5-17, 5-23, & 5-51	**Student**________ Date____	2-5-17

A

Complete the Front view using correct line types and widths. Do not erase constructon lines.

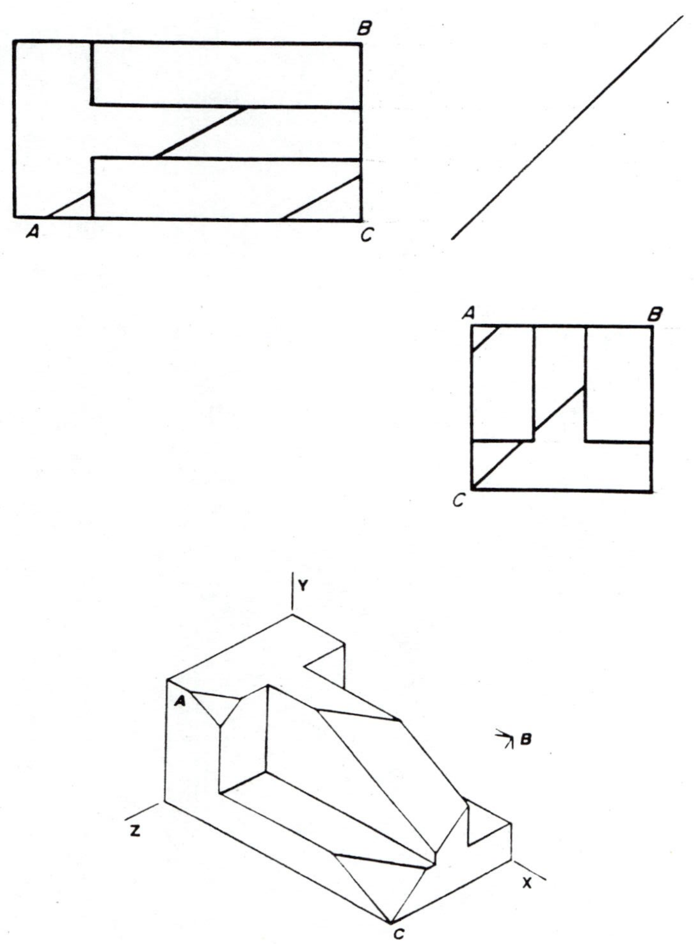

B

Complete the Right Side view using correct line types and widths. Do not erase constructon lines.

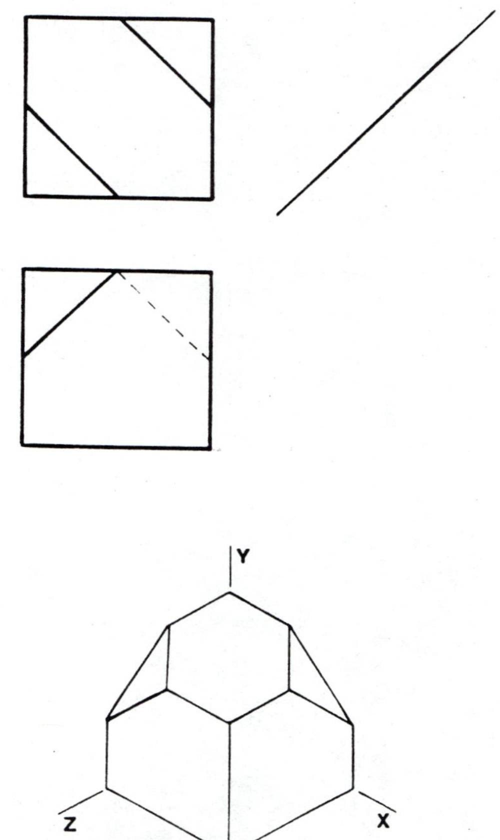

Ch5 Multiview Drawings

Section-Chapter-Problem	Orthographic views: Oblique surfaces	Course________ Section_____	Problem
2 5 18	Figures 5-15, 5-17, & 5-23	Student__________ Date_____	2-5-18

A

Complete the Right Side view using correct line types and widths. Do not erase constructon lines.

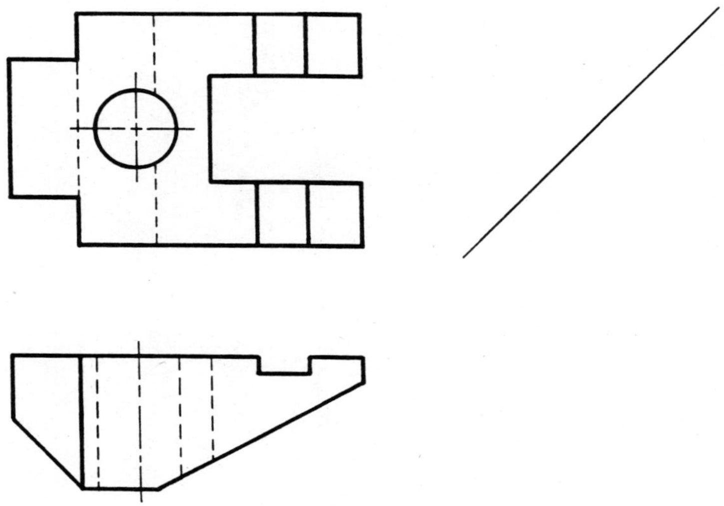

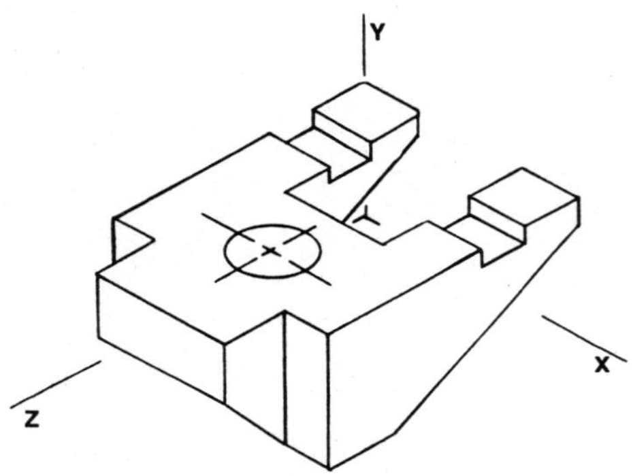

B

Complete the Top view using correct line types and widths. Do not erase constructon lines.

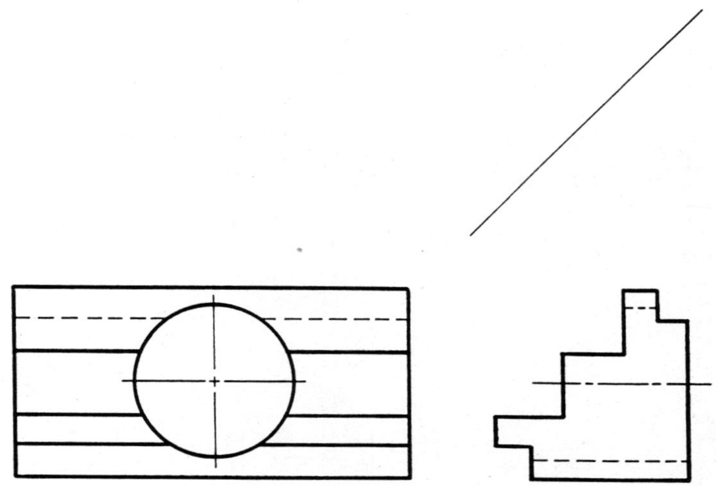

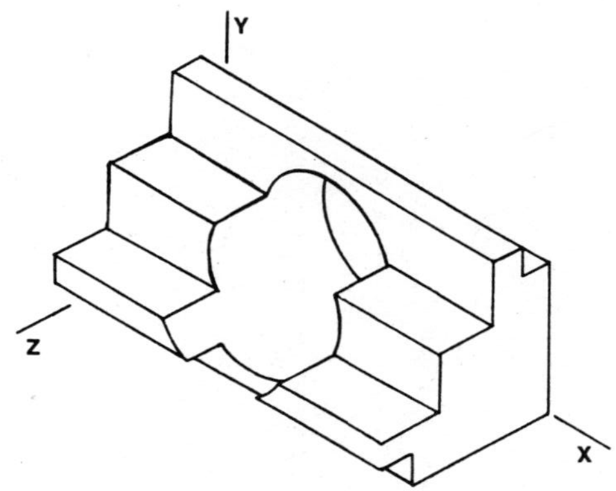

Ch5 Multiview Drawings			Problem
Section-Chapter-Problem 2 5 19	Orthographic views: Inclined and Curved surfaces Figures 5-15, 5-17, 5-23, & 5-51	Course____________Section_____ Student______________________Date______	2-5-19

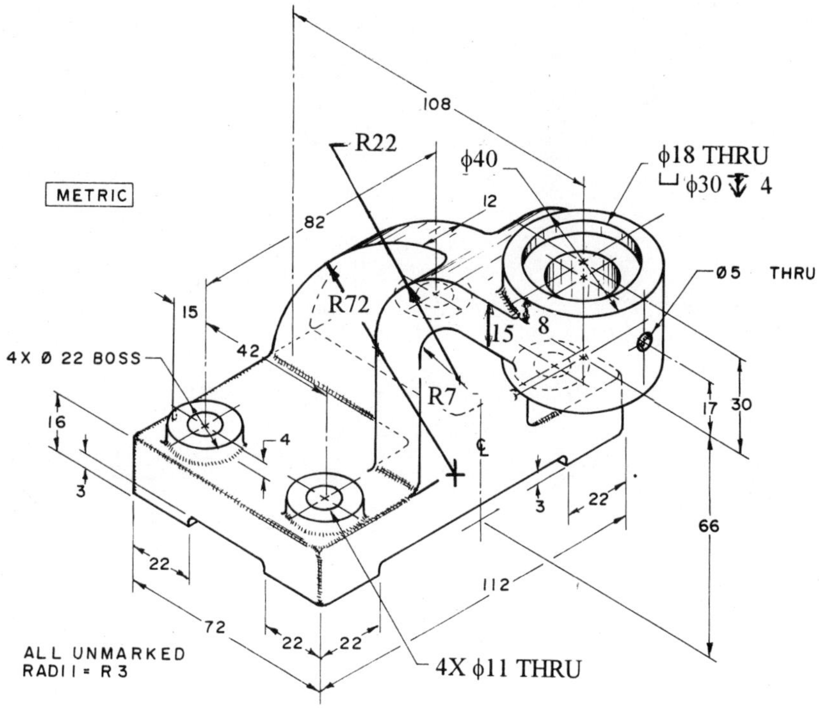

In the space provided construct the three main views, front, top, and right side of the object shown. The curve of the curved neck shows in the Front view. Center the views in the space. Leave 1" (25mm) between views. Use half scale and omit dimensions.

Ch5 Multiview Drawings			Problem
Section-Chapter-Problem	Centering orthographic views	Course____________Section_____	2-5-20
2 5 20	Figures 5-33 & 5-34	Student___________________Date______	

In the space provided construct the three main views, front, top, and right side of the object shown. View the isometric from the left for the Front view. Center the views in the space. Leave 1" between views. Use half scale and omit dimensions.

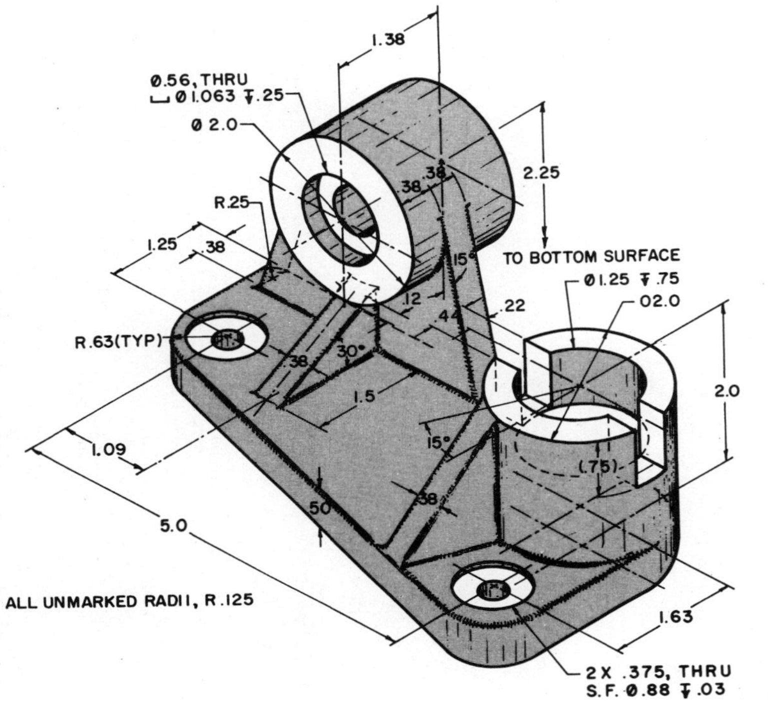

Ch5 Multiview Drawings			**Problem**
Section-Chapter-Problem 2 5 21	Centering orthographic views Figures 5-33 & 5-34	Course__________Section_____ **Student**______________Date______	**2-5-21**

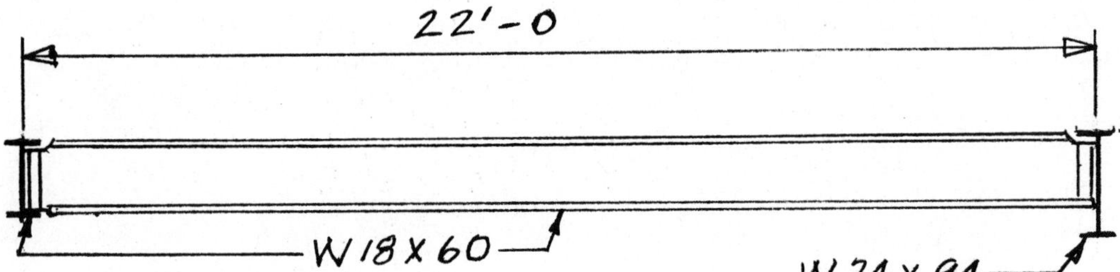

Using instruments, redraw the structural beam with the connections shown in the freehand sketch. Use the conventional Z break. Use a 1-1/2" = 1'-0 scale and include dimensions.

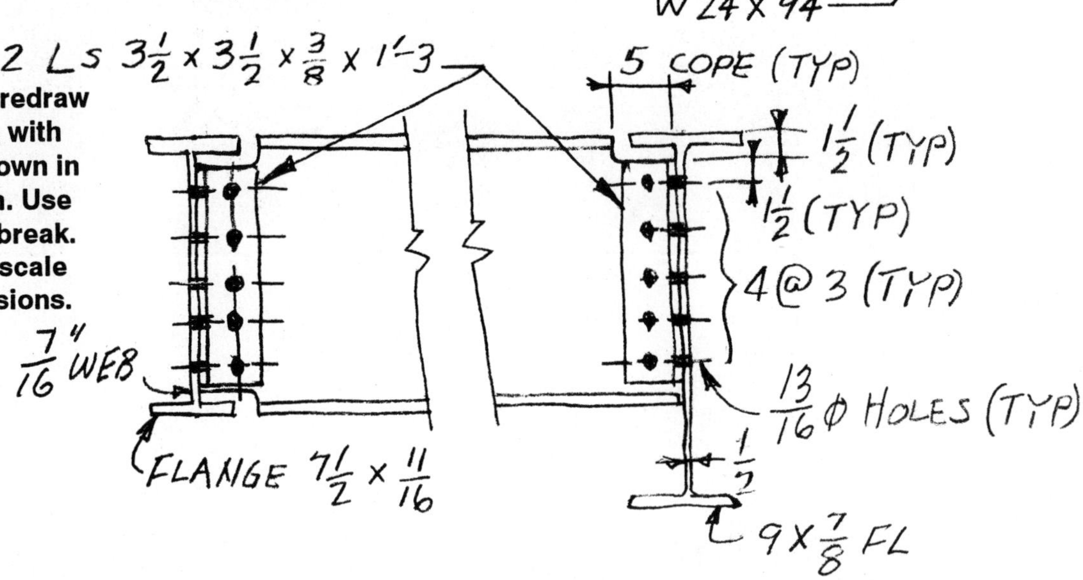

Ch5 Multiview Drawings			Problem
Section-Chapter-Problem	Structural drafting Figure 21-46	Course____________ Section_____	2-5-22
2 5 22	Figure 5-67	**Student**____________________ Date______	

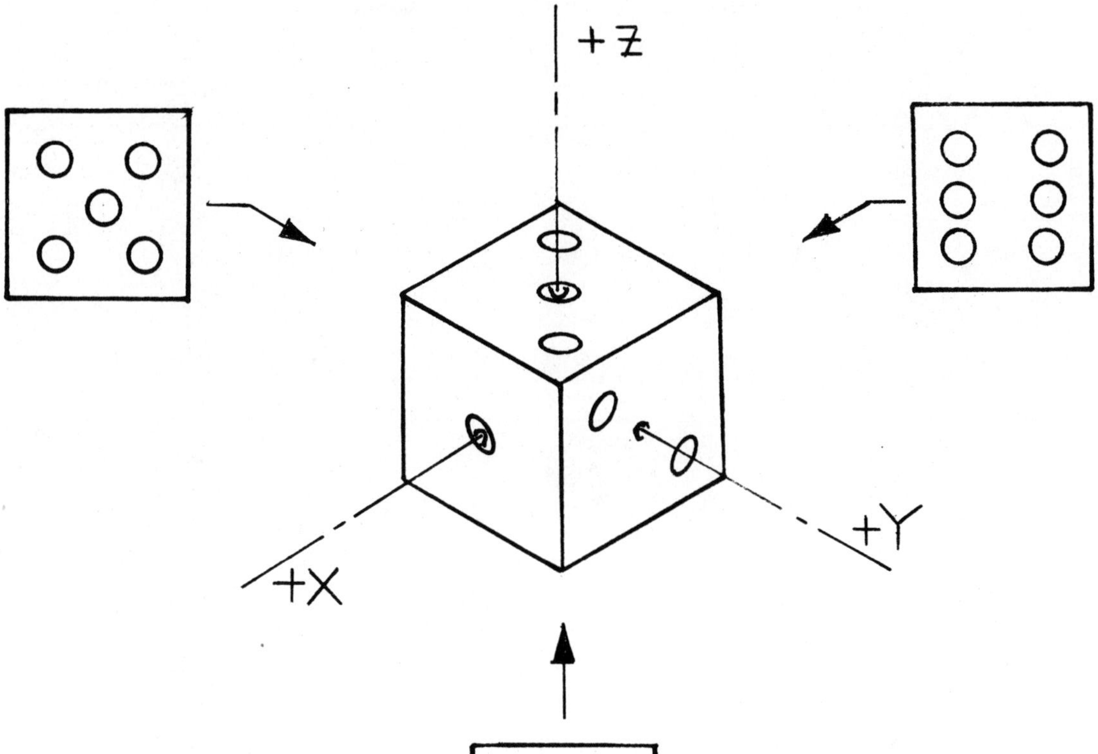

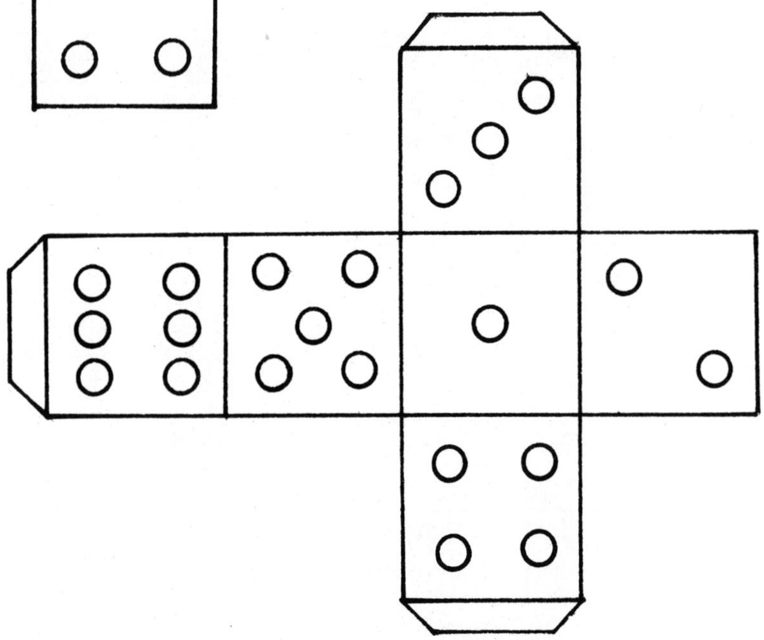

Study the die in the isometric drawing and then imagine the following rotations without looking at the isometric:

a. Rotate the die 90°CCW about the Z axis (looking toward the origin).

b. Then rotate the die 90°CW about the Y axis (looking toward the origin).

c. Finally, rotate the die 90°CCW about the X axis (looking toward the origin).

1. While facing the origin along the X axis, what number of dots do you see?
2. Which face of the die is facing up the Z axis?
3. Which face of the die is facing outward along the Y axis?

Cut out the flat pattern of the die, fold the views down 90°, tape the edges to form a cube, and then verify the result of the three rotations.

Ch5 Multiview Drawings			Problem
Section-Chapter-Problem 2 5 23	Visualization practice Figure 5-14	Course__________ Section_____ Student_______________ Date______	2-5-23

A An isometric drawing of a house is shown in drawing file "05-1.dxf. Draw the front, side, and top view of it in the space provided.

B The multiview object shown in file "05-2.dxf" is a drilling fixture used to locate the top and bottom hinge on a 6" wide by 12" long door that is on the bottom of a gun cabinet. The manufacturer is modifying the design and is therefore in need of a new fixture drawing. The new door will be 7" wide by 15" long and made of 5/8" red oak instead of ¾" pine. The hinges will therefore need to be made of a heavier gauge metal to support the heavier door. The edge of each hinge should be 1 ¼" from the top and bottom of the door respectively. Finally, the holes should be drilled slightly larger due to the fact that red oak is being used instead of pine. If necessary, visit a local hardware store and purchase a hinge that you feel would adequately support this door.

<table>
<tr><td colspan="3">Ch5 Multiview Drawings</td><td rowspan="2">Problem

2-5-24</td></tr>
<tr><td>Section-Chapter-Problem
2   5   24</td><td>House Views
Hinge Location</td><td>Course__________ Section_____
Student______________ Date_______</td></tr>
</table>

Complete the right side view making it into a full section.
Use correct line thickness.

A

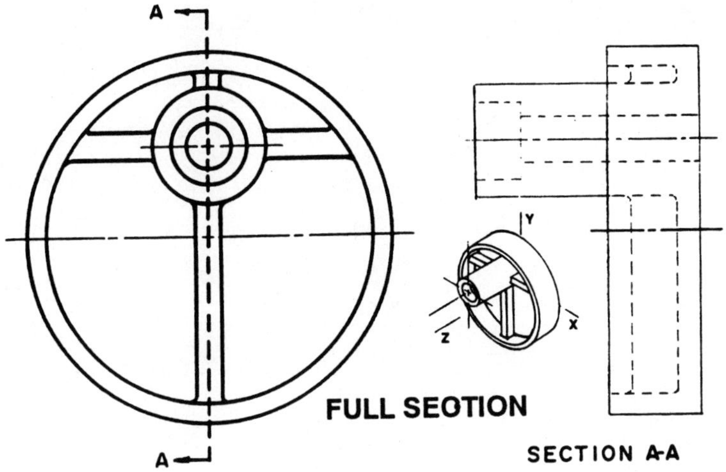

B

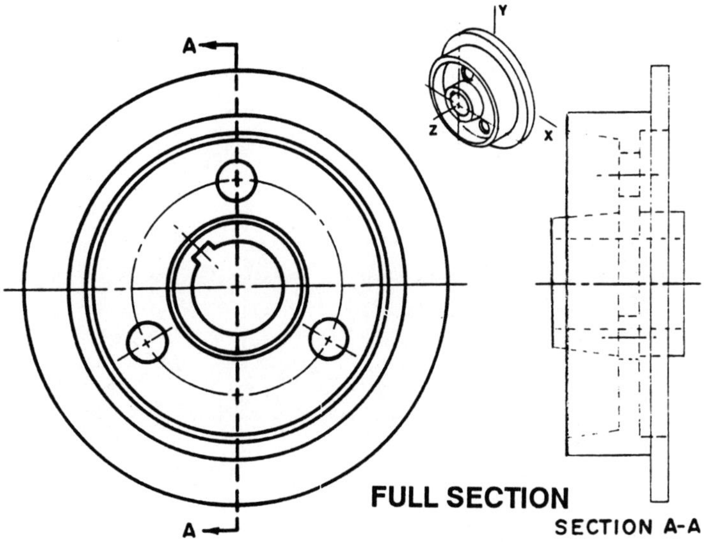

C

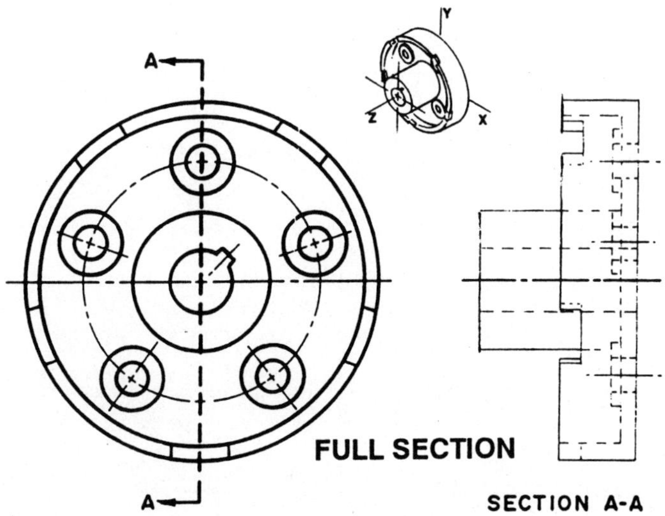

Ch6 Sectional Views			Problem
Section-Chapter-Problem 2 6 1	Full sections Figure 6-12	Course__________ Section_____ **Student**________________ Date______	**2-6-1**

A **Complete the right side view making it into a half section. Use correct line thickness.**

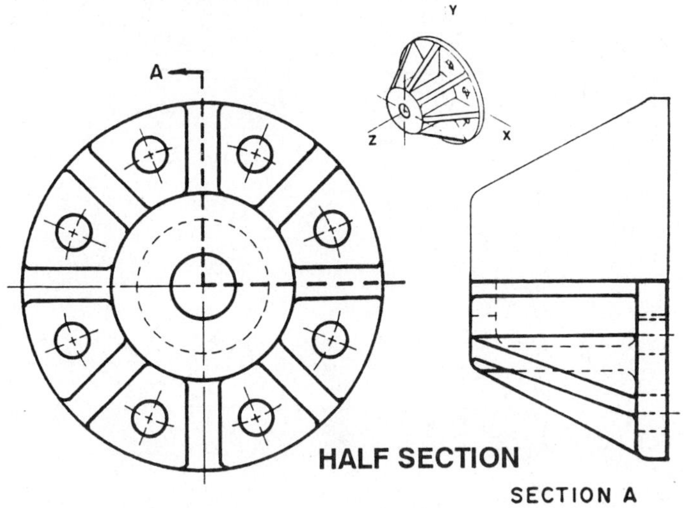

B **Complete the right side view making it into a half section. Use correct line thickness.**

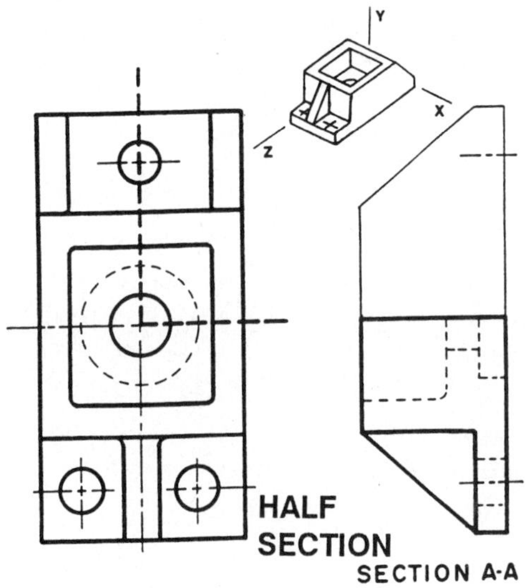

C **Make the front (top half) view into a broken-out section. Use correct line thickness.**

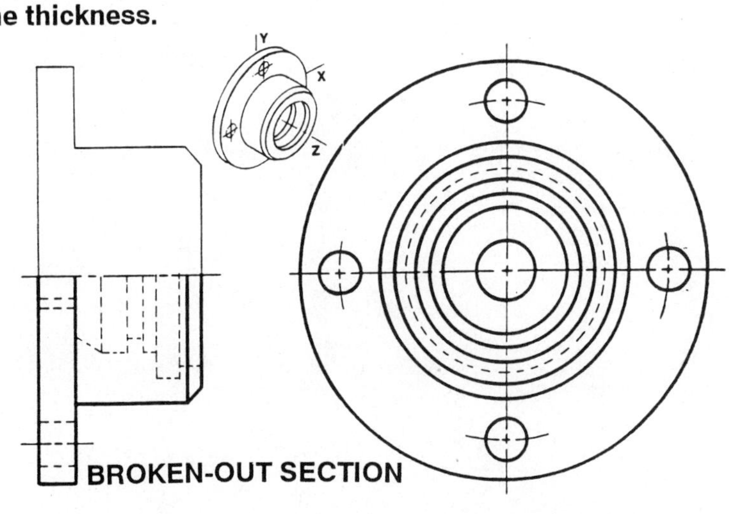

Ch6 Sectional Views			Problem
Section-Chapter-Problem 2 6 2	Half section Figures 6-17,18 Broken-out section Figure 6-20	Course__________ Section_____ **Student**______________ Date______	2-6-2

A Complete offset-removed section A-A and full section B-B.
Use correct line thickness.

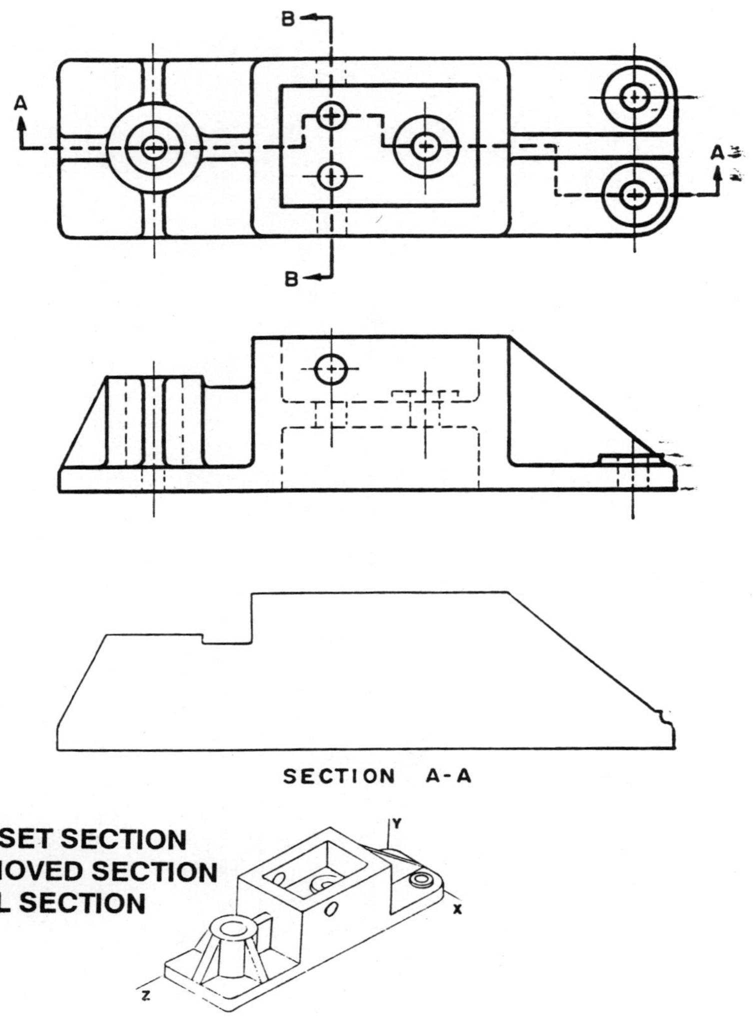

OFFSET SECTION
REMOVED SECTION
FULL SECTION

SECTION B-B

B Complete the right side view making it into a half section.
Use correct line thickness.

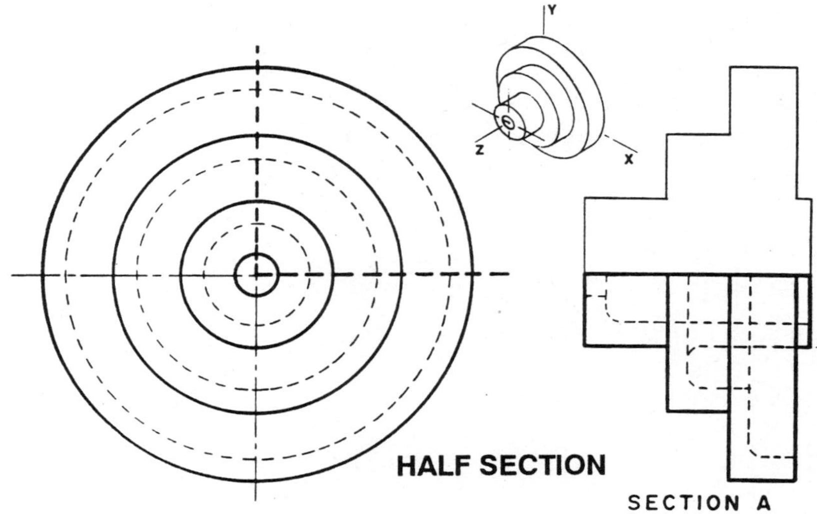

HALF SECTION

C Complete the front view making it into an offset section.
Use correct line thickness.

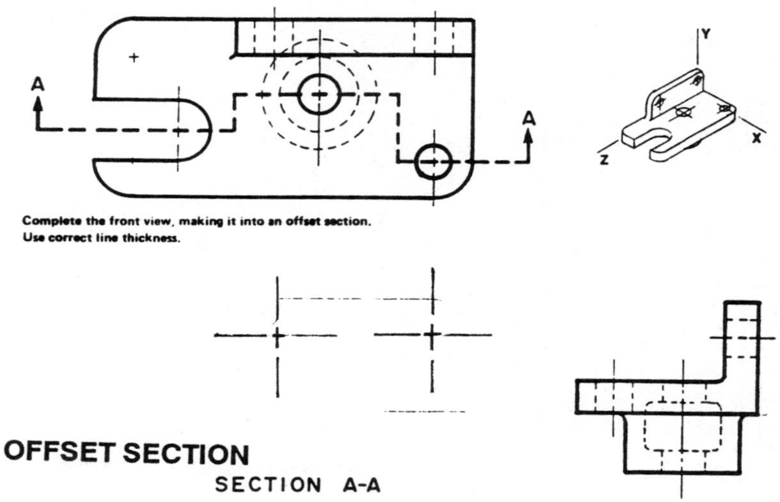

OFFSET SECTION

Ch6 Sectional Views

Section-Chapter-Problem	Offset sections Figures 6-14,15	Course______ Section______	Problem
2 6 3	Full & half sections Figures 6-12,17,18	Student______ Date______	2-6-3

A **Complete the front view making it into a full section. Use correct line thickness.**

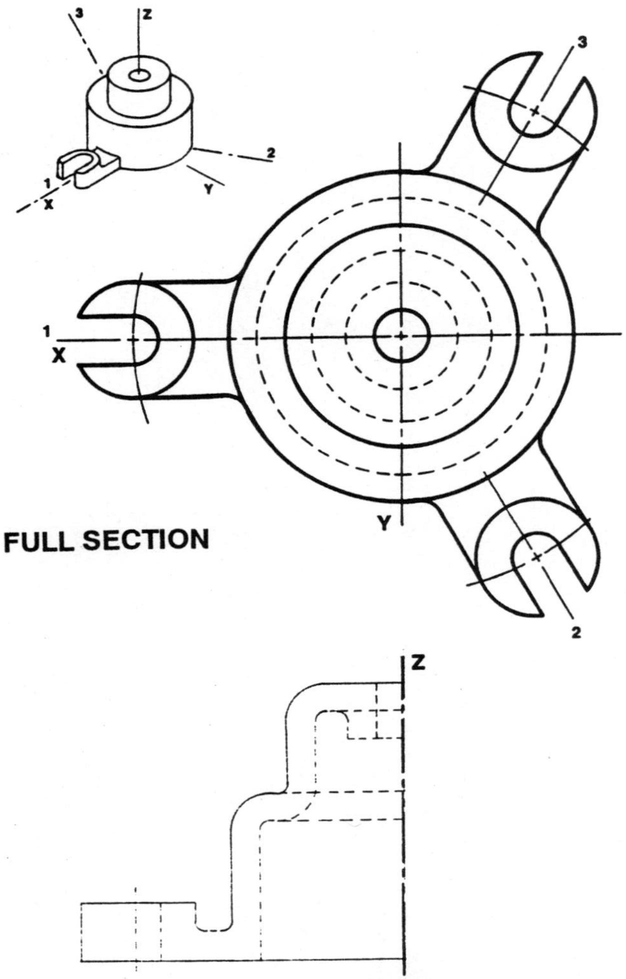

FULL SECTION

SECTION A-A

B **Complete removed sections A-A, B-B, & C-C Use correct line thickness.**

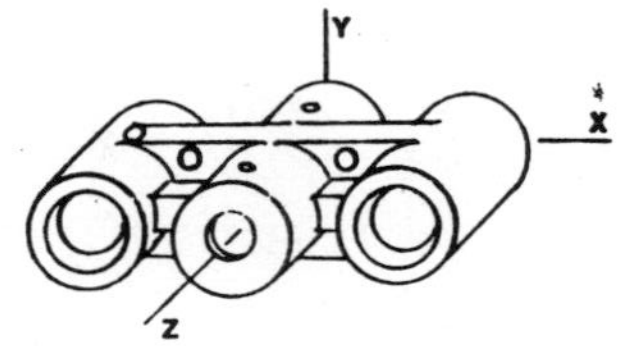

REMOVED SECTION

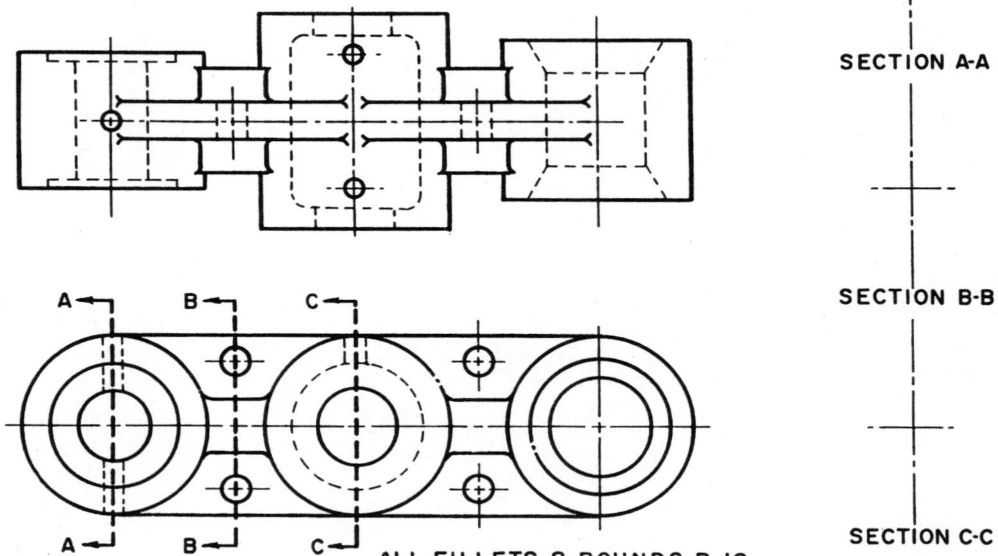

ALL FILLETS & ROUNDS R.12

Ch6 Sectional Views			Problem
Section-Chapter-Problem 2 6 4	Aligned sections Figure 6-50 Removed sectionsFigure 6-26,27	Course________ Section_____ Student__________ Date_____	2-6-4

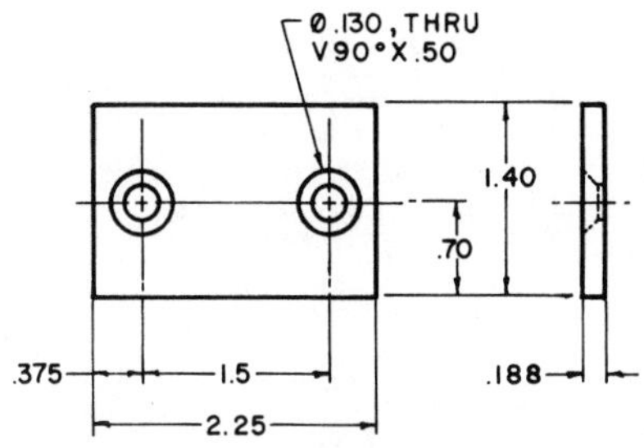

Part 6

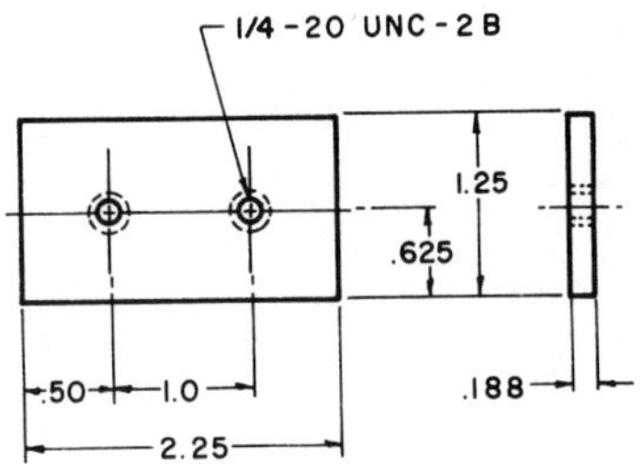

Part 7

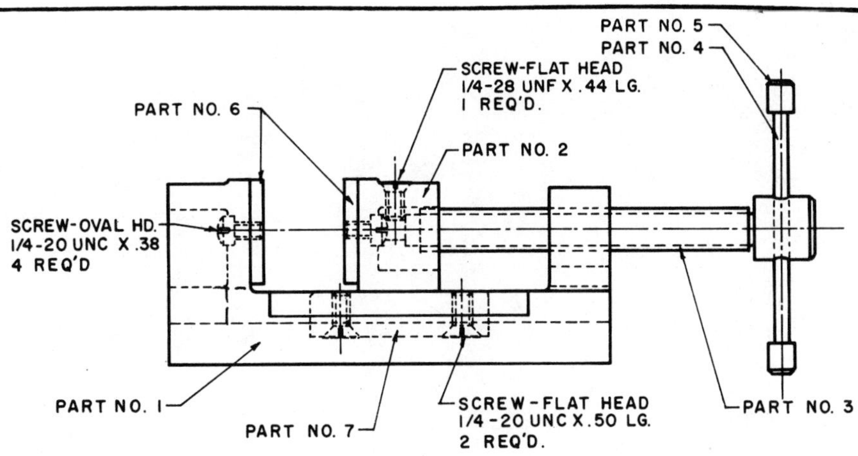

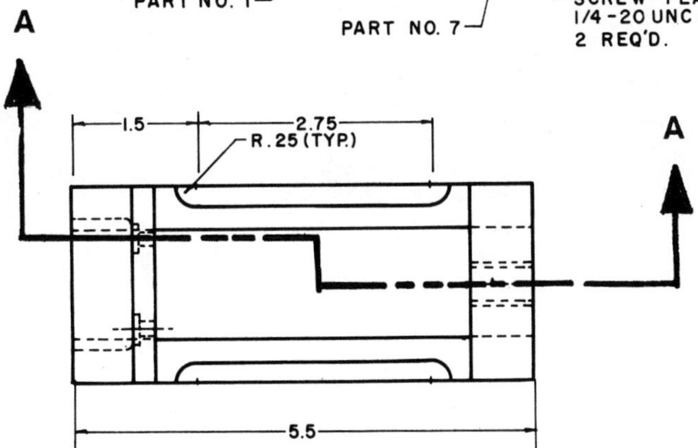

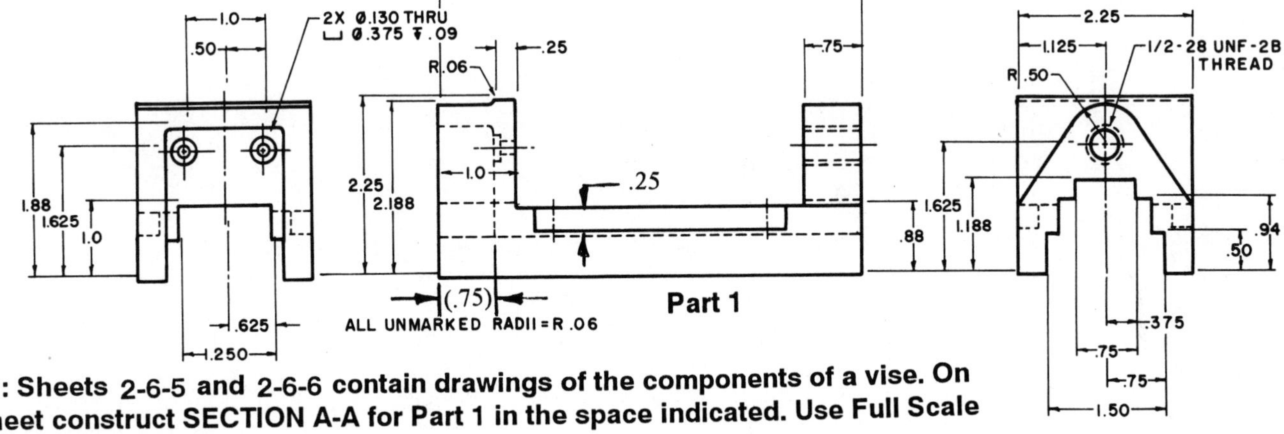

Part 1

GIVEN: Sheets 2-6-5 and 2-6-6 contain drawings of the components of a vise. On this sheet construct SECTION A-A for Part 1 in the space indicated. Use Full Scale and omit dimensions.

SECTION A-A

Ch6 Sectional Views			Problem
Section-Chapter-Problem 2 6 5	Offset section Figure 6-14	Course__________ Section_____ Student_______________ Date_____	2-6-5

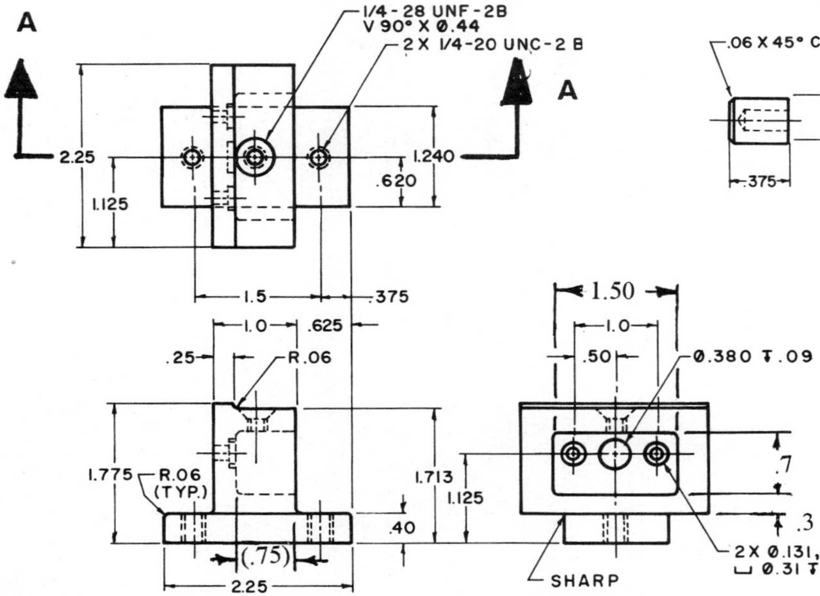

Part 2

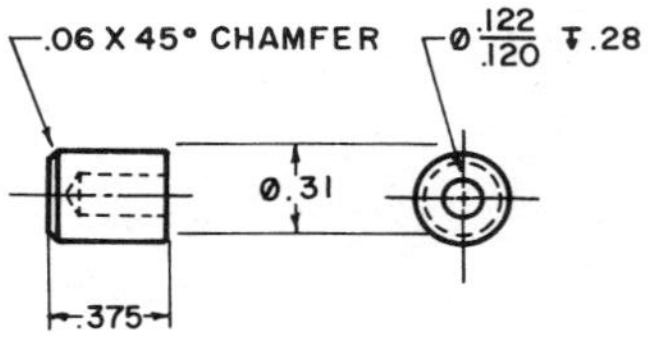

Part 5

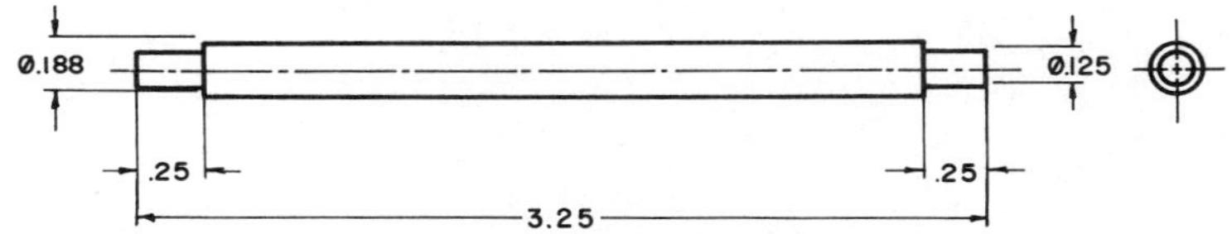

Part 4

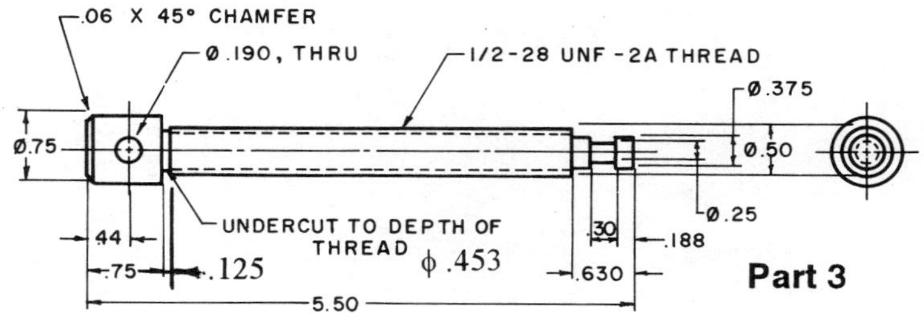

Part 3

On this sheet construct SECTION A-A for the Part 2 in the space indicated. Use Full Scale and omit dimensions.
Also construct DETAIL A to show the assembly of Parts 2 and 3. The format for Detail A is similar to a Section format ; however, a section cutting plane may or may not be used. DETAIL views are used on achitectural, civil, electrical, mechanical, and structural drawings.

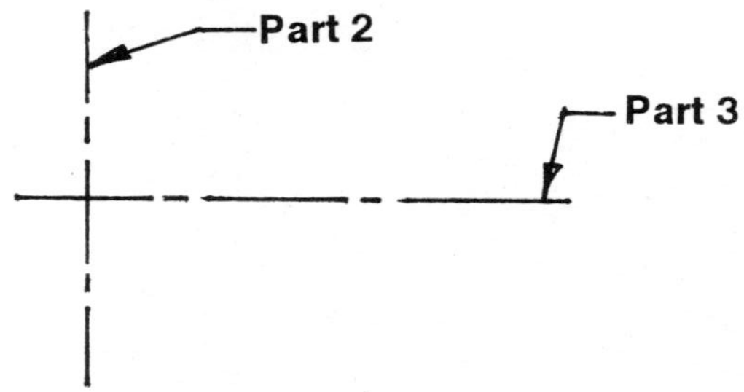

SECTION A-A

DETAIL A

Ch6 Sectional Views

Section-Chapter-Problem	Full secton Figure 6-12	Course____________Section_____	Problem
2 6 6	Detail view Figure 21-50	Student__________________Date______	2-6-6

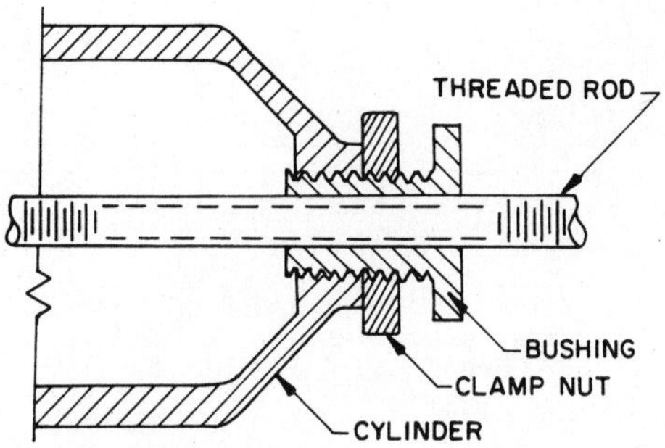

Example:
Section Assembly

In the space indicated construct to half scale a section assembly drawing of the packing gland shown in the freehand isometric sketch. Omit dimensions. Use correct line thickness. Show the metric fasteners specified.

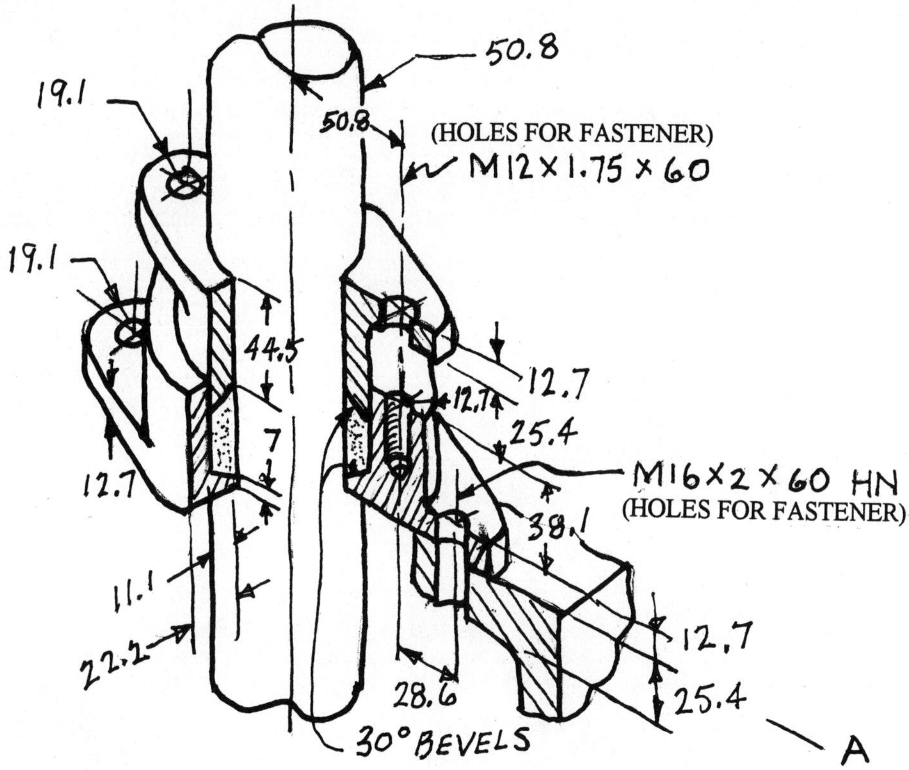

— A

Ch6 Sectional Views			Problem
Section-Chapter-Problem 2 6 7	Section Assembly Figure 6-32	Course_______ Section____ Student__________ Date____	2-6-7

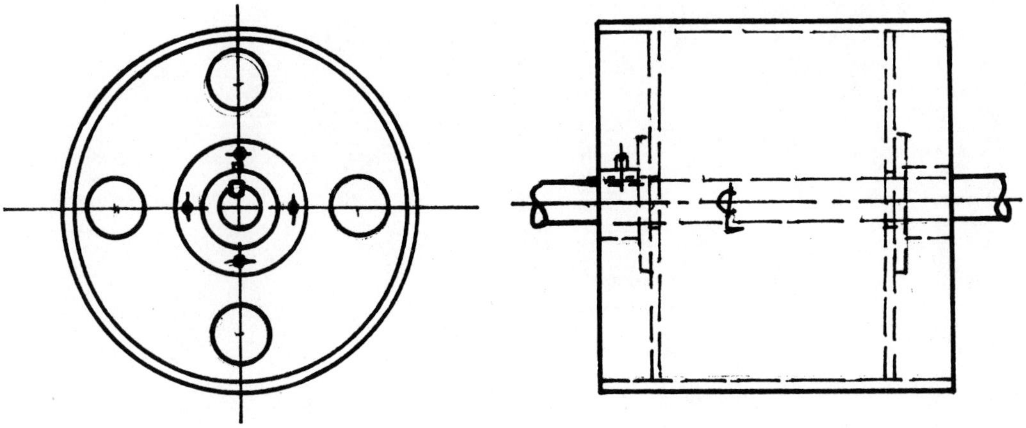

Belt Drive:
A square key held in place by a set screw transmits the turning moment from the shaft to the hub of the welded steel pulley shown in the upper two views. the hub drives the pulley through four bolts and the pulley drives a belt of a belt conveyor.

The example broken out section shown in the lower view is a similar shaft-key-hub-pulley assembly.

Prepare a broken out section 3X size of the upper pulley in the space indicated. Use the two view drawing for measurements and information. Note the access holes for assembly of the four bolts.

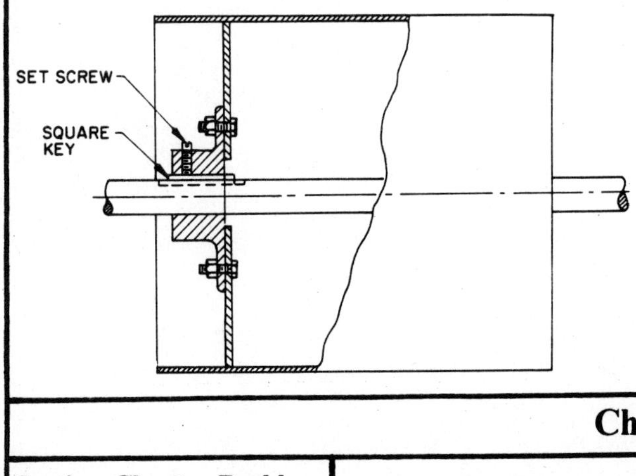

Ch6 Sectional Views			Problem
Section-Chapter-Problem 2 6 8	Broken out and Parts not sectioned Figures 6-21, 6-50, 6-51	Course__________ Section_____ Student______________ Date______	2-6-8

The freehand sketch below specifies edge radius requirements at three stations (cross sections) of a blow-torch nozzle. At each station along the nozzle the cross sectional area must be the same for best flow conditions. The cross sectional area at station 1 is:

$$\frac{\Pi(.8)^2}{4} = .503 sq.inches$$

Show the removed cross sections for each station (Similar to the example drawing.). Show inside dimensions for each cross section. Use full size. Show your calculations.

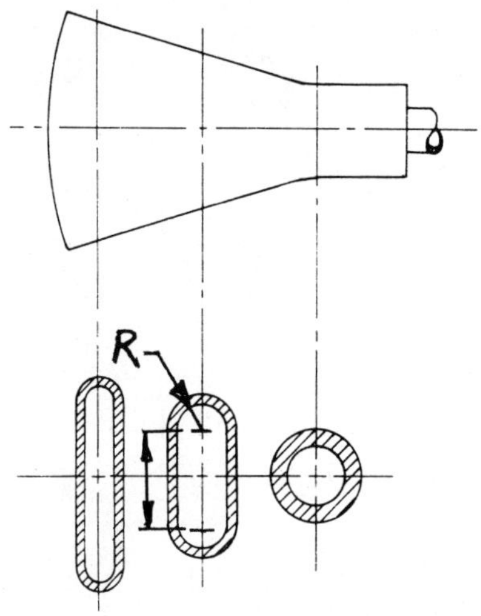

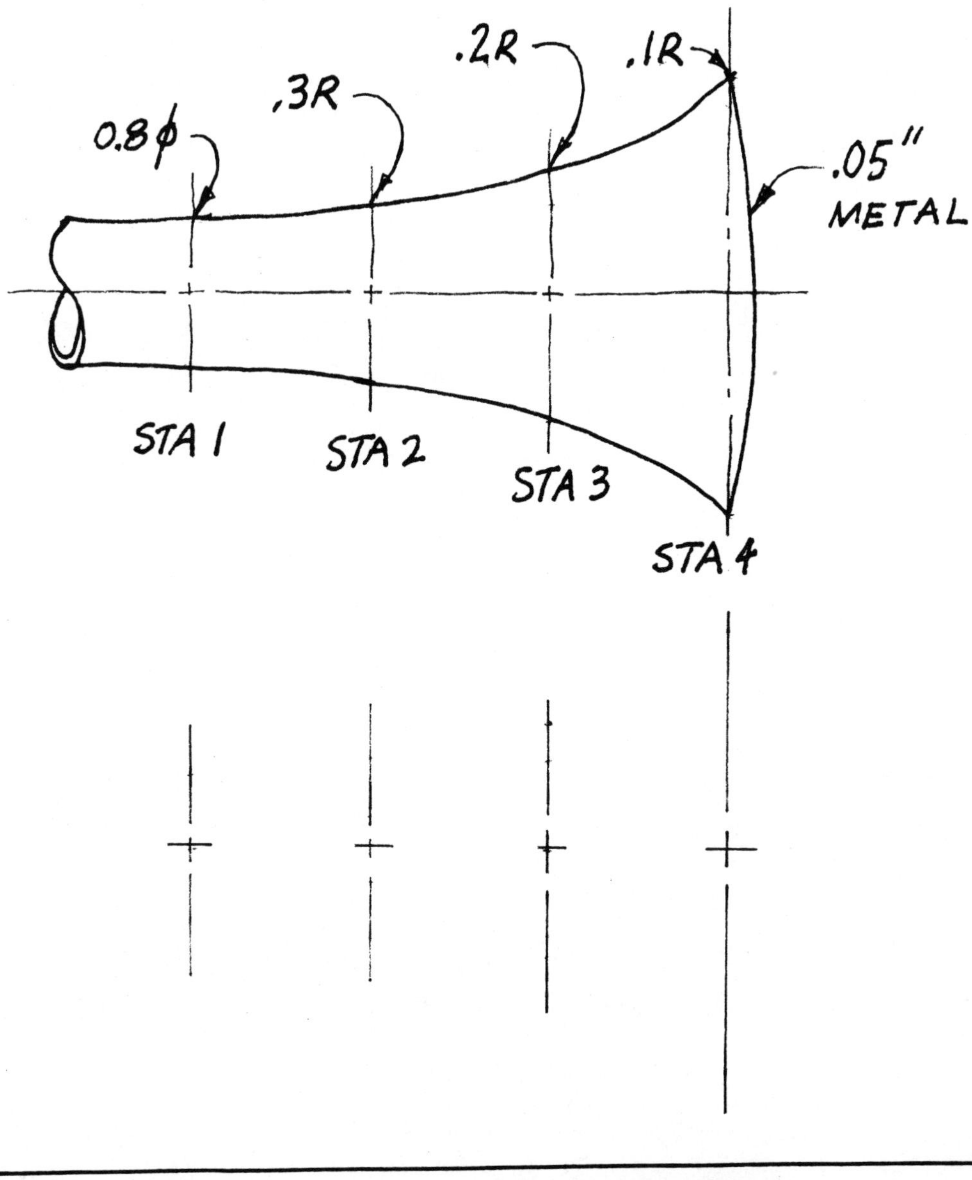

Ch6 Sectional Views			Problem 2-6-9
Section-Chapter-Problem 2 6 9	Removed sectons Figures 6-27	Course_______ Section____ Student_________ Date_____	

A

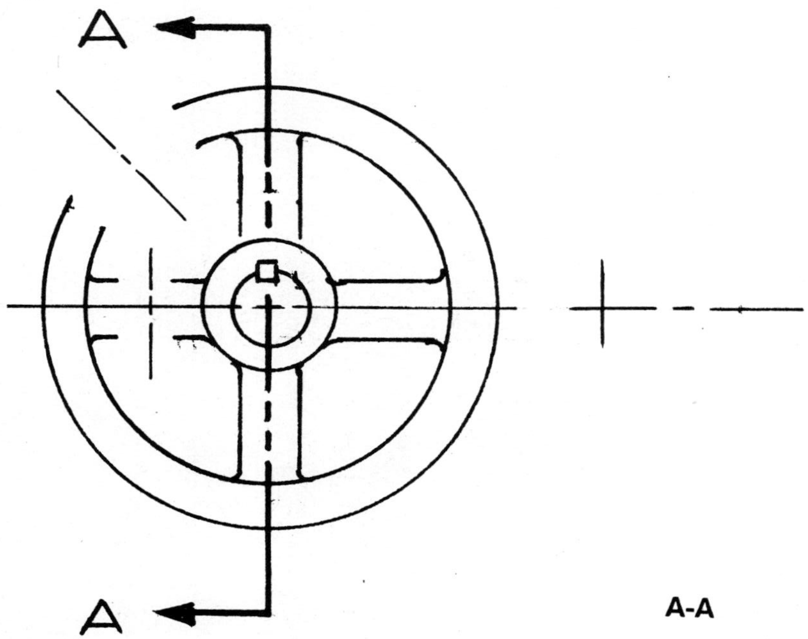

Handwheel Specifications

Rim: circular cross section
Spokes: Major Dia. 3 times Minor Dia.
Hub length = Hub dia.
Hub bore = shaft dia.
Key =1/8 of shaft dia.; square key

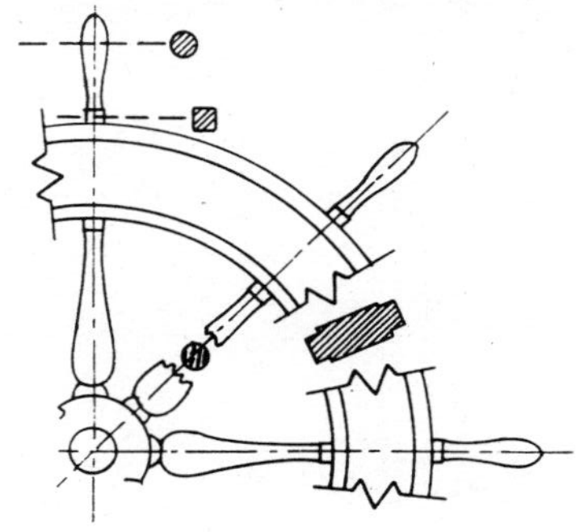

Examples

Scale the handwheel shown and do the following in the spaces indicated:

a. Construct section A-A.

b. Do a revolved section in the handwheel rim.

c. Do a revolved section in a spoke.

A-A

B

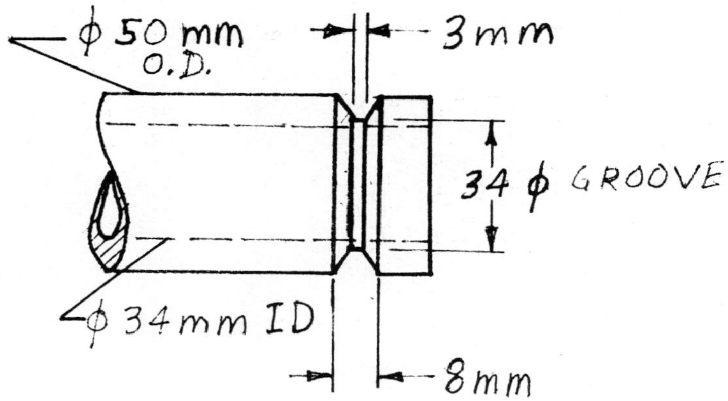

Use the information on the drawing of the grooved tube and prepare a 4X enlarged removed-section of the groove.

Ch6 Sectional Views			Problem
Section-Chapter-Problem 2 6 10	Aligned sections Figure 6-48 & 6-50 Revolved sections Figure 6-26 Enlarged removed-sections Figure 6-28	Course______ Section____ Student______ Date____	2-6-10

A **Construct as indicated auxiliary section A-A of the wave guide with a pressure transducer (centered on the narrow side). Do not dimension. Use correct lines.**

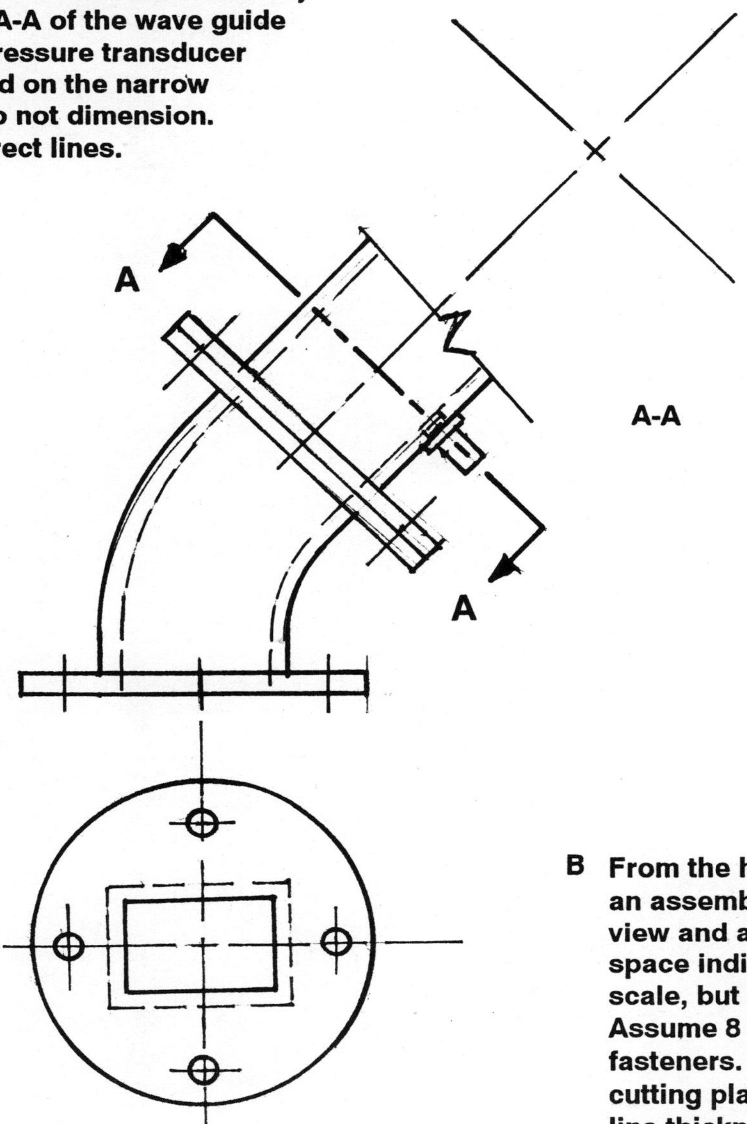

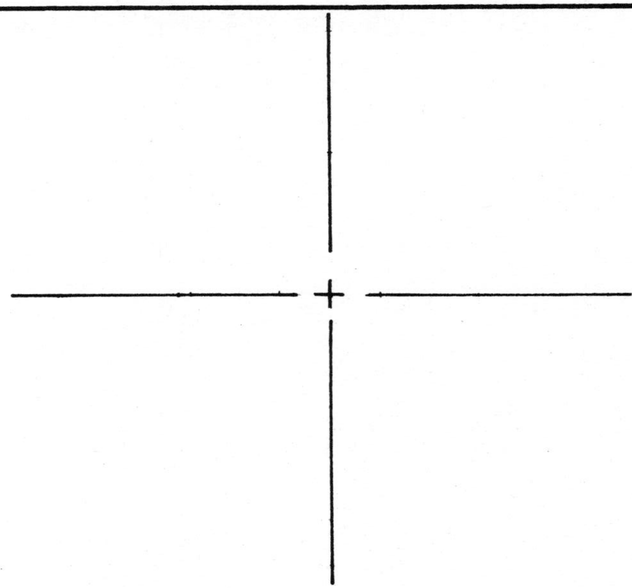

B **From the half section view of an assembly construct a top view and a front view in the space indicated. Use the same scale, but do not dimension. Assume 8 equally spaced fasteners. Show the sectioning cutting plane. Use correct line thicknesses.**

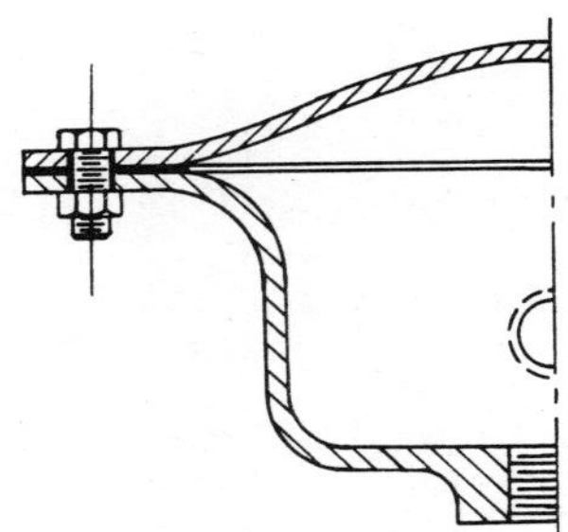

Ch6 Sectional Views

Section-Chapter-Problem		Course____________Section_____	Problem
2 6 11	Auxiliary section Figure 6-29Thinwall section Figure 6-30 Half section Figure 6-17	Student__________________Date______	2-6-11

A A linear drive follower is shown in drawing file "06-1.dxf". Draw a full section according to the placement of the cutting plane line shown in the top view. The linear drive follower is made of brass.

B The pivot base is shown in drawing file "06-2.dxf". The machinist is requesting an offset section view of it, with two design changes. First, the shaft that goes in the 5/8" hole has been changed to 7/8" and now requires a ¼" x ¼" x 1" key for extra support. In addition, the hole that supports the shaft is now 1" deep instead of 3/4". This object will be cast iron.

<table>
<tr><td colspan="3">Ch6 Sectional Views</td><td rowspan="2">Problem

2-6-12</td></tr>
<tr><td>Section-Chapter-Problem
2 6 12</td><td>Full Section
Offset Section</td><td>Course__________ Section_____
Student__________________ Date_______</td></tr>
</table>

Refer to the views shown and answer the following TRUE-FALSE questions.

____1. To be able to draw an auxiliary view of a plane surface, the plane surface must first appear as an edge .

____2. To prepare an auxiliary view of a given plane which appears as an edge, projection lines must be perpendicular to the given plane.

___3. All the lines in a sectional drawing (except for specific conventions such as bolts and shafts) are in true size because the section cutting plane shows as an edge.

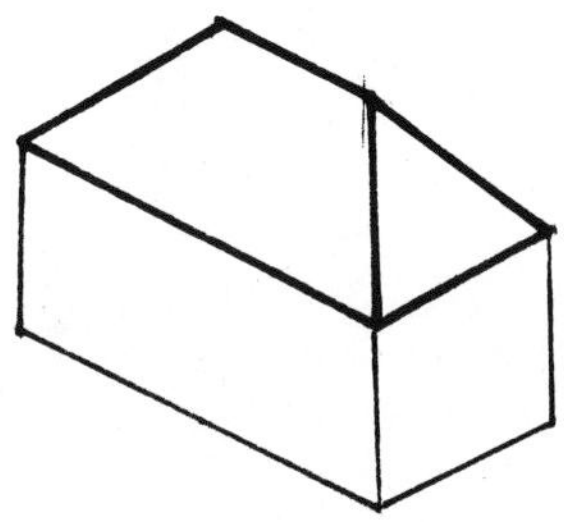

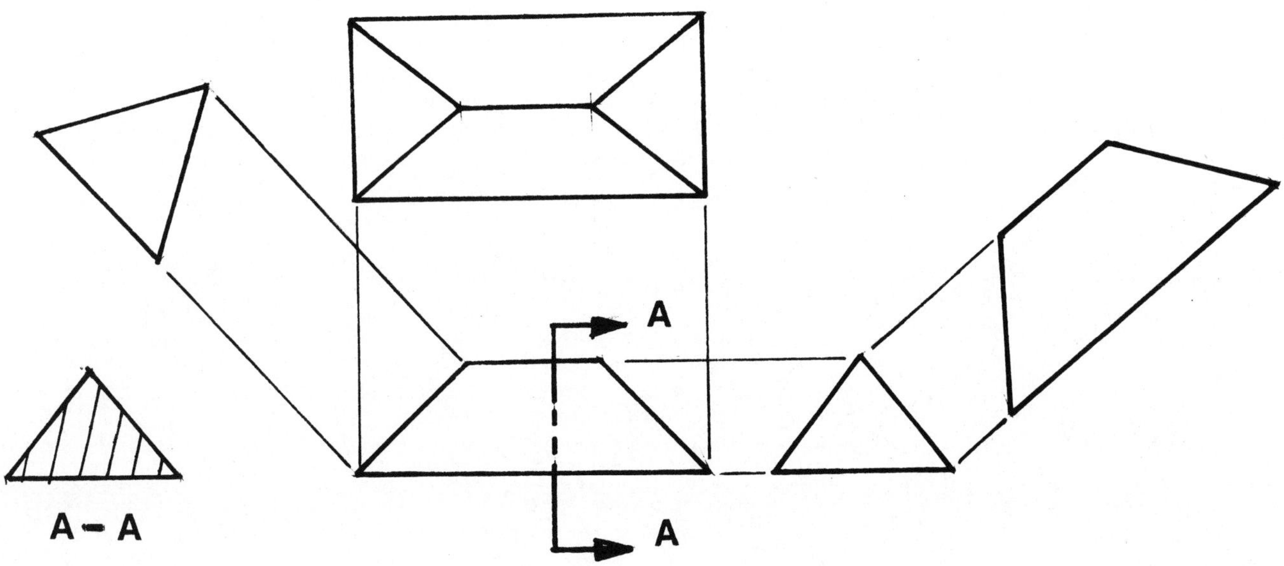

Ch7 Auxiliary Views			Problem
Section-Chapter-Problem 2 7 1	Figures 7-1 through 7-6 Figure 8-17	Course__________ Section_____ Student_______________ Date______	2-7-1

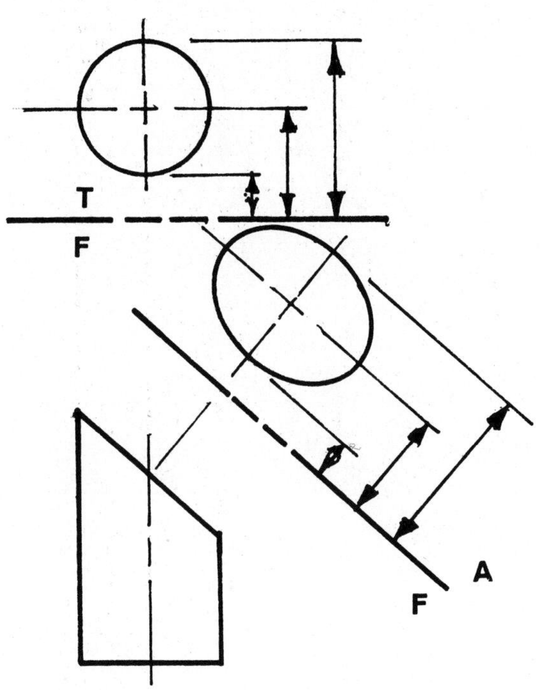

EXAMPLE: FOLD LINE METHOD

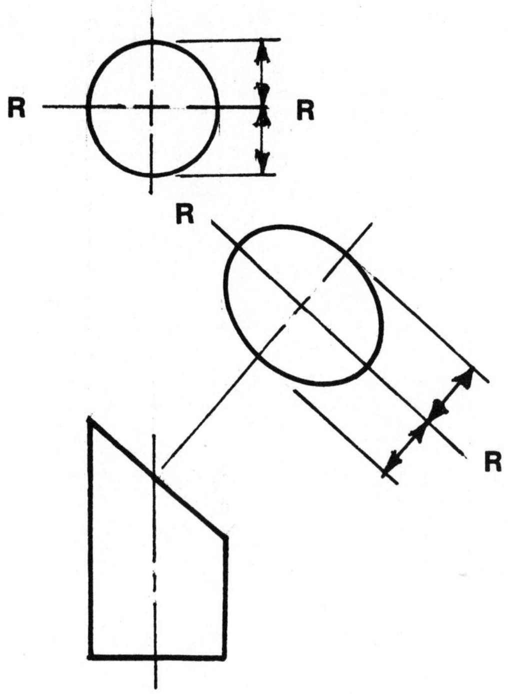

EXAMPLE: REFERENCE LINE METHOD

Construct an auxiliary view of the inclined portion of the object using Fold Line A/S. Construct an auxiliary view of the inclined portion of the object using Reference line R-R. Do not erase construction lines. Use correct lines. Use a "Z" break.

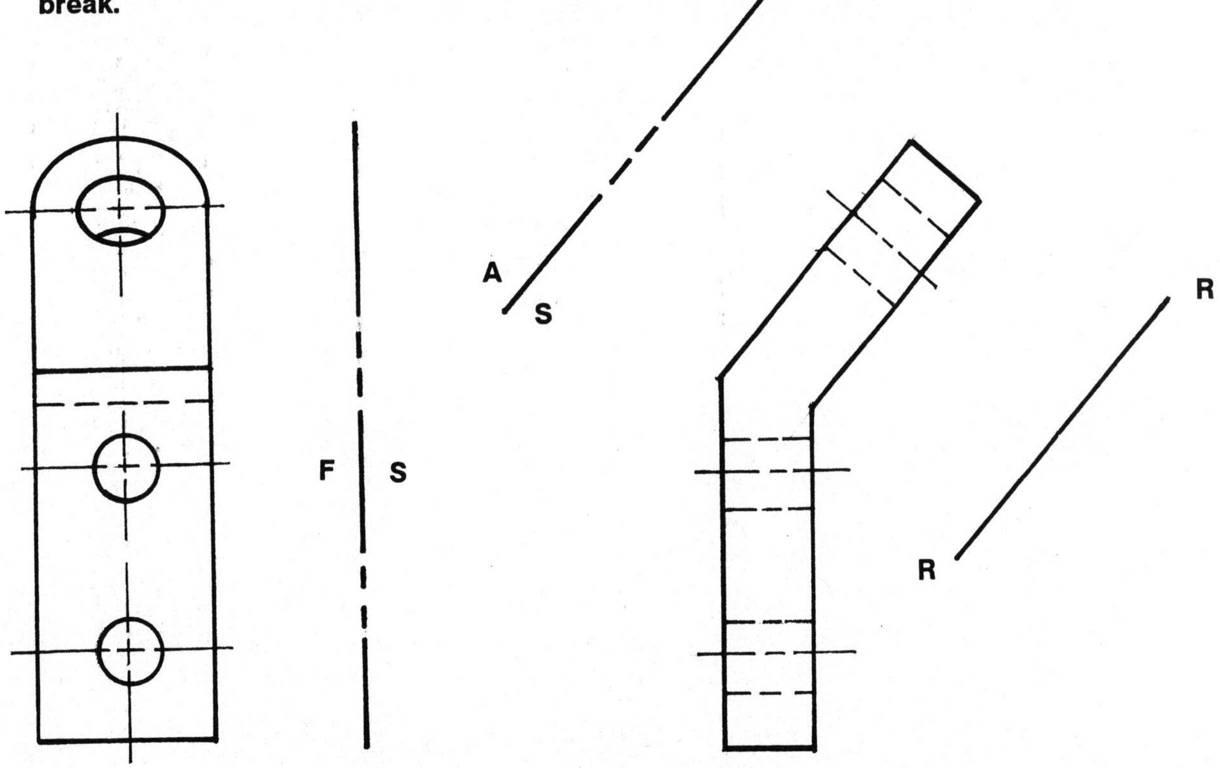

Ch7 Auxiliary Views			Problem 2-7-2
Section-Chapter-Problem 2 7 2	Reference Line method Figure 7-6 Fold Line method Figure 7-8	Course________ Section_____ **Student**__________ Date_____	

A

Prepare auxiliary views of the inclined planes using the reference plane method. Do not erase construction lines. Use correct line thickness.

B

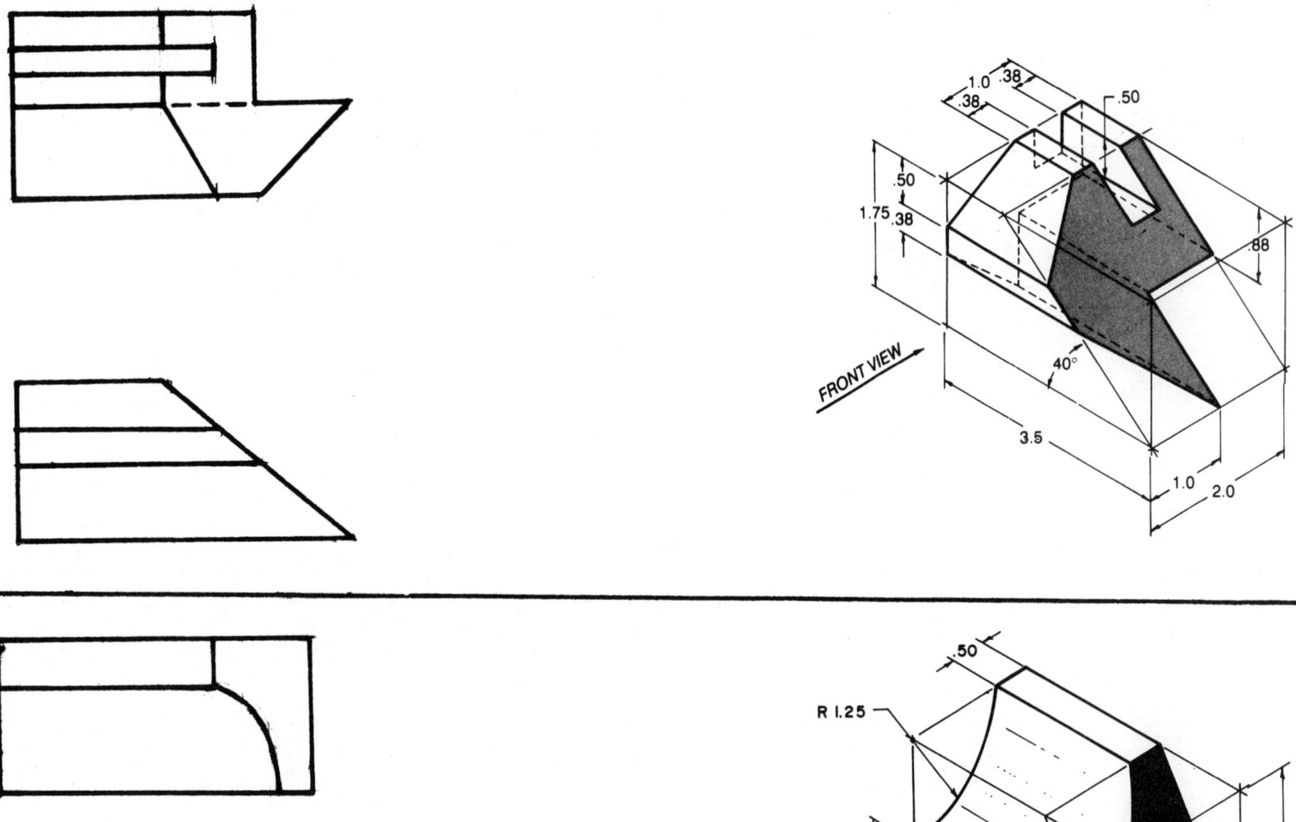

C

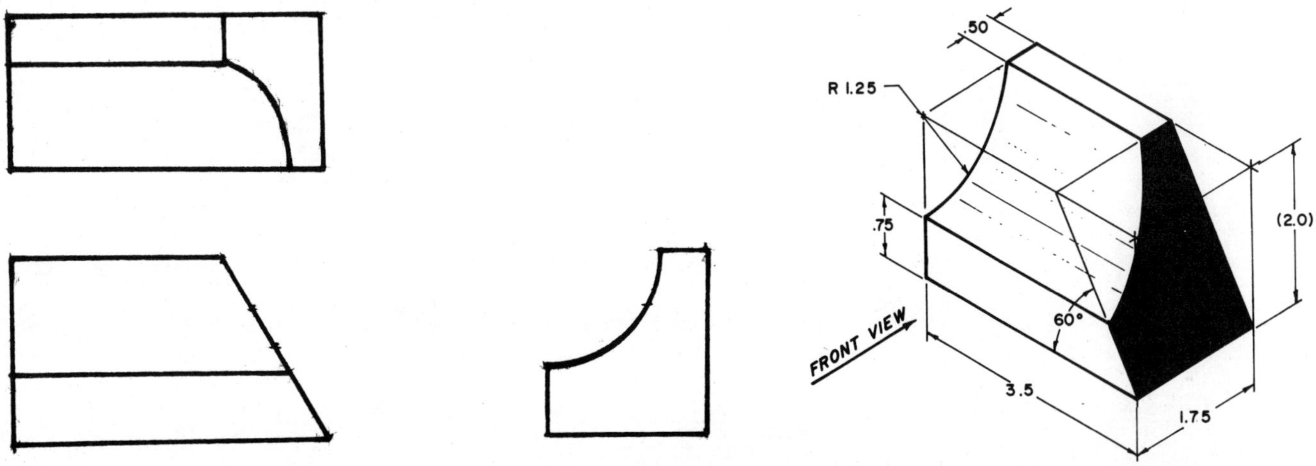

Ch7 Auxiliary Views			Problem
Section-Chapter-Problem 2 7 3	Reference Line method Figure 7-6	Course__________ Section_____ Student______________ Date______	2-7-3

A

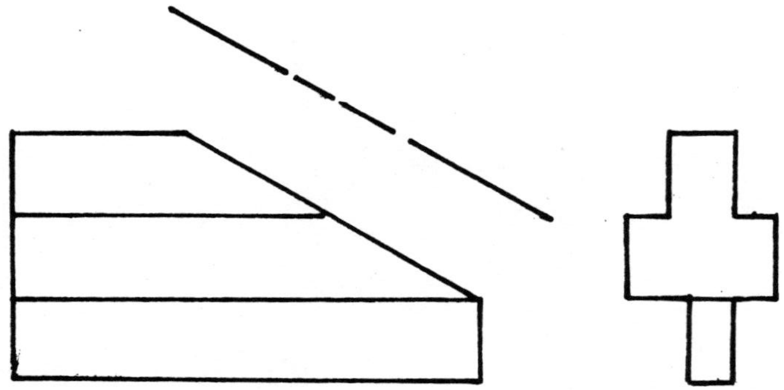

B

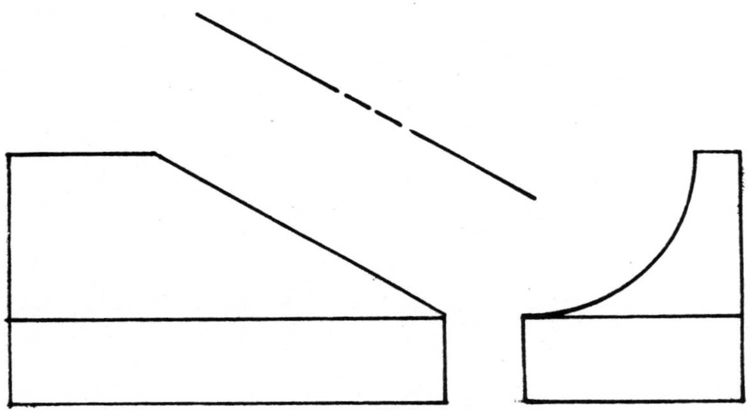

C

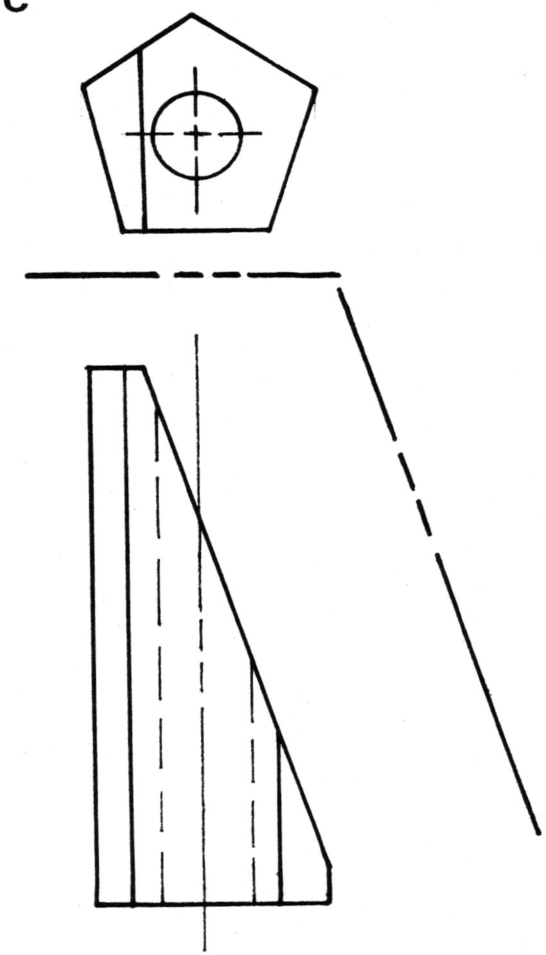

Prepare auxiliary views of the inclined planes using the fold line method. Do not erase construction lines. Use correct line thickness.

Ch7 Auxiliary Views			Problem
Section-Chapter-Problem 2 7 4	Fold Line method Figure 7-8	Course________ Section_____ Student____________ Date_____	2-7-4

A

Prepare an auxiliary view of the inclined plane using the fold line method.

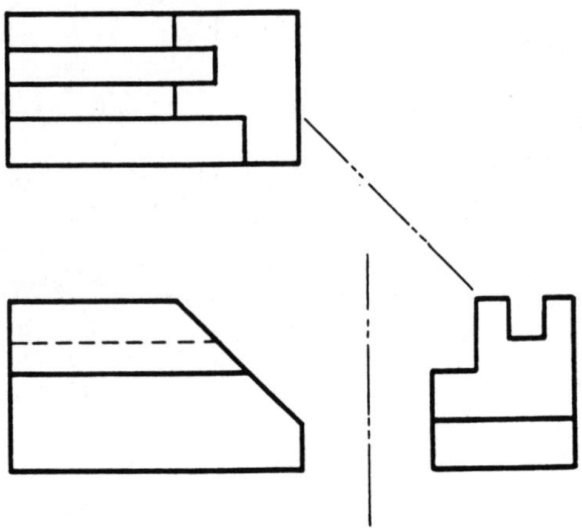

B

Complete the front view. Do not erase construction lines. Use correct line thickness.

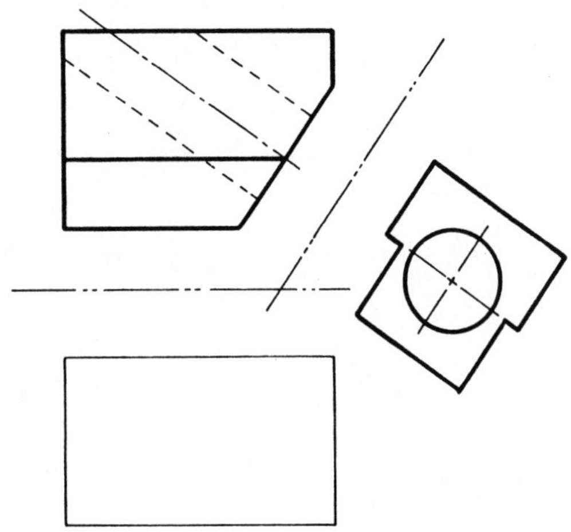

C

Complete the front and right side views. Do not erase construction lines. Use correct line thickness.

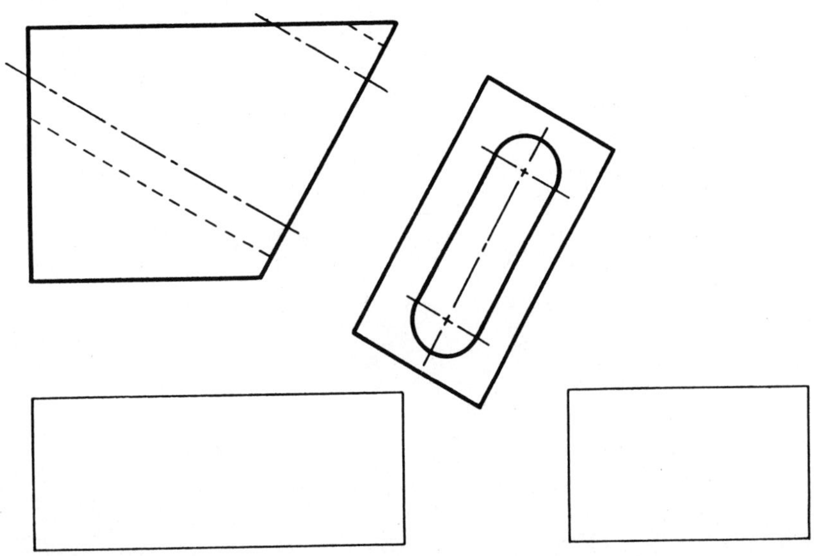

Ch7 Auxiliary Views			Problem
Section-Chapter-Problem 2 7 5	Fold Line method Figure 7-8	Course____________ Section______ Student__________________ Date______	2-7-5

A

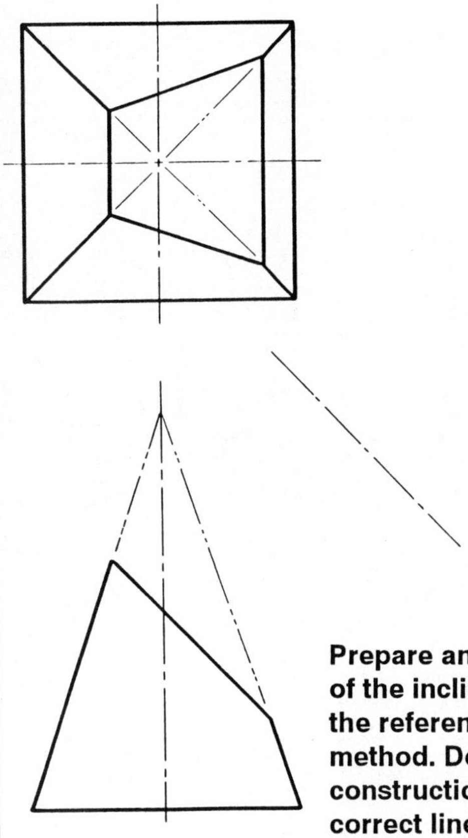

Prepare an auxiliary view of the inclined plane using the reference plane method. Do not erase construction lines. Use correct line thickness.

B

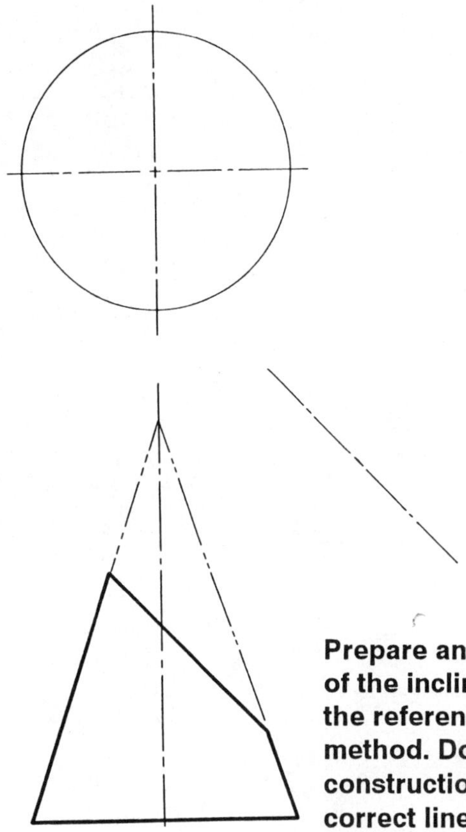

Prepare an auxiliary view of the inclined plane using the reference plane method. Do not erase construction lines. Use correct line thickness.

C

Prepare an auxiliary view of the inclined plane using the reference plane method and complete the front view.

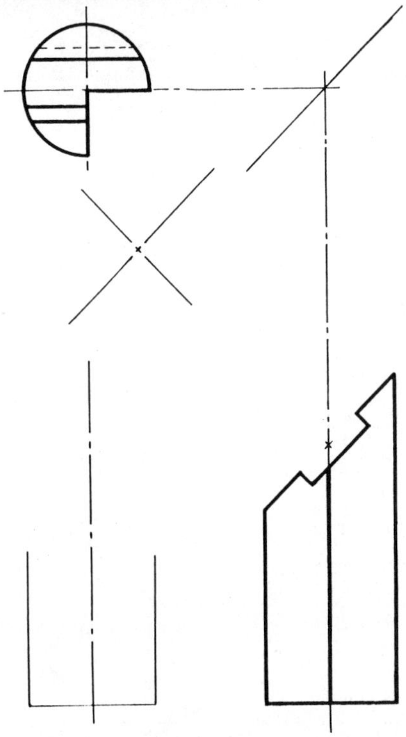

D

Prepare an auxiliary view of the inclined plane using the fold line method and complete the right side view.

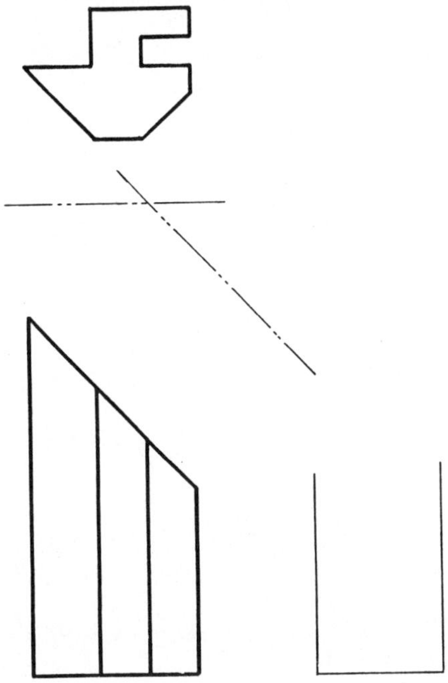

Ch7 Auxiliary Views

Section-Chapter-Problem	Reference Line method Figure 7-6	Course____________ Section_____	Problem
2 7 6	Fold Line method Figure 7-8	Student__________________ Date______	2-7-6

A

Prepare an auxiliary view of the inclined plane using the reference plane method and complete the right side view. Do not erase construction lines. Use correct line thickness.

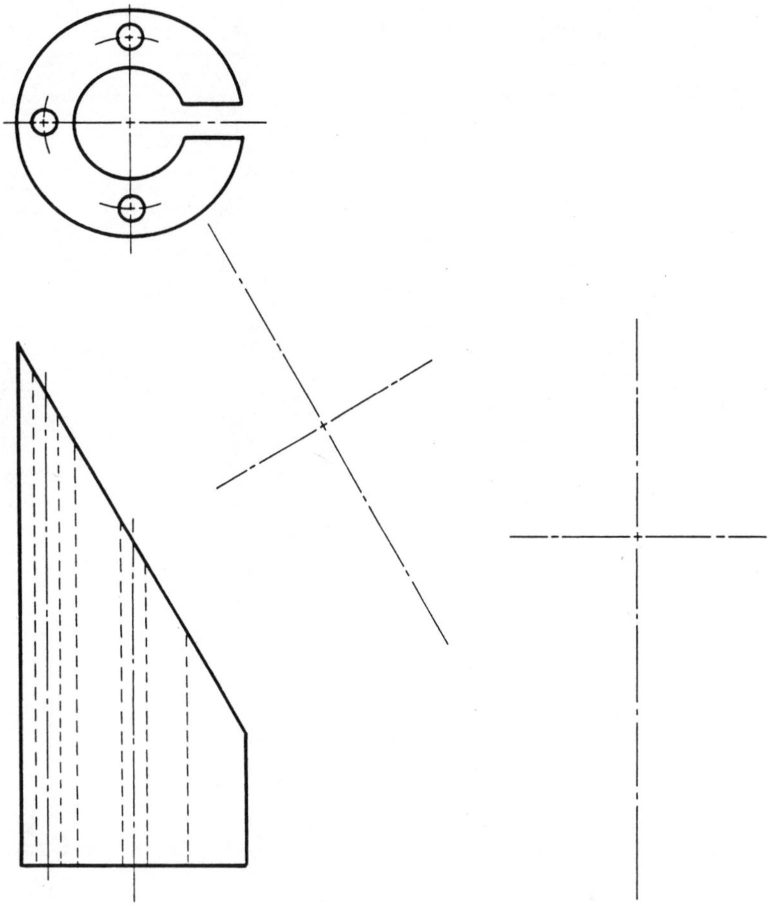

B Prepare auxiliary views of the inclined planes using the reference plane or the fold line method. Complete the right side view. Do not erase construction lines. Use correct line thickness.

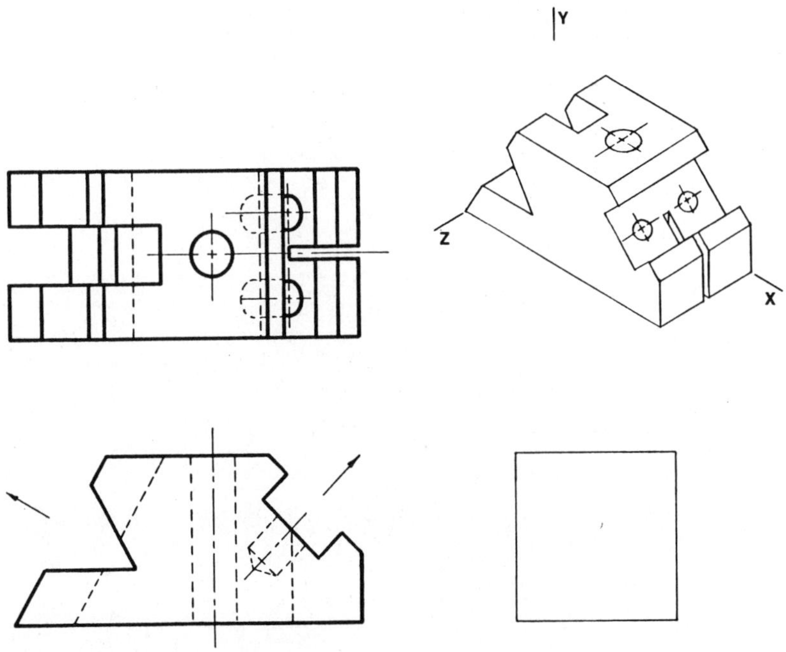

Ch7 Auxiliary Views			Problem
Section-Chapter-Problem 2 7 7	Reference Line method Figure 7-6 Fold Line method Figure 7-8	Course__________ Section_____ Student____________ Date______	2-7-7

A Complete the top and front views. Prepare an auxiliary view of the inclined plane using the reference plane method. Do not erase construction lines. Use correct line thickness.

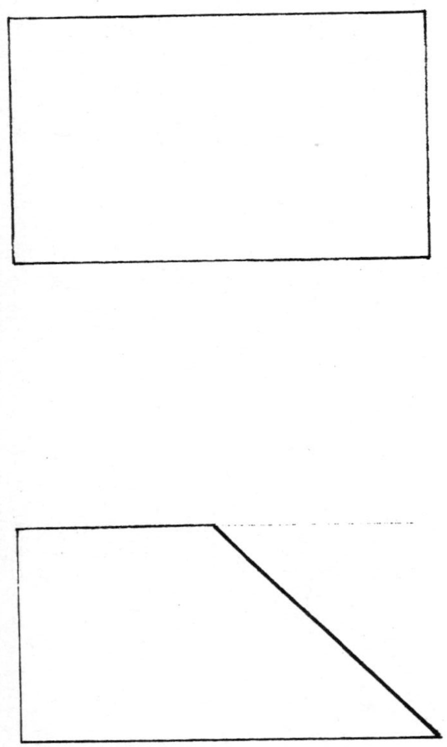

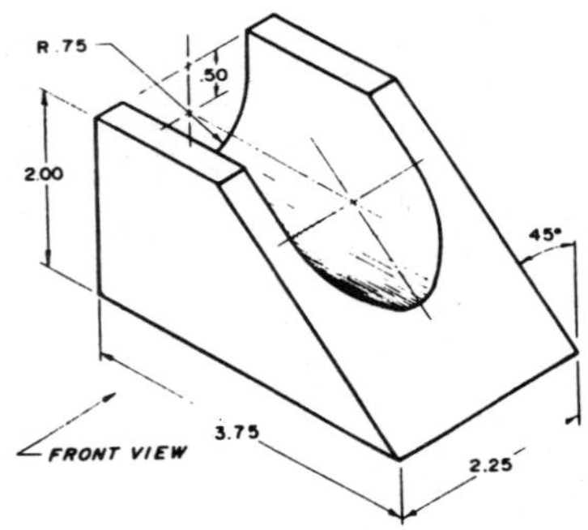

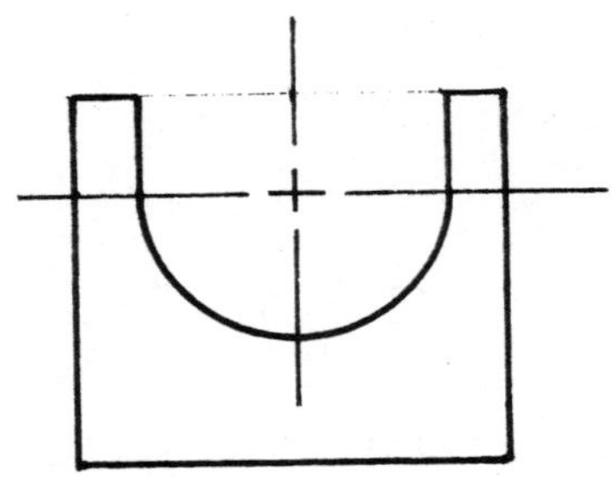

B

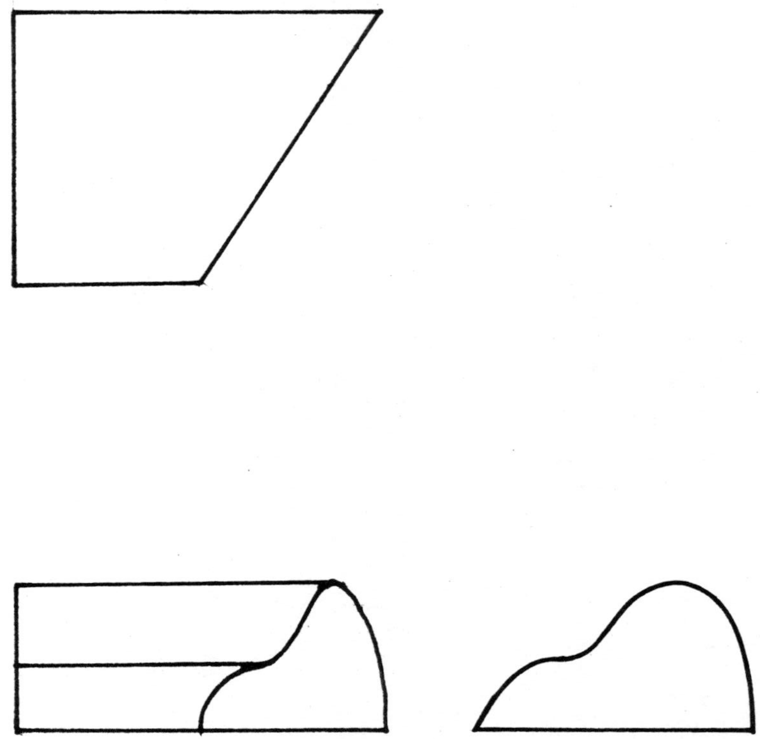

Prepare an auxiliary view of the inclined plane using the fold line method. Is the top view complete as drawn? Do not erase construction lines. Use correct line thickness.

Ch7 Auxiliary Views

Section-Chapter-Problem	Reference Line method Figure 7-6 Fold Line method Figure 7-8	Course________ Section____	Problem
2 7 8	Intersecting surfaces Figure 5-47	Student________ Date____	2-7-8

A

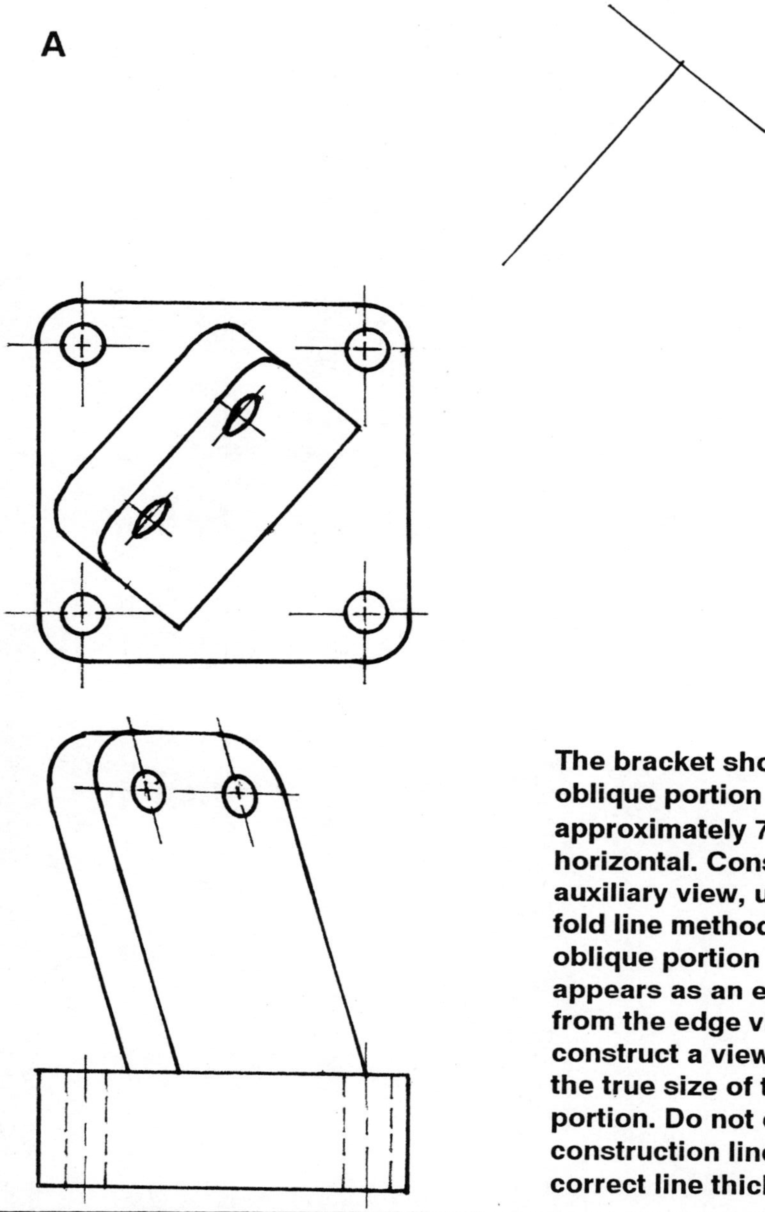

The bracket shown has an oblique portion at approximately 70° to the horizontal. Construct an auxiliary view, using the fold line method, of the oblique portion so it appears as an edge. Then from the edge view construct a view showing the true size of the oblique portion. Do not erase construction lines. Use correct line thickness.

B

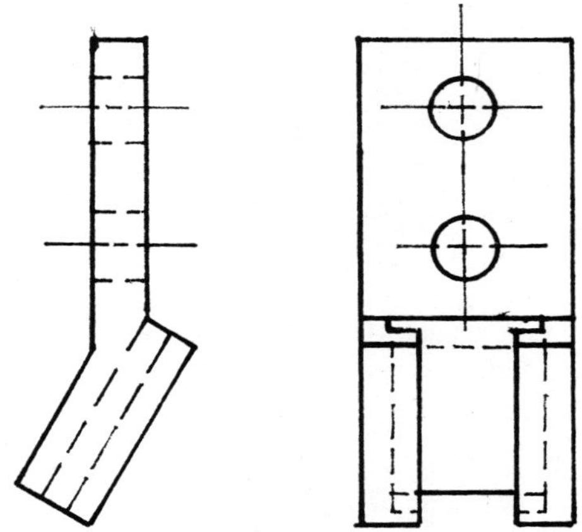

Prepare an auxiliary view of the inclined slot using the reference plane method. Then using the auxiliary view prepare a secondary view showing an end view of the slot. Do not erase construction lines. Use correct line thickness.

Ch7 Auxiliary Views			Problem
Section-Chapter-Problem 2 7 9	Reference Line method Figure 7-6 Fold Line method Figures 7-8, 19A	Course______ Section____ Student________ Date____	2-7-9

The top view T, a partially completed front view F, and an auxiliary view A are shown in the drawing.

a.With the given information complete the front view F.

b. Assume that the surface in the auxiliary view is drawn to a scale of 1m=10mm and determine an approximate area of the surface. Show your calculations. (Suggestion: Divide the area into familiar shapes such as triangles, rectangles, and trapezoids.)

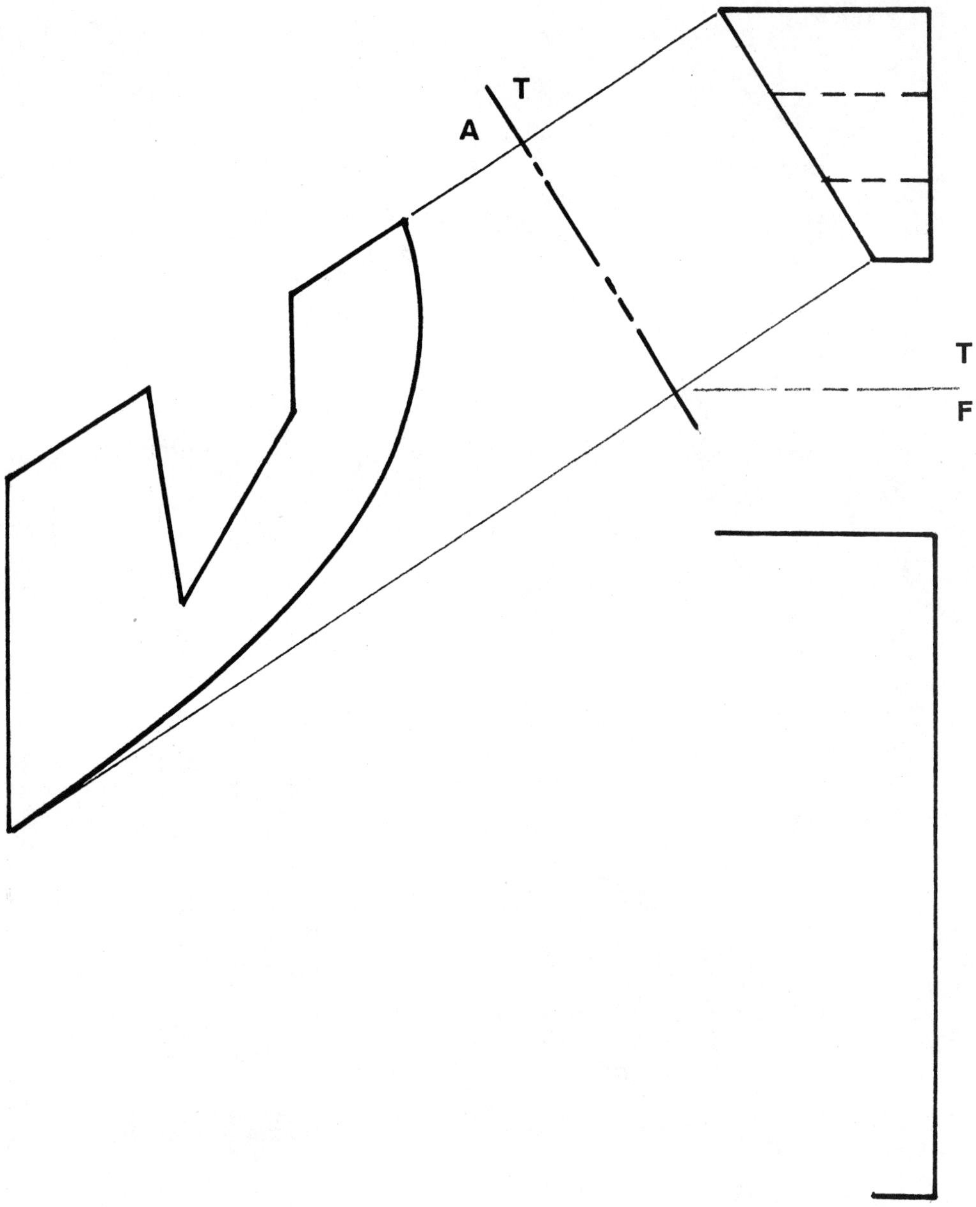

Ch7 Auxiliary Views			Problem
Section-Chapter-Problem 2 7 10	Fold Line method Figure 7-8	Course____________ Section_____ Student________________ Date_____	2-7-10

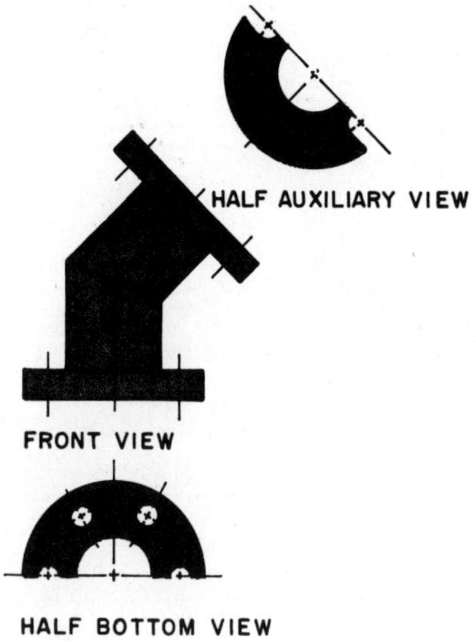

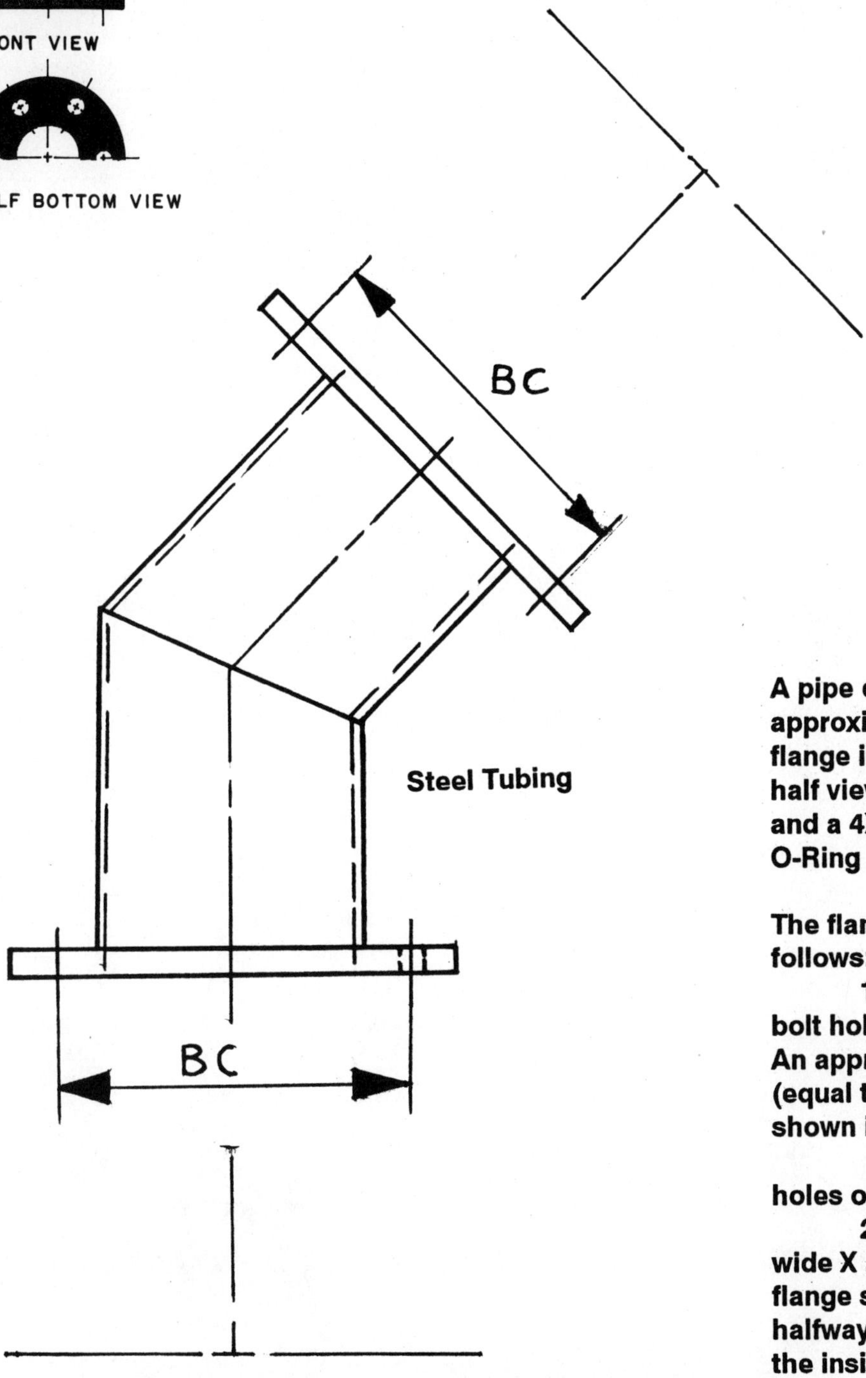

A pipe elbow drawing (Shown approximately to scale.) needs to have flange information added to it. Prepare half views at upper and lower flanges, and a 4X detail cross section showing an O-Ring groove.

The flange specifications are to be as follows:

1. The lower flange requires eight bolt holes on bolt circle BC as indicated. An approximate bolt hole diameter (equal to the flange thickness .38") is shown in the edge view of the flange.

The upper flange needs six bolt holes of the same size.

2. A circular O-Ring groove , .13" wide X .10" deep is required in each flange so that the groove is located halfway between the bolt circle BC and the inside diameter of the pipe.

Ch7 Auxiliary Views			Problem
Section-Chapter-Problem 2 7 11	Reference Line method Figure 7-6	Course__________ Section____ Student________________ Date_____	2-7-11

A In file “07-1.dxf” you will find a multiview drawing of a paper airplane. Draw the appropriate auxiliary view of the main wing and then calculate the surface area of it.

B Drawing file 07-2.dxf” is divided into four quadrants. Create drawings in each quadrant that would illustrate the main steps to follow when creating an auxiliary view. Be sure to add text to describe what procedure to follow. The main steps have been identified in the appropriate quadrants.

Ch7 Auxiliary Views			Problem
Section-Chapter-Problem **2 7 12**	Paper Airplane Auxiliary Views	Course__________ Section_____ **Student**________________ Date______	**2-7-12**

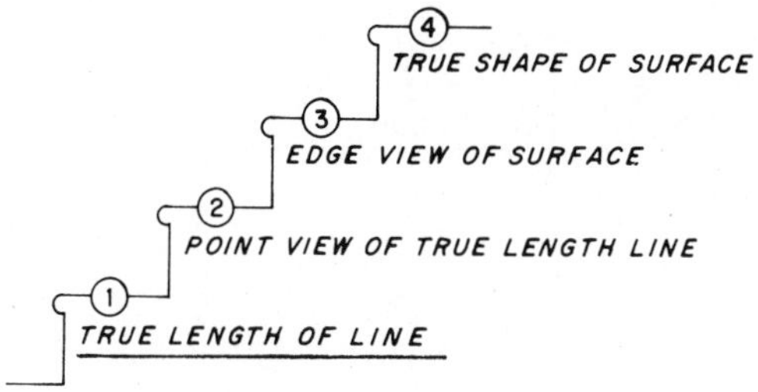

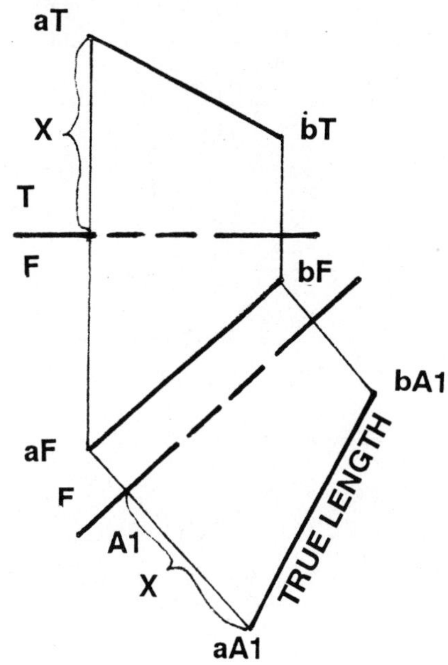

A

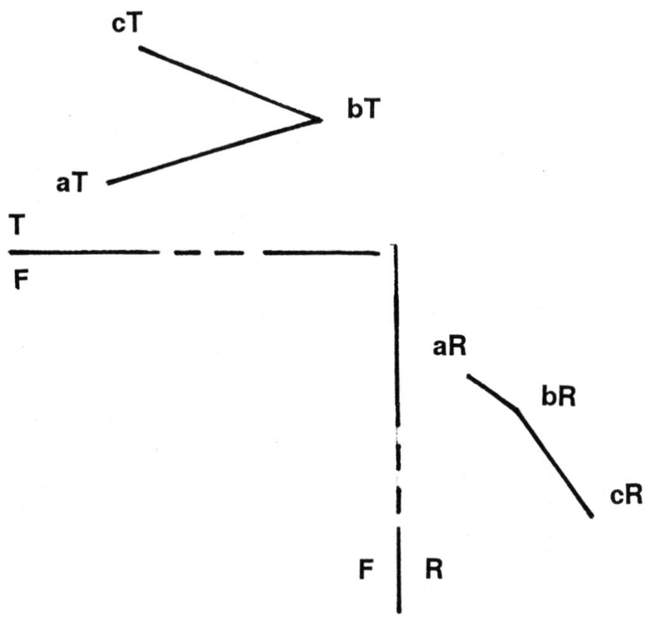

Locate a-b and b-c in the F view. Then determine the true length of a-b from the F view. Do not erase construction lines. Lable your answers.

B

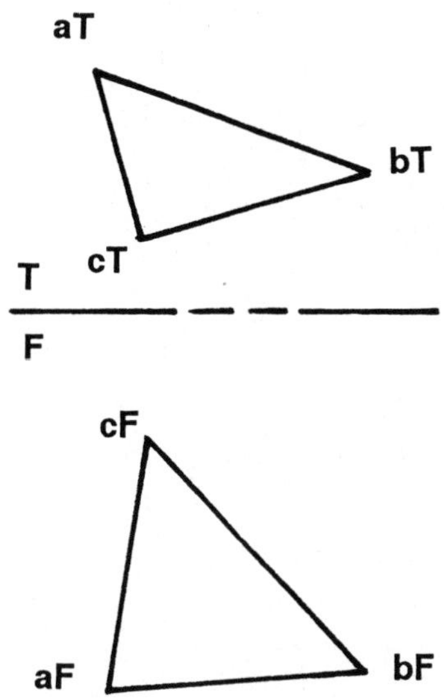

Determine the true lengths of a-b and b-c from the F view. Do not erase construction lines. Lable your answers.

Ch8 Descriptive Geometry			Problem 2-8-1
Section-Chapter-Problem 2 8 1	Finding True length of a line Figure 8-11	Course________ Section____ Student__________ Date_____	

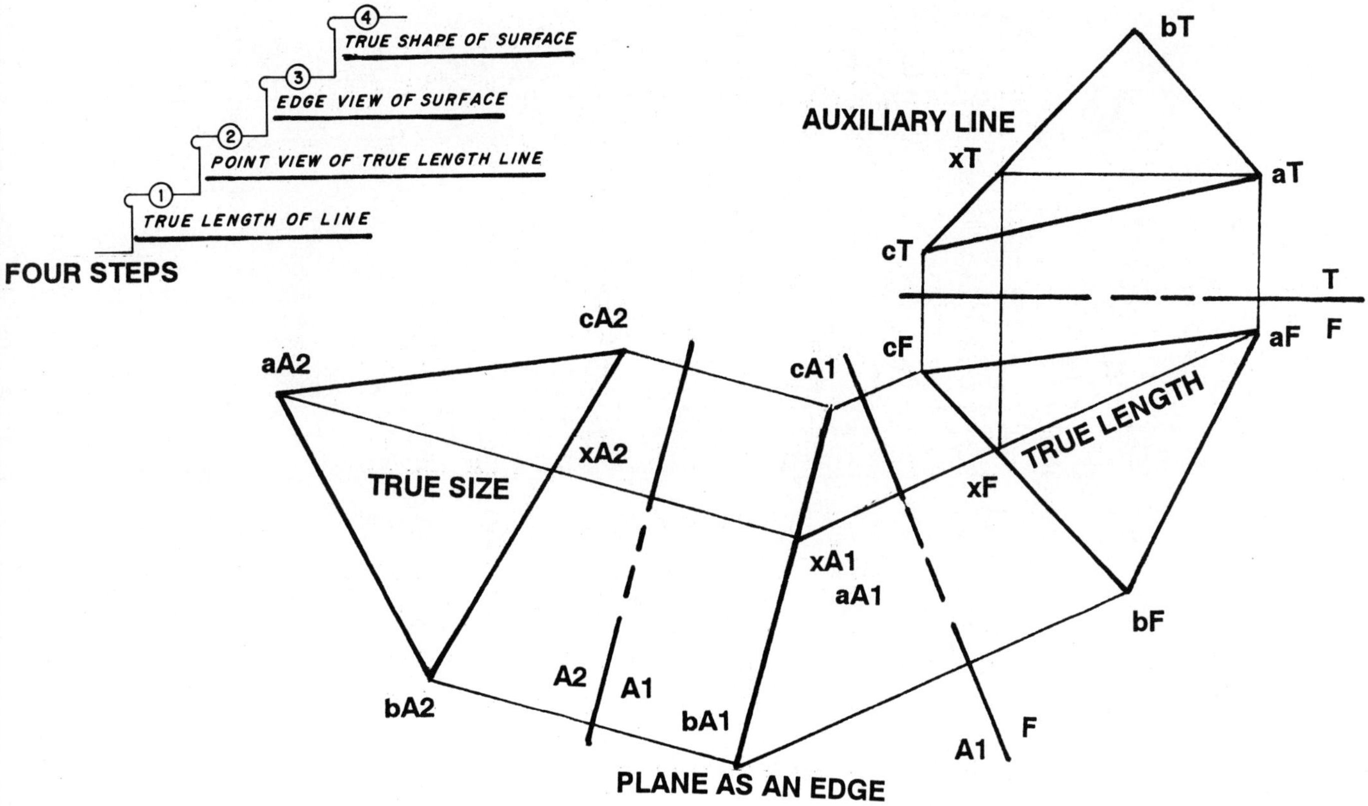

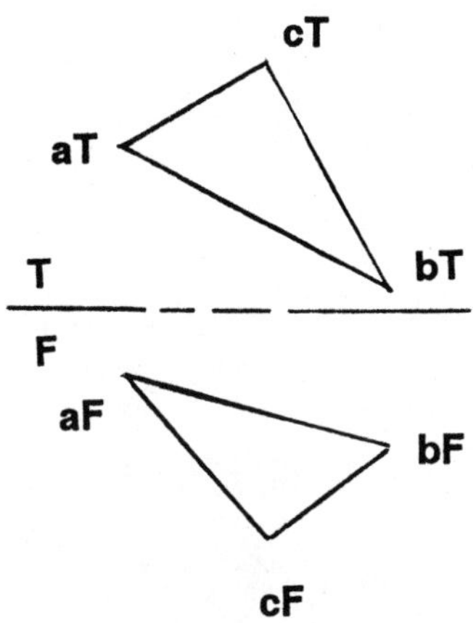

Determine the true size of plane a-b-c.
Do not erase construction lines. Lable all points in each view.

Ch8 Descriptive Geometry			Problem
Section-Chapter-Problem 2 8 2		Course__________ Section_____ Student__________________ Date______	2-8-2

A

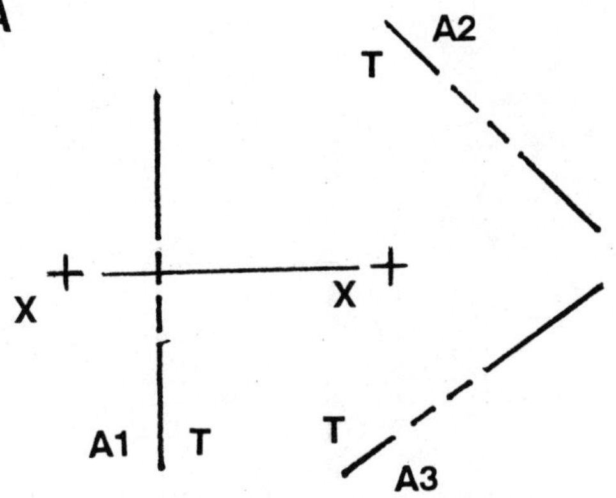

In problems A,B,C, and D do not erase construction lines. Lable all points in each view. (Note that measurements should originate from the centers of lines.)

A. Locate point x in the auxiliary views A2 and A3.

B

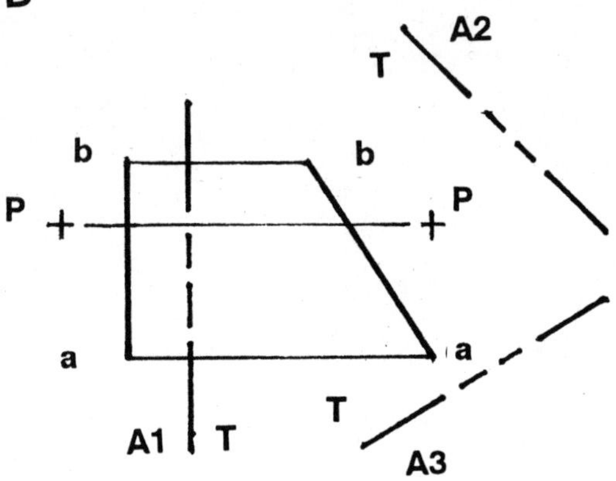

B. Project line a-b and point P into the auxiliary views A2 and A3. Why does the shortest distance from point P to line a-b appear in one of the auxiliary views?

C

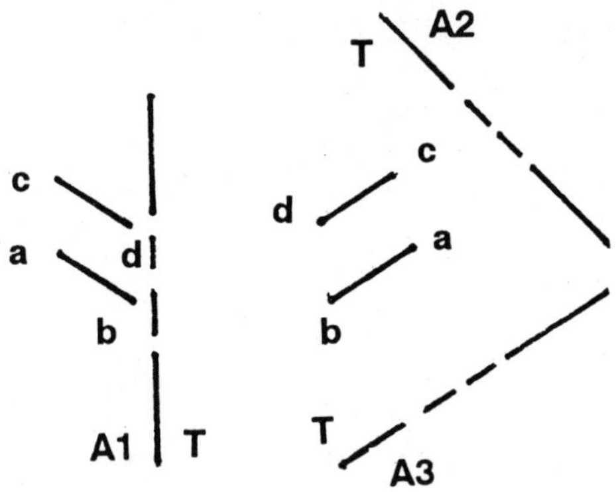

C. Project parallel lines a-b and c-d into the auxiliary views A2 and A3. Then project from one of the auxiliary views to determine the shortest distance between the two lines. (Note that parallel lines are parallel in all views.)

D

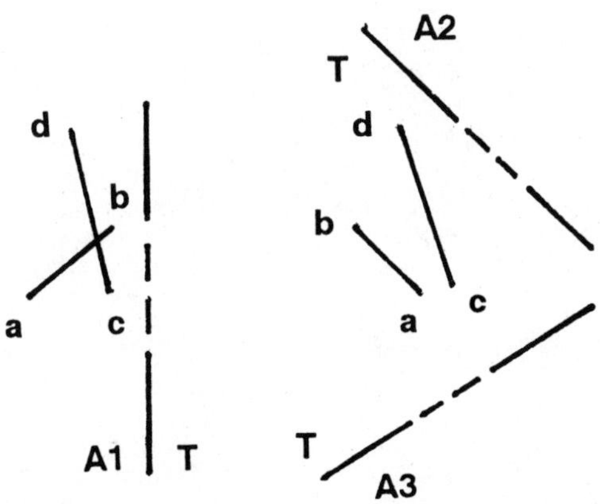

D. Project lines a-b and c-d into the auxiliary views A2 and A3. Then project from one of the auxiliary views to determine the shortest distance between the two lines.

Ch8 Descriptive Geometry			Problem 2-8-3
Section-Chapter-Problem 2 8 3	Shortest distances Figures 8-14 & 8-15	Course__________ Section_____ Student________________ Date______	

A

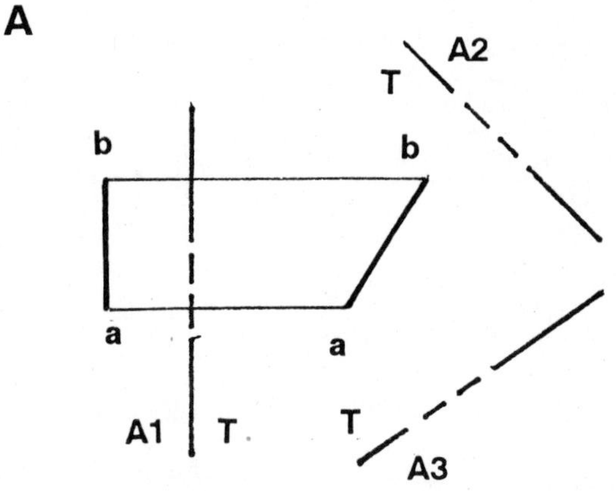

In problems A,B,C, and D do not erase construction lines. Lable all points in each view. (Note that measurements should originate from the centers of lines.)

A. Project line a-b from the T view into the auxiliary views A2 and A3. Explain why these projections seem to be identical.

B

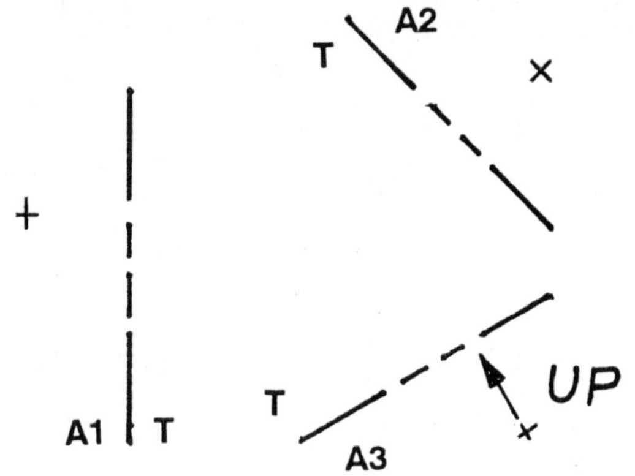

B.View A3 shows a vector indicating the "up" direction with respect to the T view. Show a vector indicating the "up" direction in the A1 and A2 views.

C

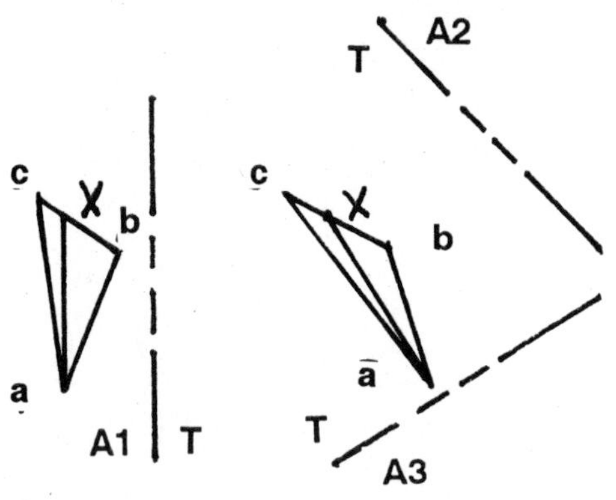

C. In the A1 view, in triangle a-b-c, line a-x is parallel to the fold line A1/T. Project t.riangle a-b-c and a-x from the T view into the A2 and A3 views. Comment on the projections.

D

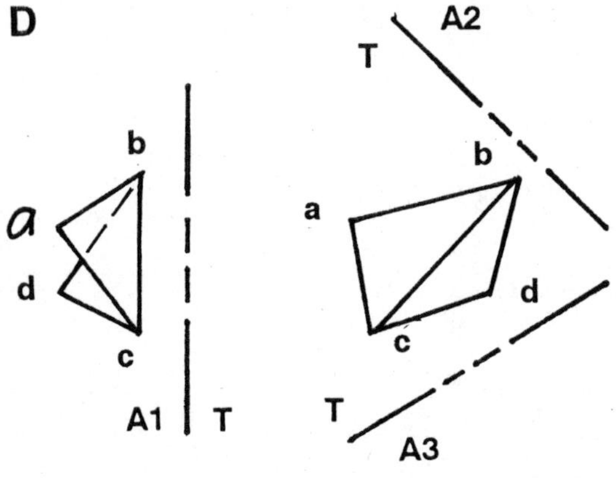

D. Project triangles a-b-c and b-c-d from the A1 view into the T view. Then project the triangles into the auxiliary views A2 and A3. Comment on the projections.

Ch8 Descriptive Geometry			Problem 2-8-4
Section-Chapter-Problem 2 8 4	Projections Figures 8-8,9,10 Dihedral angles Figures 8-19	Course__________ Section_____ Student________________ Date______	

A Determine the shortest distance between lines a-b and c-d. Project from the front view F. Do not erase construction lines. Lable all points in each view. Lable your answer.

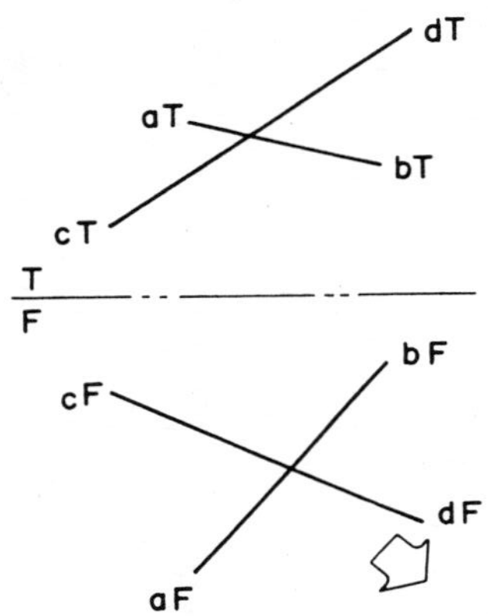

B Determine the shortest distance between point x and the plane a-b-c-d. Project from the front view F. Do not erase construction lines. Lable all points in each view. Lable your answer.

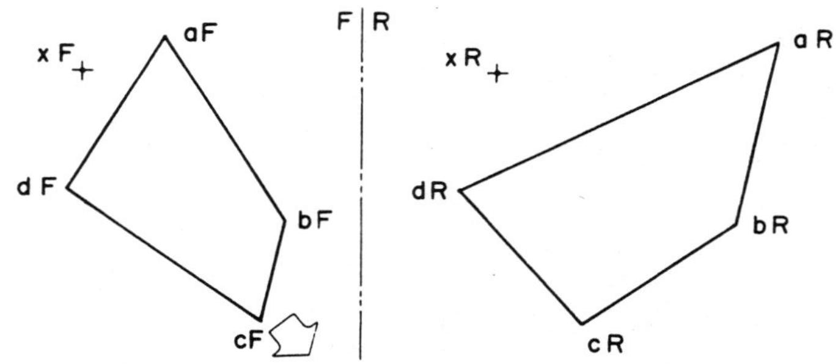

Ch8 Descriptive Geometry

Section-Chapter-Problem		Course_____ Section_____	Problem
2 8 5	Shortest distance between two non-parallel lines Figure 8-15 Shortest distance between a plane and a point not on the plane Figure 8-18	**Student**_____ Date_____	2-8-5

A A. Find a view which shows line a-b as a point. Project from the F view.

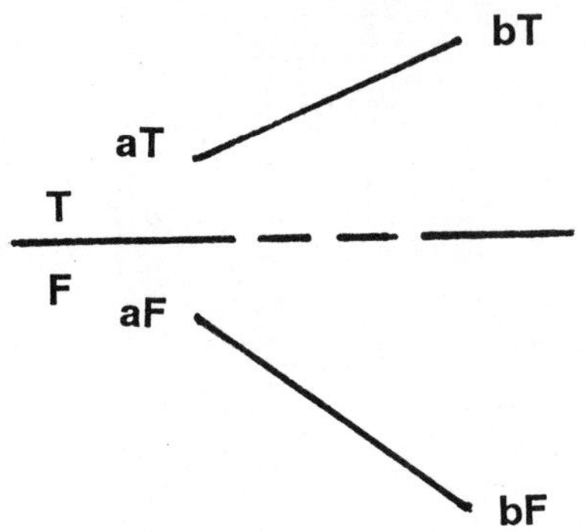

POINT VIEW OF TRUE LENGTH LINE

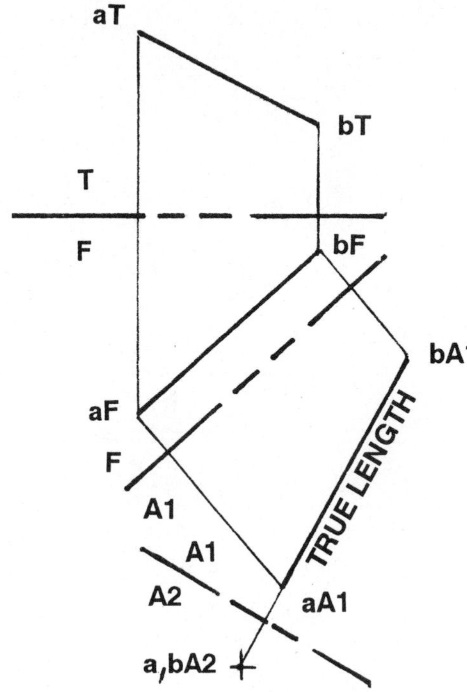

B

In problems A,B, and C do not erase construction lines. Lable all points in each view. (Note that measurements should originate from the centers of lines.)

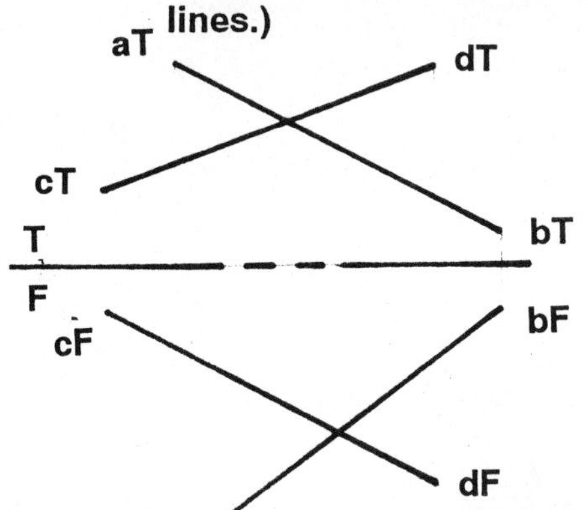

B. Determine the shortest distance between line a-b and line c-d. Project from the F view. Lable your answer in mm.

C

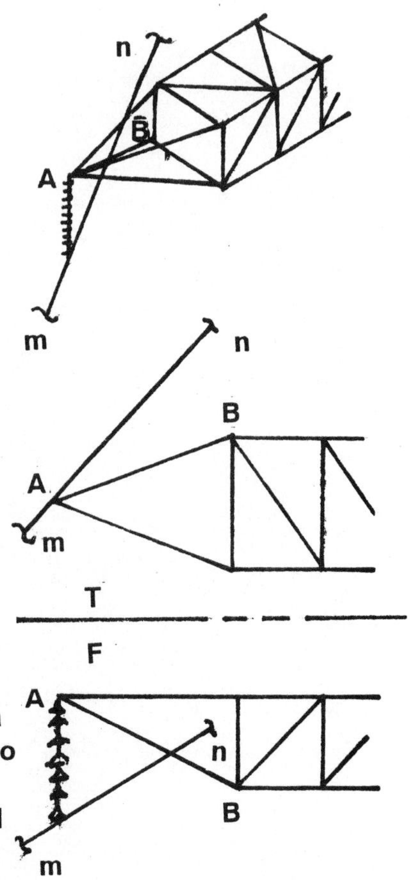

C. Assume a scale of 1"=10' and determine the shortest distance a spark would have to jump between power line m-n and member A-B of the truss.

Ch8 Descriptive Geometry			Problem
Section-Chapter-Problem 2 8 6	Shortest distance between two non-parallel lines Figure 8-15	Course________ Section_____ Student________ Date_____	2-8-6

Determine the true size of plane A and of plane B. Project from the top view T. Do not erase construction lines. Lable all points in each view.

A

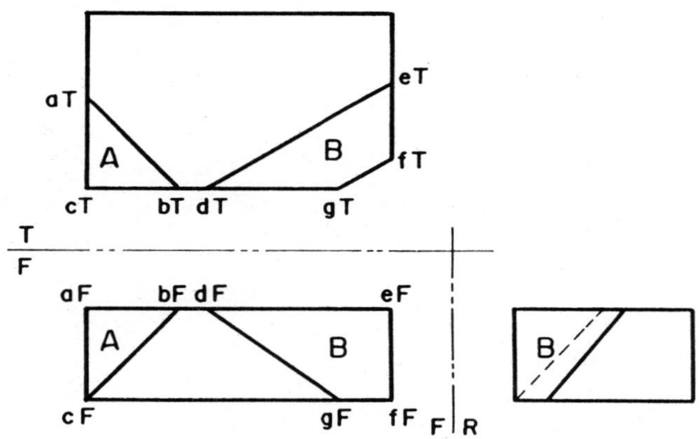

B

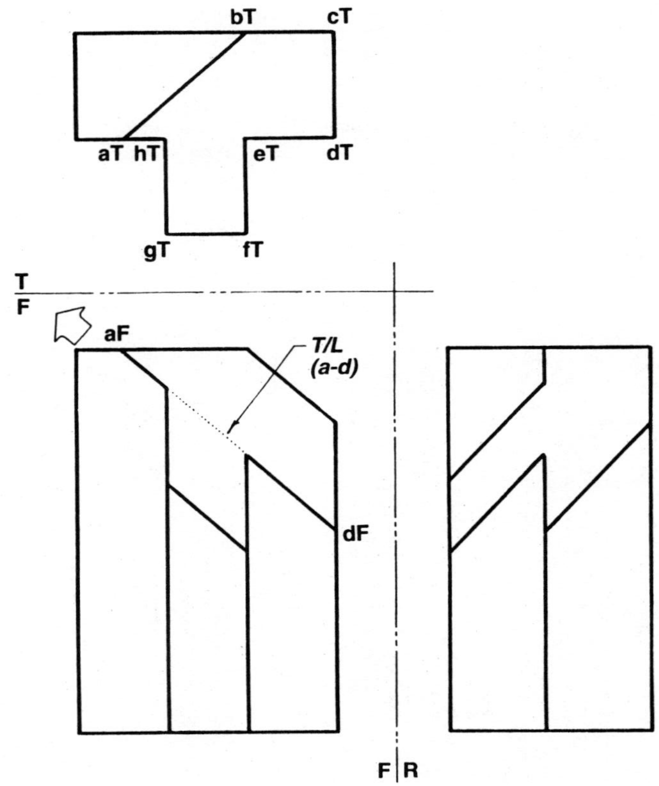

Determine the true size of plane a-b-c-d-e-f-g-h. Project from the front view F. Do not erase construction lines. Lable all points in each view.

Ch8 Descriptive Geometry			Problem
Section-Chapter-Problem 2 8 7	True size of a plane Figure 8-17	Course________ Section_____ Student____________ Date_____	2-8-7

A

Determine the true size of plane a-b-c-d. Do not erase construction lines. Lable all points in each view.

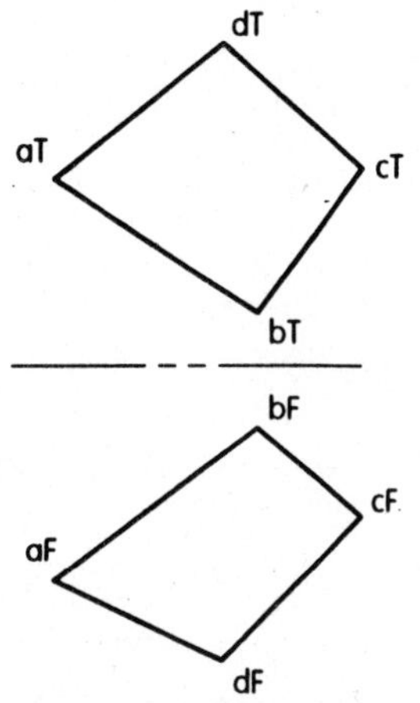

B

Determine the true angle between lines a-b and b-c. Do not erase construction lines. Lable all points in each view.

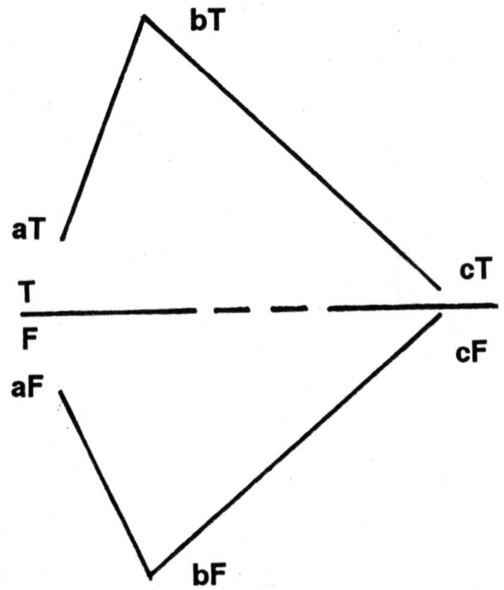

C

Determine the true size of plane a-b-c-d. Do not erase construction lines. Lable all points in each view.

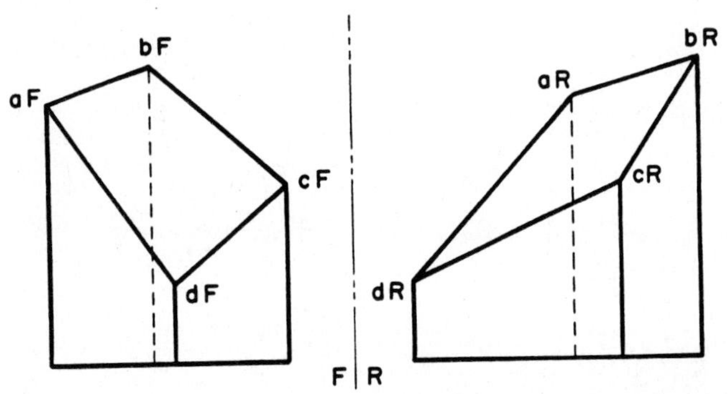

Ch8 Descriptive Geometry			Problem 2-8-8
Section-Chapter-Problem 2 8 8	True size of a plane Figure 8-17	Course__________ Section_____ Student________________ Date______	

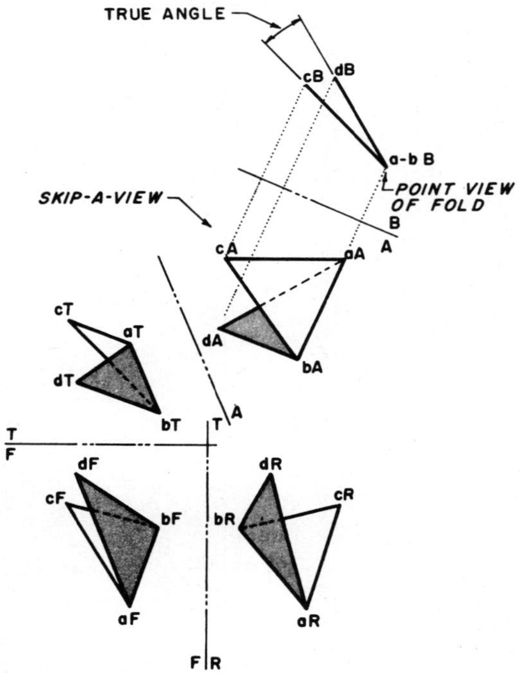

EXAMPLE: DIHEDRAL ANGLE CONSTRUCTION

A

Determine the true angle between the. windshield planes. Do not erase construction lines

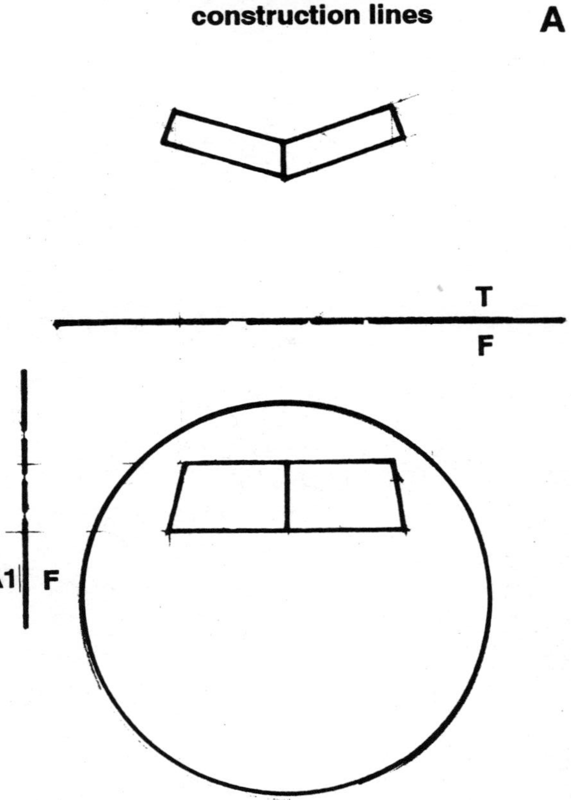

B

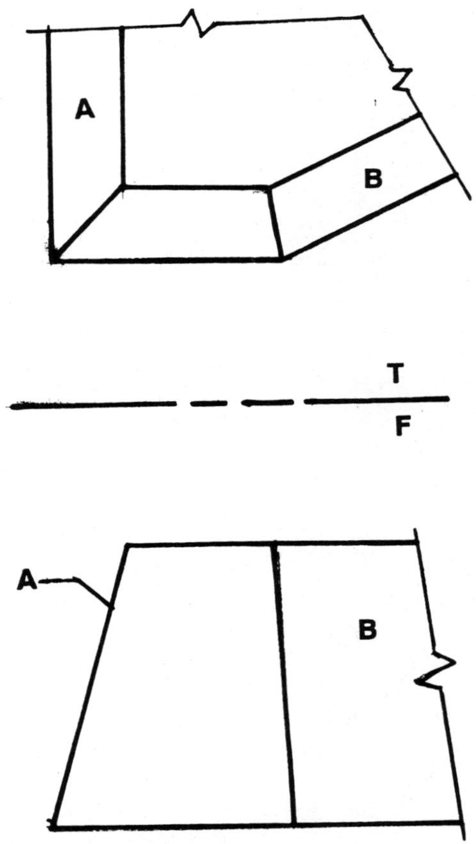

Determine the true angle between planes A and B for the bulwork shown. Do not erase construction lines. (Hint:: Extend plane B until it intersects with plane A.)

Ch8 Descriptive Geometry			Problem
Section-Chapter-Problem 2 8 9	True angle between two plane surfaces Figure 8-19	Course__________ Section_____ Student________________ Date______	2-8-9

Determine the true angle between plane a-b-c and plane b-c-d. Also, determine the shortest connector between plane b-c-d and point E. Show the connector in all views. Do not erase construction lines. Lable all points in each view.

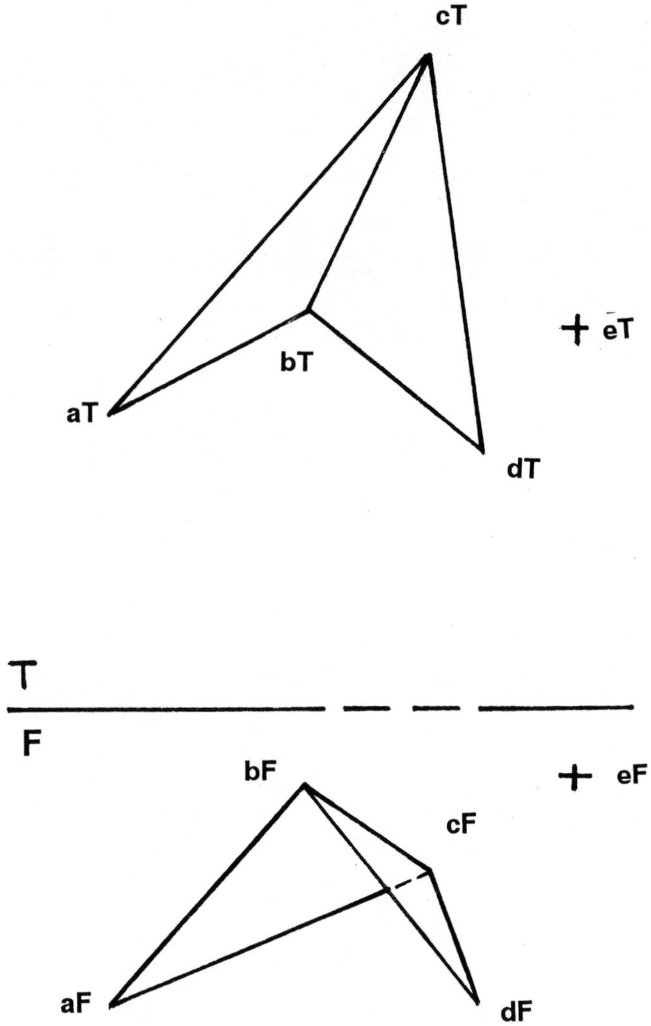

Ch8 Descriptive Geometry			Problem
Section-Chapter-Problem 2 8 10	True angle between two plane surfaces Figure 8-19	Course__________ Section_____ Student______________ Date______	2-8-10

A

Use the line projection method to determine the intersecton of the plane and the cylinder. Do not erase construction lines.

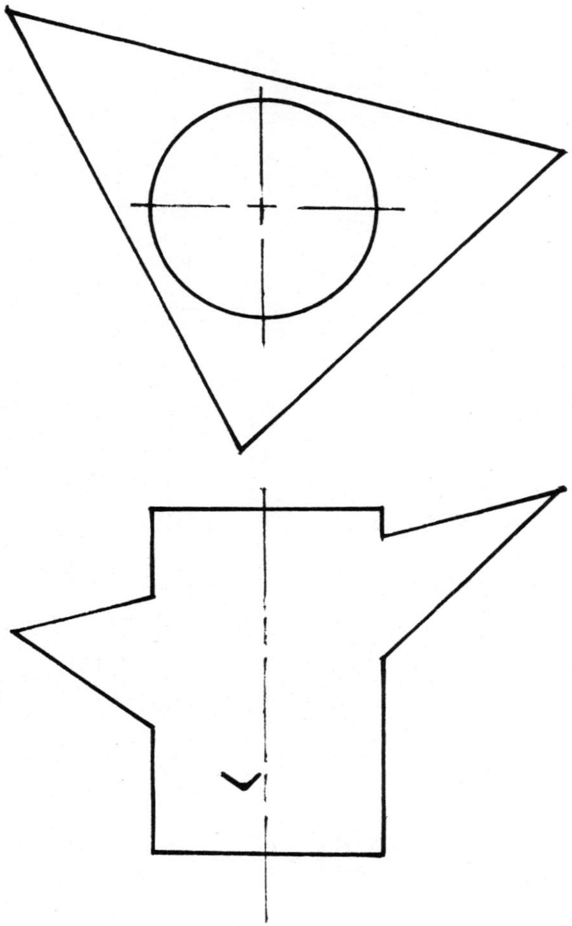

B

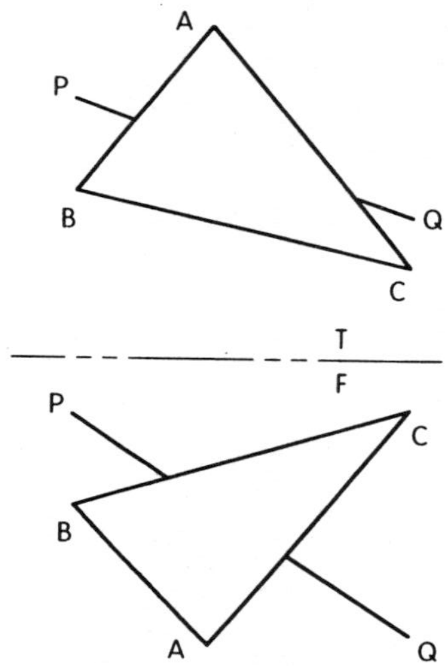

Use two different methods to determine the point where lline P-Q pierces plane A-B-C: the projection (edge view) method and the line projection (cutting-plane) method. Do not erase construction lines. Lable all points in each view.

C

Complete the front view F. Do not erase construction lines.

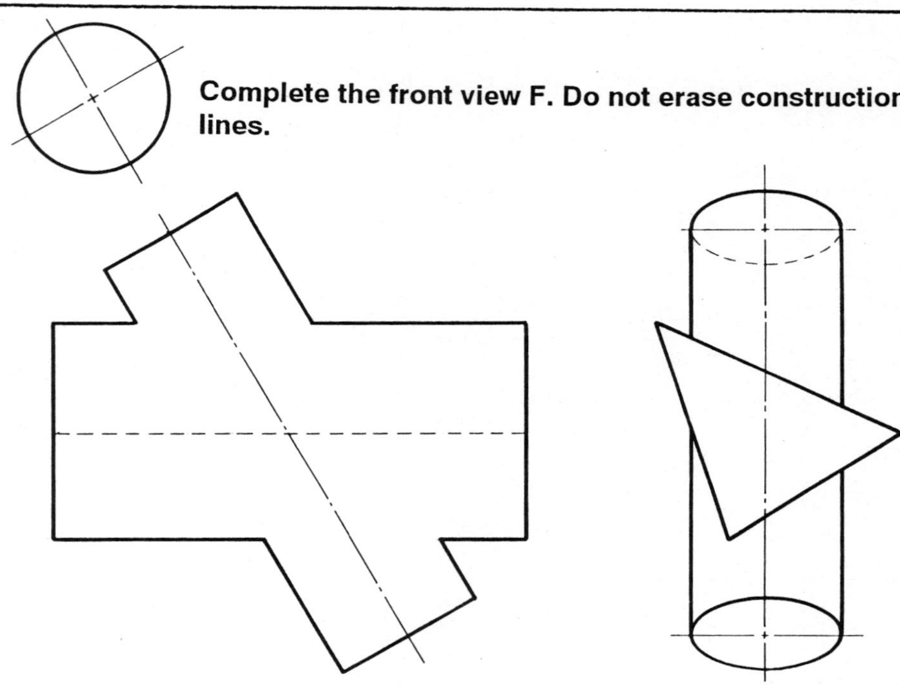

Ch8 Descriptive Geometry

Section-Chapter-Problem	Intersections of planes, and prisms Figures 8-23 & 8-30	Course__________ Section_____	Problem
2 8 11	Intersections of lines and planes Figures 8-23 & 8-30	**Student**______________ Date_____	2-8-11

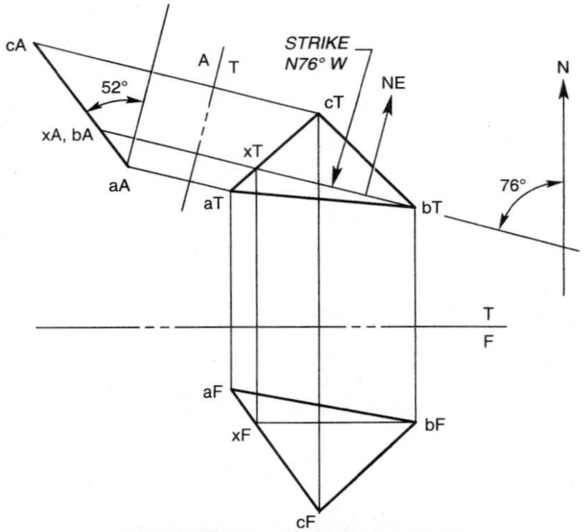

EXAMPLE: STRIKE AND DIP CONSTRUCTION

A

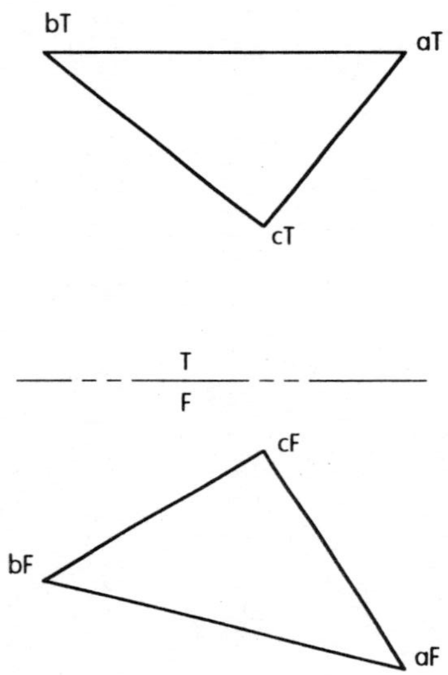

Determine the strike and dip of plane a-b-c shown. Show and lable them in the T view. Do not erase construction lines.

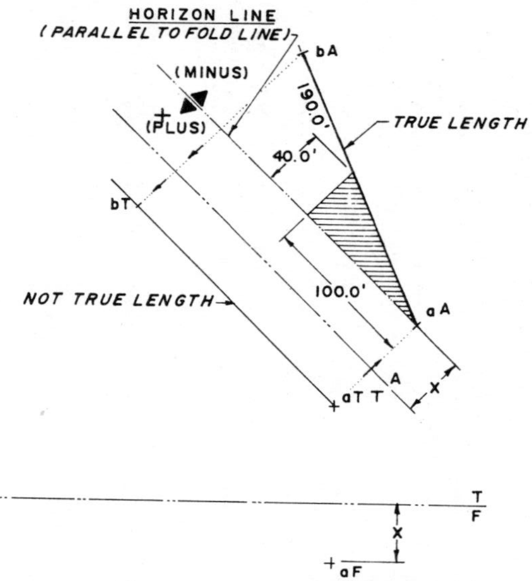

EXAMPLE: LINE LAYOUT

B

Construct a line N80°W at -20° slope from point "a" shown in the T and F views, 1000' long. Show the line in the T and F views. Do not erase construction lines.Scale 1"=500'.

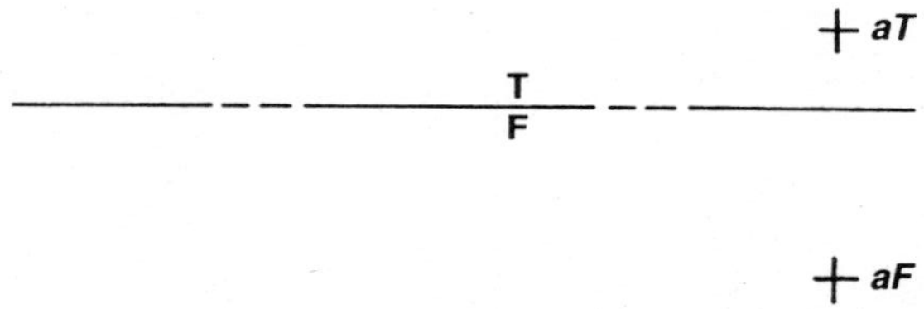

Ch8 Descriptive Geometry			
Section-Chapter-Problem 2 8 12	Strike & Dip Figure 8-59 Specified bearing, grade, & length Figure 8-53	Course________ Section____ Student____________ Date____	Problem 2-8-12

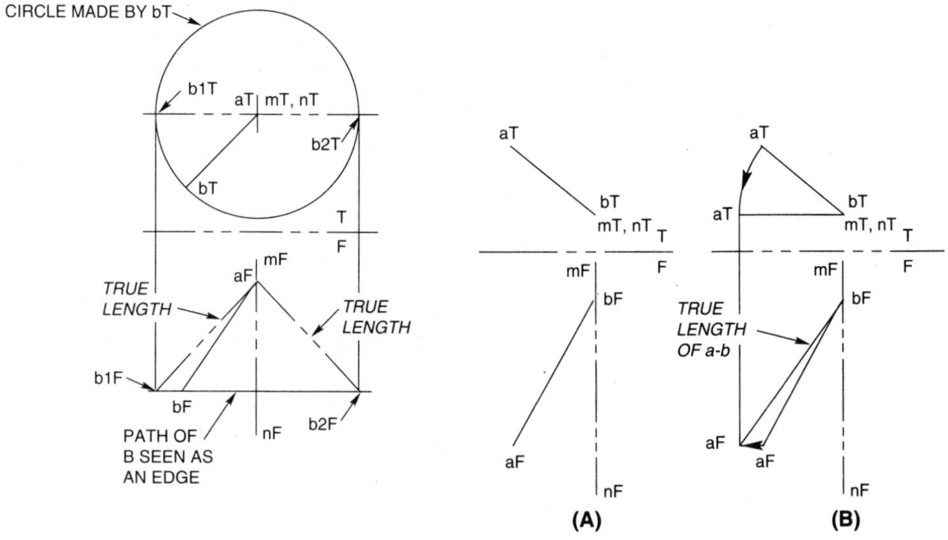

EXAMPLE: REVOLUTION METHOD

A

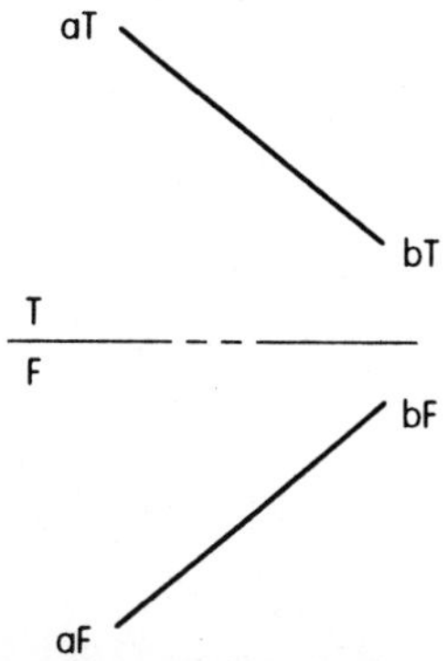

Deterrmine the length of line a-b by the method of revolution. Rotate point aT about point bT and then show the true length of a-b in view F. Do not erase construction lines.

B

Use revolution methods to determine the true lengths of the edges of the plates of the hopper shown. Use centerlines m-n and j-k to rotate lines a-b and c-d respectively. Show true lengths in the F view. Do not erase construction lines.

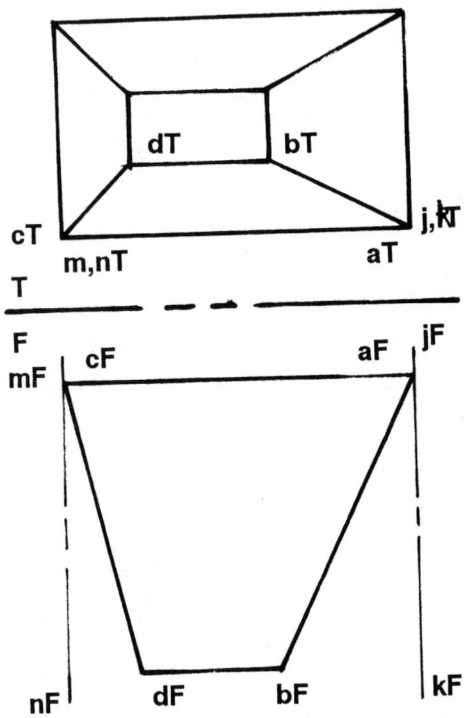

C

Use revolution methods to determine the true lengths of the guy wires a-o, b-o, & c-o supporting pole o-p. Use centerline o-p to rotate

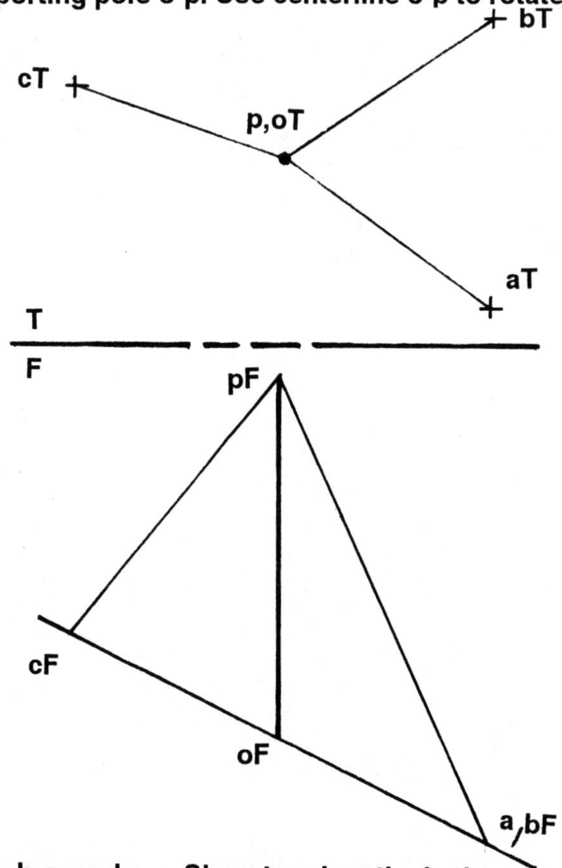

lines a-o, b-o, and c-o. Show true lengths in the F view. Do not erase construction lines.

Ch8 Descriptive Geometry			Problem 2-8-13
Section-Chapter-Problem 2 8 13	True length of a line byRevolution Figure 8-84	Course______ Section____ Student______ Date____	

The proposed roadway shown is to be horizontal at an elevation of 80 meters.Plot the cut and fill areas at a slopes of 1:1. Lable the cut and fill areas. The contour lines are at 5 meter intervals. Use the suggested elevations and the location of the roadway shown. Do not erase construction lines.

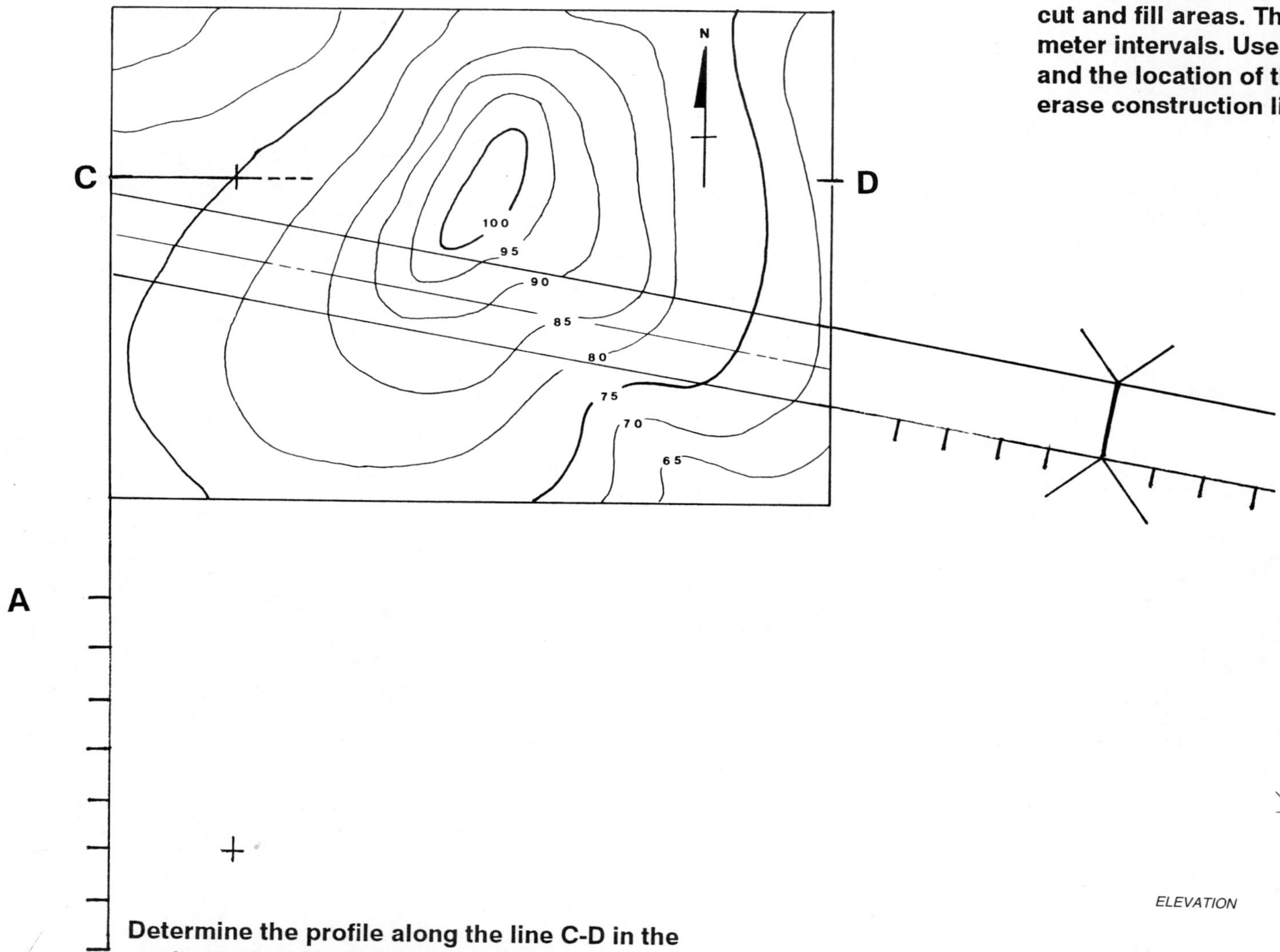

Determine the profile along the line C-D in the contour map above. One point on the profile plot is given. Do not erase construction lines.

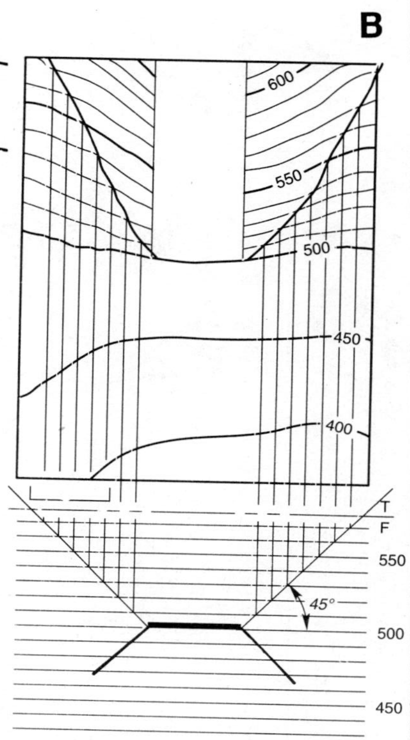

EXAMPLE: CUTS CONSTRUCTION (CUTS AND FILLS)

Ch8 Descriptive Geometry

Section-Chapter-Problem	Contour lines and Profiles Figure 8-55	Course______ Section_____	Problem
2 8 14	Cuts and Fills Figures 8-56 & 8-57	Student______ Date_____	2-8-14

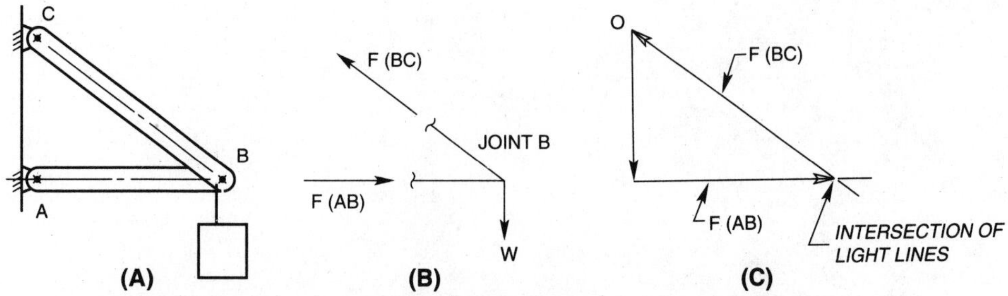

EXAMPLE: VECTOR GRAPHICS WITH FORCE DIAGRAM

A

Determine the magnitude and type of force (tensile or compressive) in members AB and BC due to the applied load of 500 kN. Members AB and BC are drawn to scale. Show your answers in a force diagram. Do not erase construction lines.

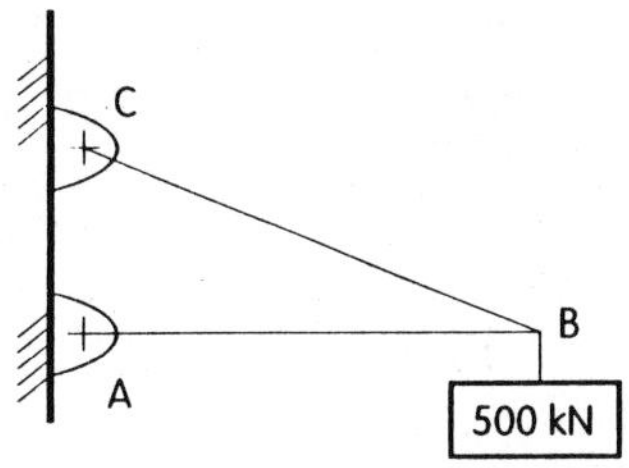

B

Determine the magnitude and type of force (tensile or compressive) in members RS and ST due to an applied force F of 10,000 pounds. Members RS and ST are drawn to scale. Show your answers in a force diagram. Do not erase construction lines.

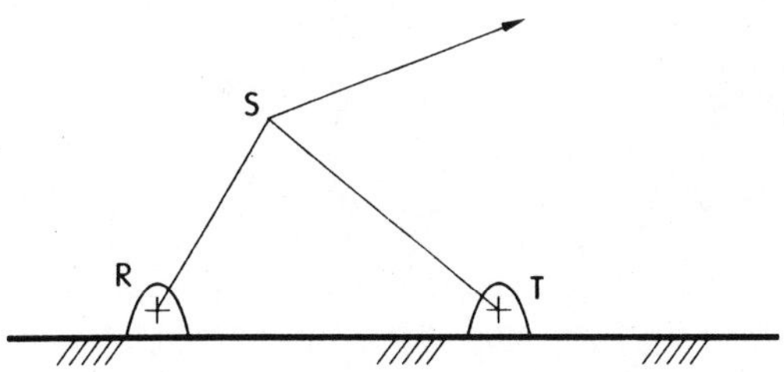

Ch8 Descriptive Geometry			Problem
Section-Chapter-Problem 2 8 15	Vector graphics Figure 8-73	Course________ Section_____ Student____________ Date_____	2-8-15

1. The “object name” shown in drawing file “08-1.dxf” is not complete. It shows the top and front view of two objects that intersect. Draw the intersection lines in red and on a separate layer.
2. A contour map is shown is drawing file “08-2.dxf”. An architect is planning to design an “L” shaped house on the property. There will need to be some excavating done in order for the house to be located on the property. Redesign the contour for the house and a driveway and show a before and after excavation profile drawing at the center of the house.

Ch8 Descriptive Geometry			Problem
Section-Chapter-Problem 2 8 16	Intersections Topography	Course__________ Section_____ **Student**________________ Date_____	**2-8-16**

From the development of the "box" given do the following:

a. Determine the shape and construct the three orthographic views F, T & S of the box. Put the small square side in front. Assume that all of the fold lines require an up bend. Scale the figure for dimensions.

b. Determine the number of tabs required for this object. TABS= ________

c. Prepare an isometric drawing of the box with the small square side at the left front.

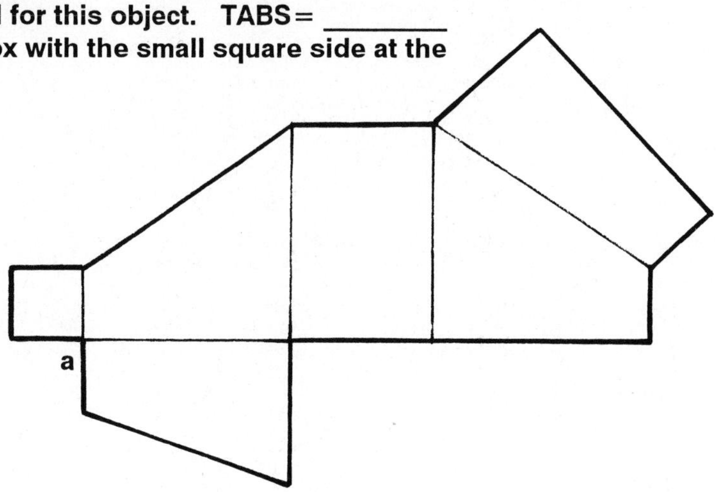

Ch9 Patterns and Developments			Problem
Section-Chapter-Problem 2 9 1	Parallel lines Figure 9-8	Course__________ Section_____ Student__________________ Date______	2-9-1

A drafter's files revealed the development drawing shown. Add the missing tabs and determine the object drawing used by the drafter to prepare the development. Show your answer in three appropriate views F, T, & S.

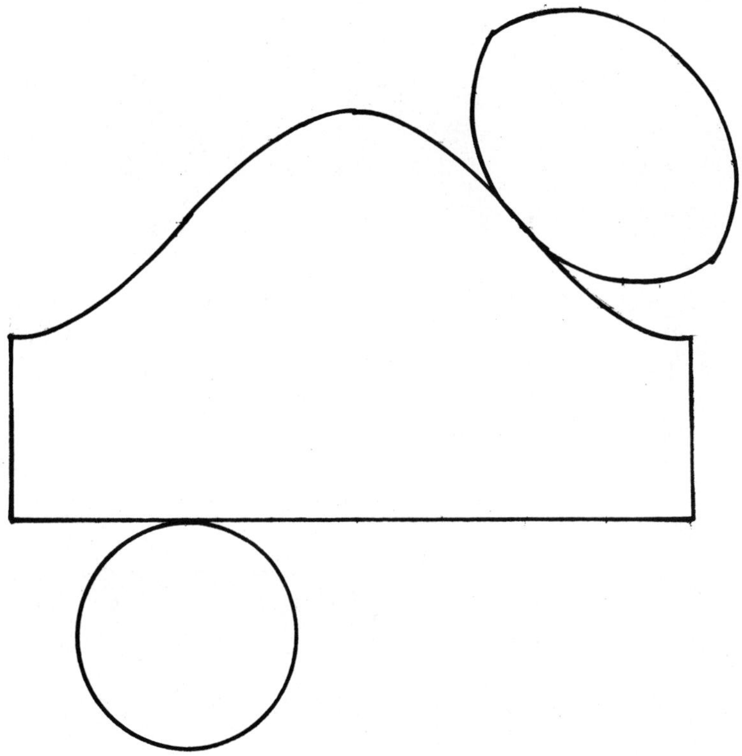

Ch9 Patterns and Developments			Problem
Section-Chapter-Problem 2 9 2	Parallel lines Figure 9-9	Course__________ Section_____ Student__________________ Date______	2-9-2

Using the parallel line development method, develop the objects shown starting from the seam indicated. Add the required tabs. Use half scale.

A

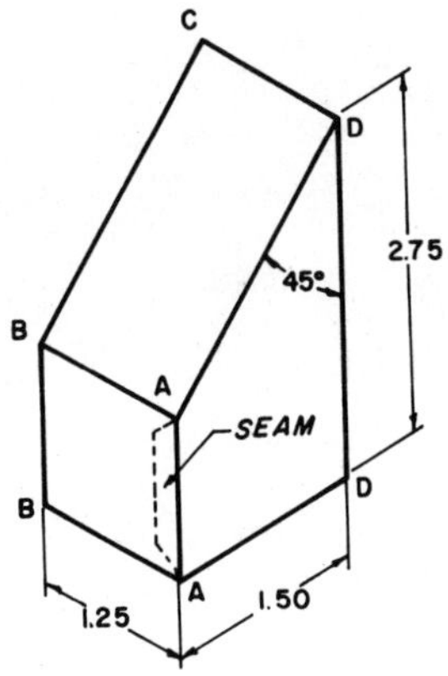

A

A

B

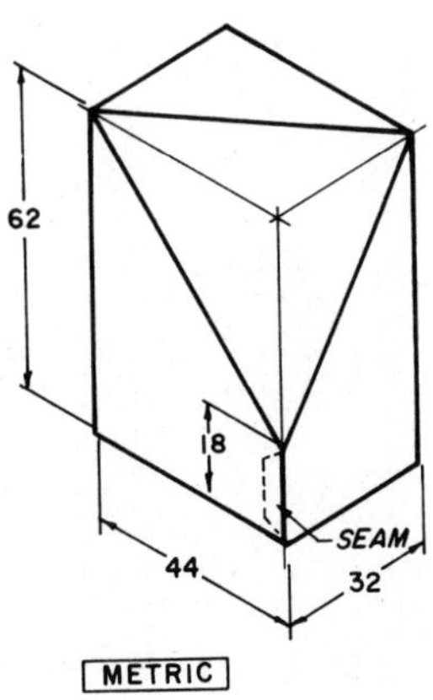

Ch9 Patterns and Developments			Problem
Section-Chapter-Problem 2 9 3	Parallel lines Figure 9-8	Course________ Section_____ Student________ Date_____	2-9-3

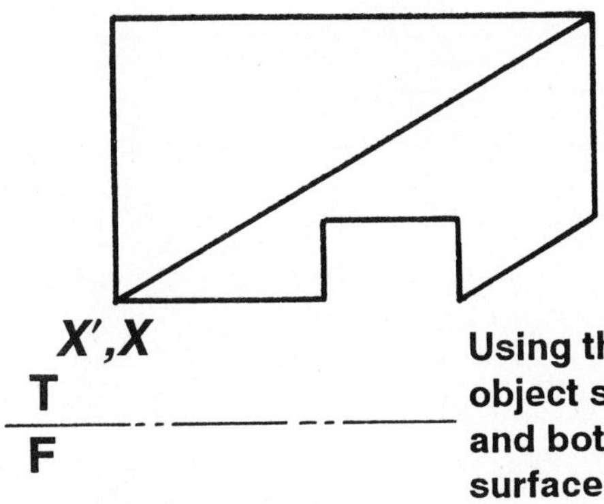

Using the parallel line development method, develop the hollow object shown starting from the seam indicated. Add tabs to the top and bottom of each vertical side.. Determine the true size of the top surface before starting the development. Add a tab along edge X'-X.

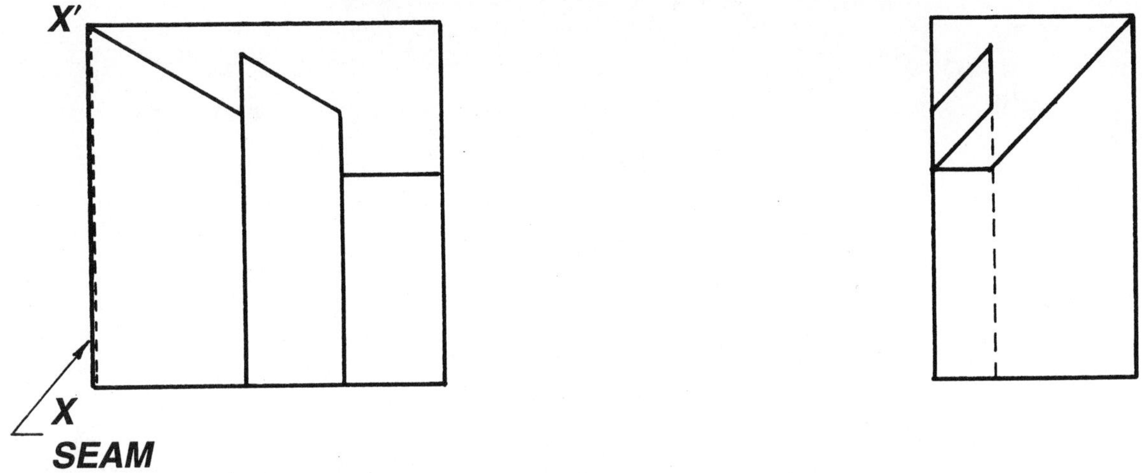

Ch9 Patterns and Developments			Problem
Section-Chapter-Problem 2 9 4	Parallel lines Figure 9-8	Course__________ Section_____ Student_______________ Date______	2-9-4

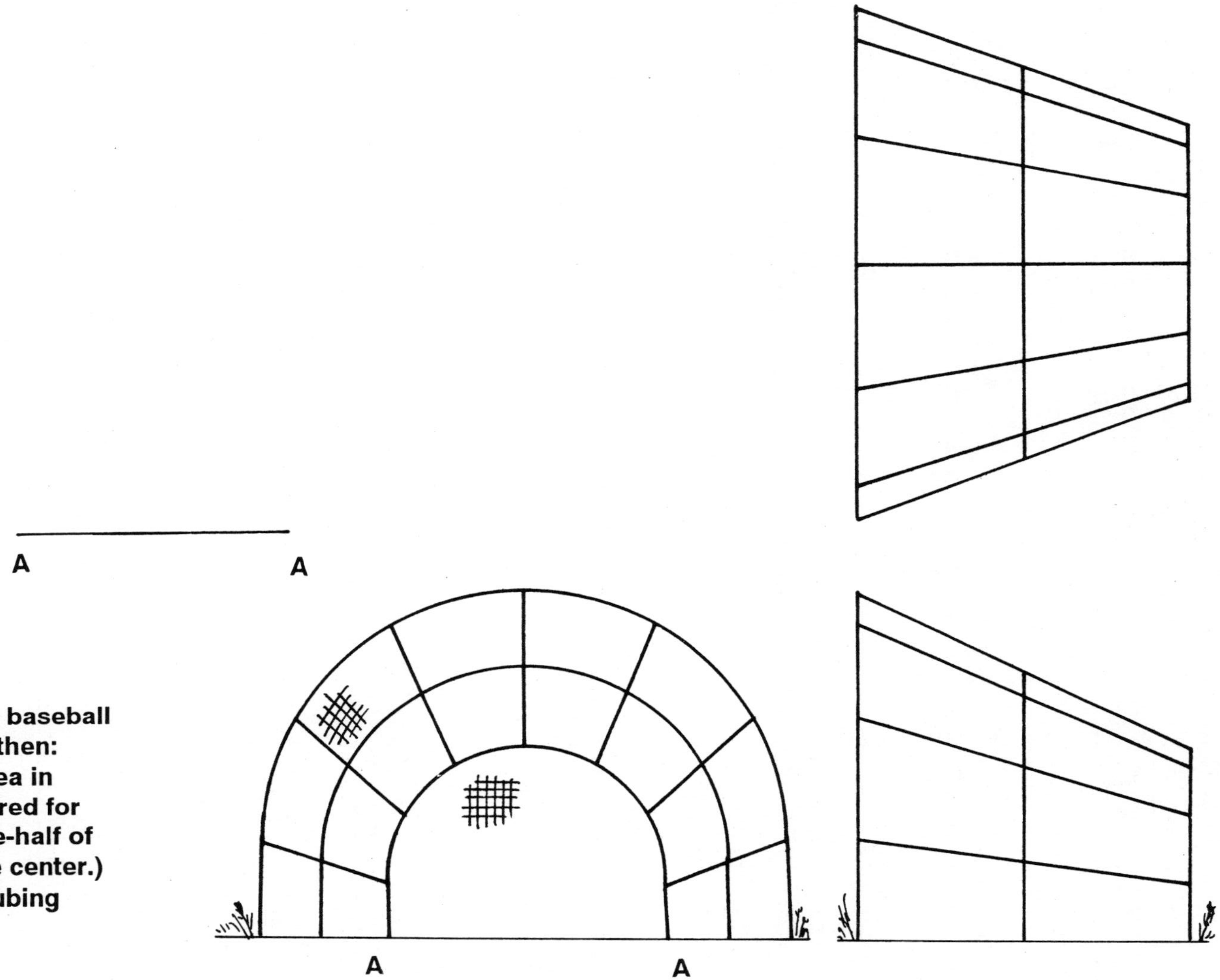

Use the three views of a conial-shaped baseball backstop to develop a flat pattern and then:

a. Determine an approximate area in square meters of the wire screen required for the backstop. (Suggestion: do only one-half of the conical development starting at the center.)

b. Estimate the length of steel tubing required to support the screen. Scale 1mm=100mm.

Ch9 Patterns and Developments			Problem 2-9-5
Section-Chapter-Problem 2 9 5	Triangulation Figure 9-17	Course__________ Section_____ Student______________ Date_____	

A

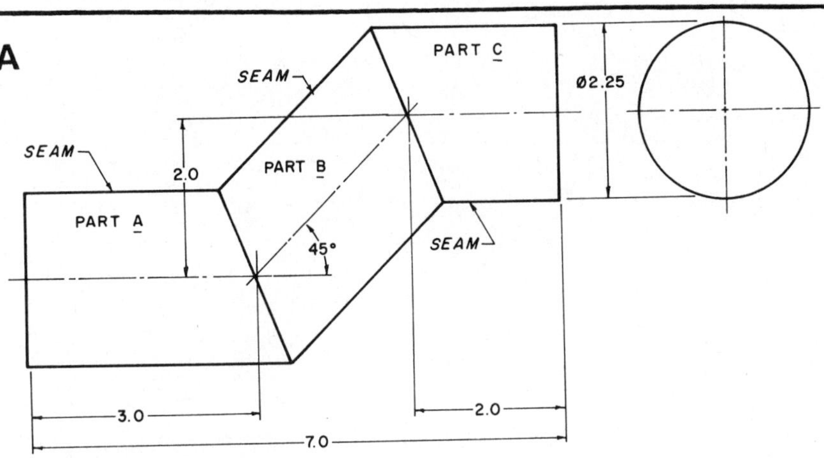

Use the parallel line development method to develop the two transition sections shown. Scales: If using separate A size sheets use full scale. For using this sheet scale the figures as drawn.

B

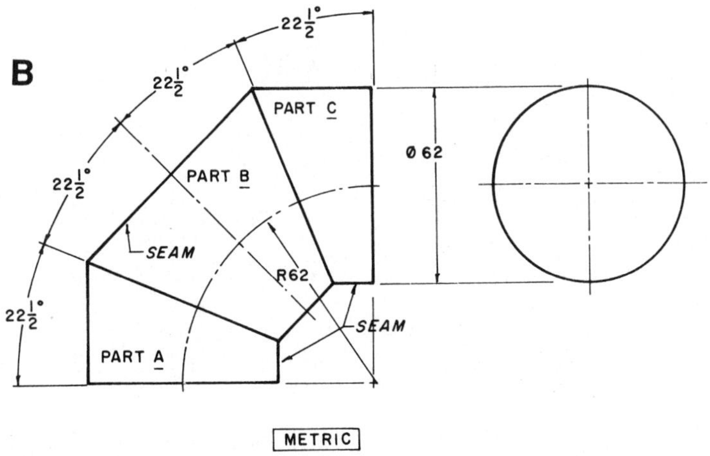

Ch9 Patterns and Developments			Problem
Section-Chapter-Problem 2 9 6	Parallel lines Figure 9-9	Course______ Section_____ Student______ Date_____	2-9-6

If using separate A size sheets use full scale. For using this sheet scale the hollow figures as drawn and complete the intersections. Then do developments of the objects and add the required tabs.

A

ALL DIMENSIONS ARE TO OUTSIDE LIMITS OF PARTS

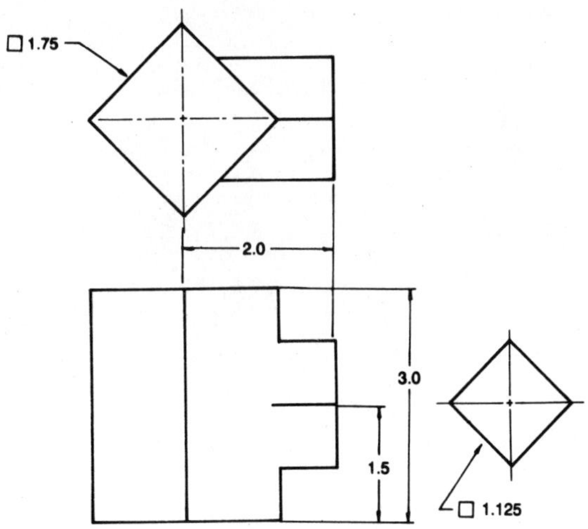

B

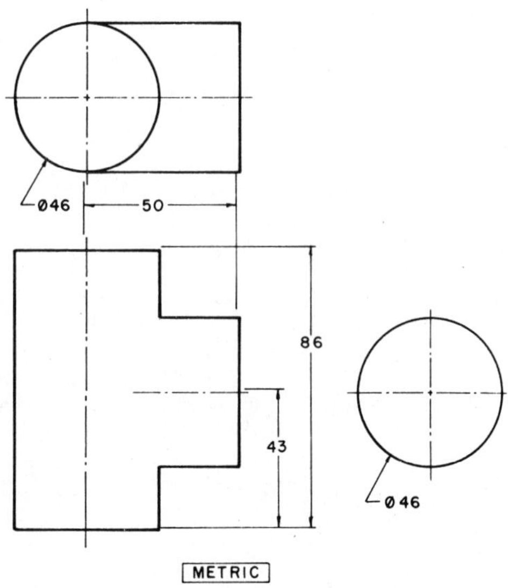

Ch9 Patterns and Developments			Problem
Section-Chapter-Problem 2 9 7	Parallel lines Figures 9-8 & 9-9 Figures 5-56 & 8-32	Course__________ Section_____ Student________________ Date______	2-9-7

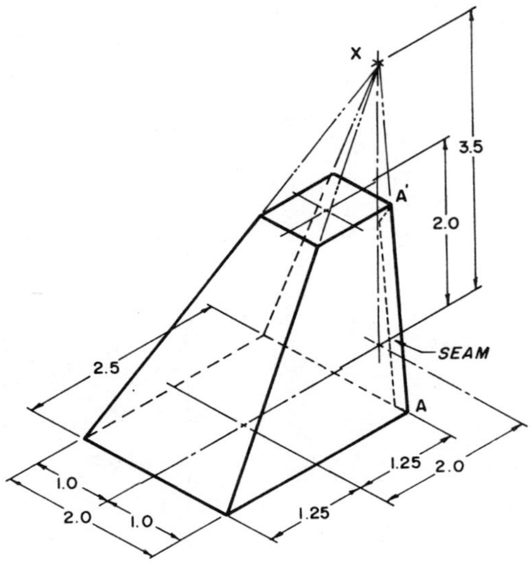

Use the radial line development method to develop the closed top and bottom, truncated, pyramid shown. Add the required tabs. If using separate A size sheets use full scale. For using this sheet use half scale. Prepare orthographic views as needed.

X

xT

xF

Ch9 Patterns and Developments			Problem
Section-Chapter-Problem 2 9 8	Radial lines Figure 9-13	Course______ Section______ Student______ Date______	2-9-8

Use the radial line development method to develop the closed top and bottom, truncated, cone. shown. Add the required tabs. Use full scale. Prepare orthographic views as needed.

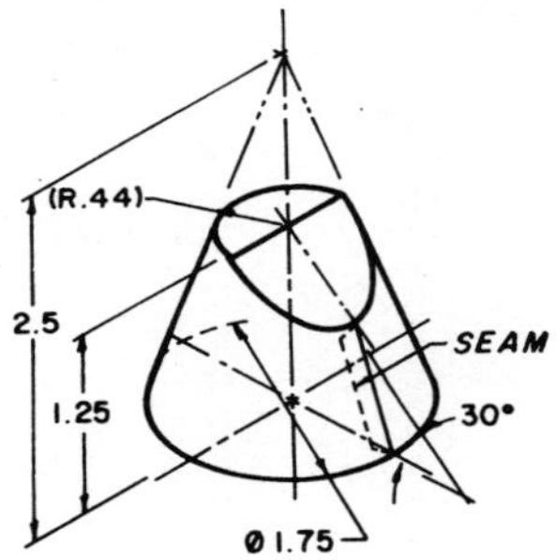

Ch9 Patterns and Developments			Problem
Section-Chapter-Problem 2 9 9	Radial lines Figure 9 -14	Course____________Section_____ Student______________________Date_____	2-9-9

B **Use the triangulation method to develop the open top and bottom, hopper shown. Add the required tabs. Scale the figure as drawn.**

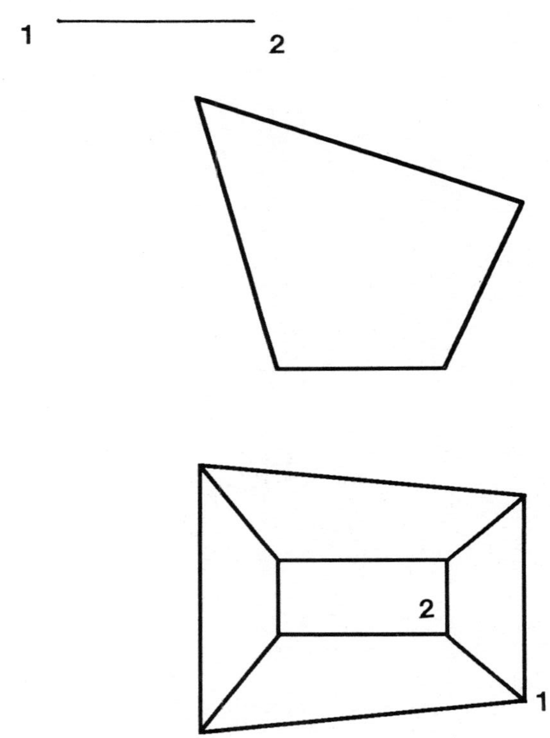

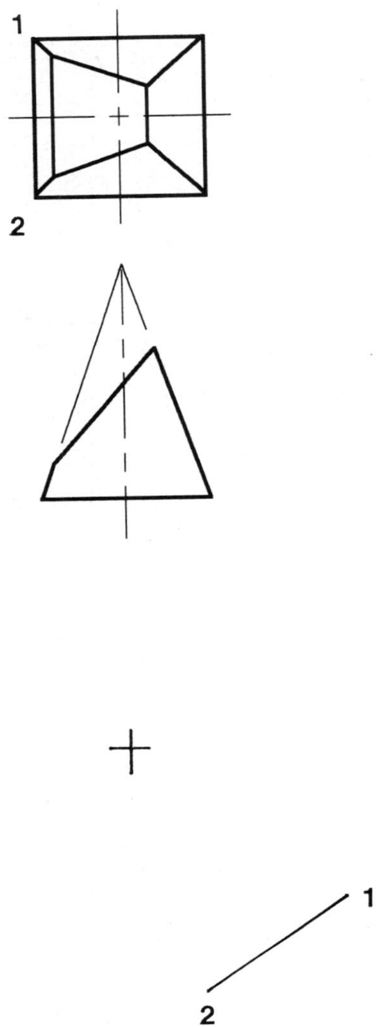

A

Use the radial line development method to develop the closed top and bottom, truncated, pyramid shown. Add the required tabs. Scale the figure as drawn.

Ch9 Patterns and Developments			Problem
Section-Chapter-Problem 2 9 10	Radial lines Figure 9-13 Triangulation Figure 9-15	Course______ Section______ Student______ Date______	2-9-10

Prepare a development of the open top and bottom, transition duct shown. (Hint: Use revolution and a true length diagram.) Add any required tab(s).

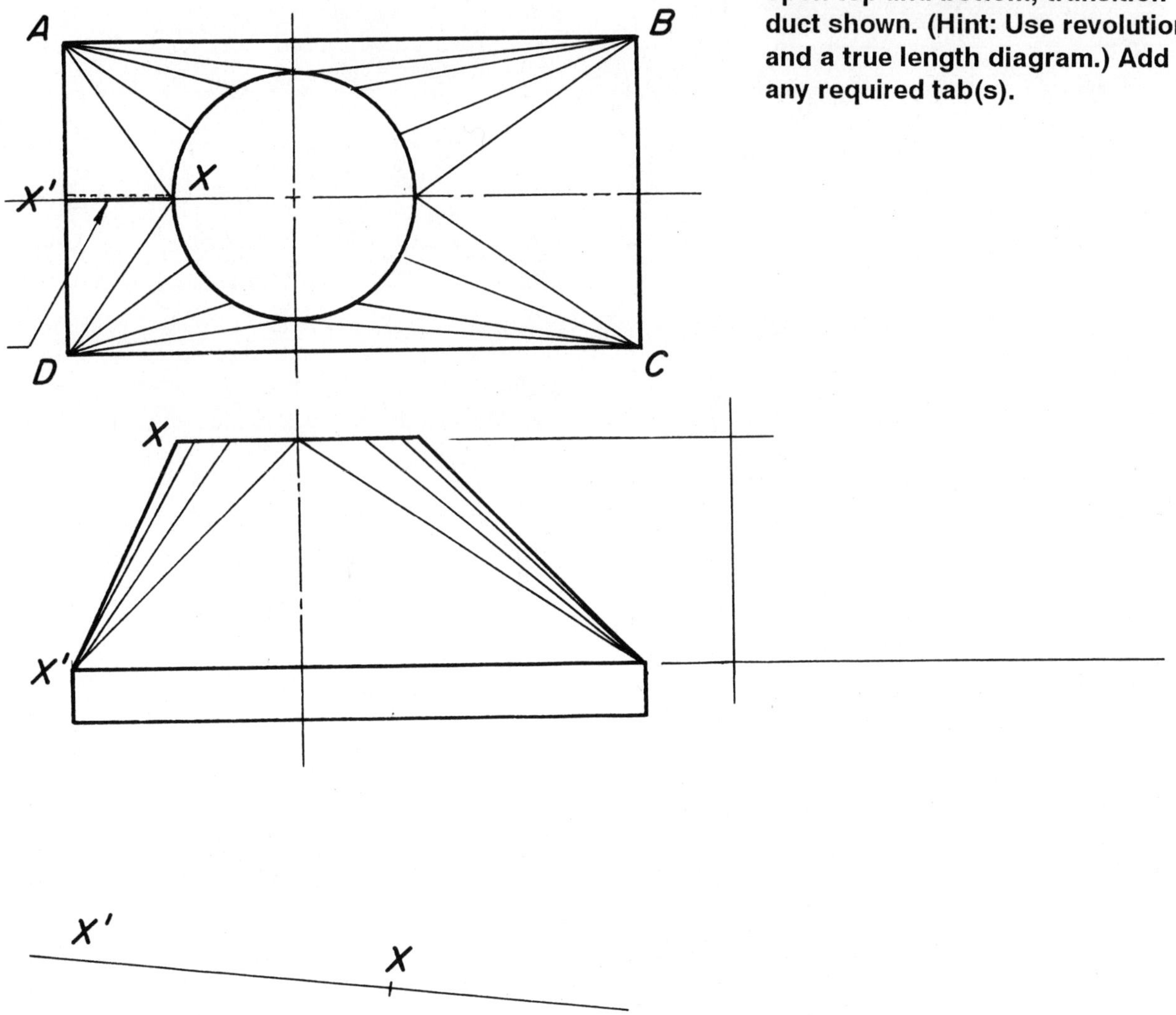

Ch9 Patterns and Developments			Problem
Section-Chapter-Problem 2 9 11	Triangulation Figure 9-16	Course__________ Section_____ Student______________ Date______	2-9-11

Prepare an isometric drawing of the development shown. The development (approximate detail) is a cardboard fold-up package used commercially. No top or bottom is employed; however, once the "box" is assembled and a product included, the "box" is totally enclosed in clear plastic.

All fold lines are down except the two indicated. For better visualization trace the development, cut it out, do the folds, and tape the tab end inside the far left panel.

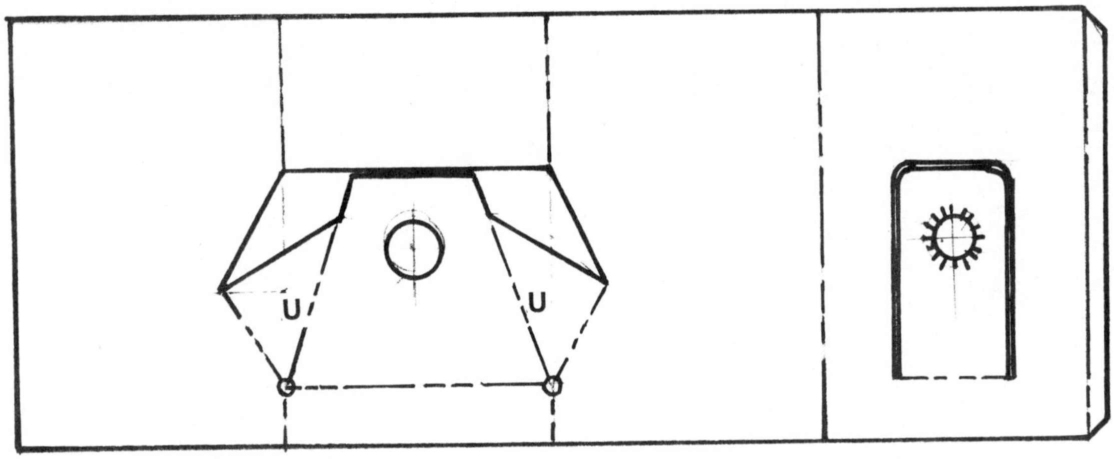

Ch9 Patterns and Developments			Problem
Section-Chapter-Problem 2 9 12	Parallel lines Figure 9-8	Course______ Section____ Student______ Date____	2-9-12

Using development charts calculate the length of each object shown. Round-off answers to three places. Show your work.

A

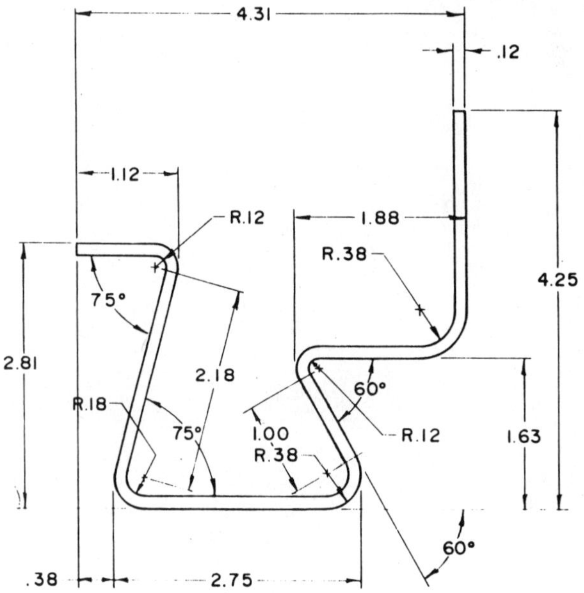

B

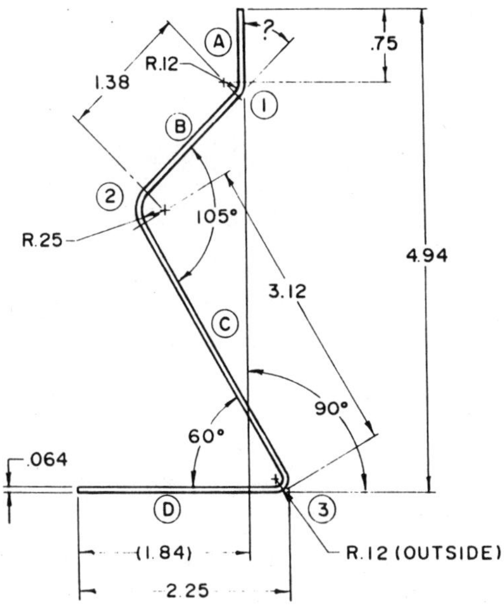

C

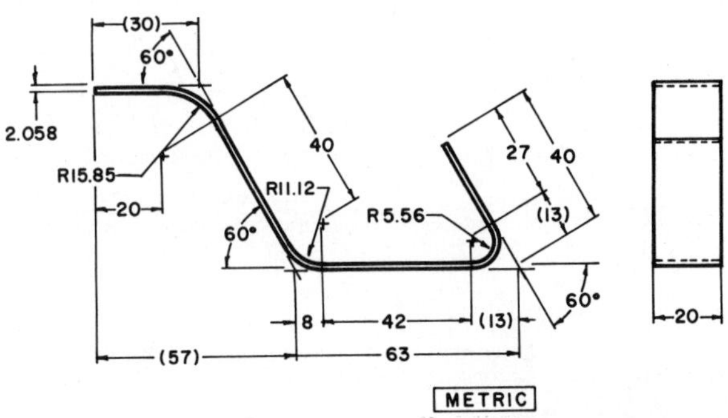

Ch9 Patterns and Developments			Problem 2-9-13
Section-Chapter-Problem 2 9 13	Bends Figure 9-20	Course____________Section______ Student____________________Date________	

A A heart-shaped chocolate tin is shown in drawing "09-1.dxf". Draw a pattern development of the base. Note that there are two dividers within the box. No more than two pieces of metal should be used for the divider and the base. The bottom will be fastened separately.

B Design a new side order container for a fast food restaurant. It must hold both French fries and onion rings, but they must not come in contact with each other. Draw the pattern development for this container. A sample French fry holder pattern development is shown in drawing file "09-2.dxf."

Ch9 Patterns and Developments			Problem
Section-Chapter-Problem 2 9 14	Flat Patterns for Food Containers	Course__________ Section_____ **Student**________________ Date______	**2-9-14**

Measure and dimension, unidirectionally, only the features noted in the captions. Full size. Use accepted practices, decimal inches to two places, and decimal angles to two places. Measure to the centers of lines.

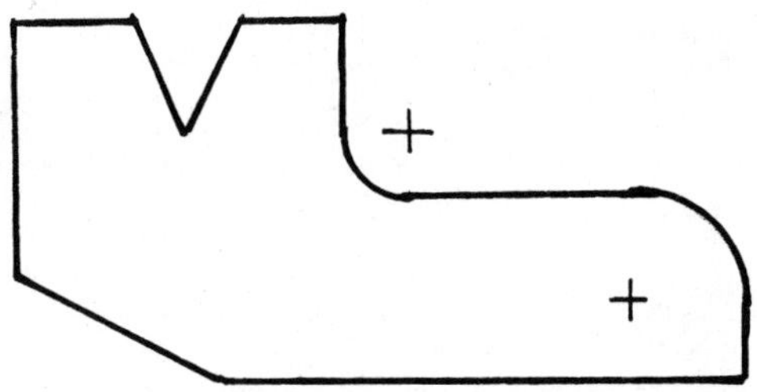

Angles and Rounds, Figures 10-33--10-36

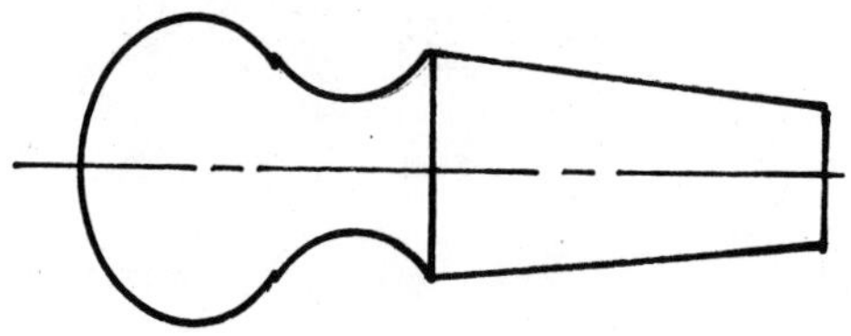

Offsets of Round Object, Figure 10-45

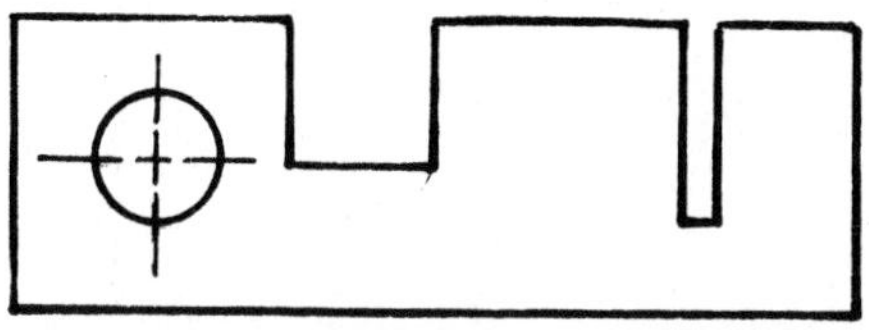

Hole and Grooves, Figures 10-13, 10-26

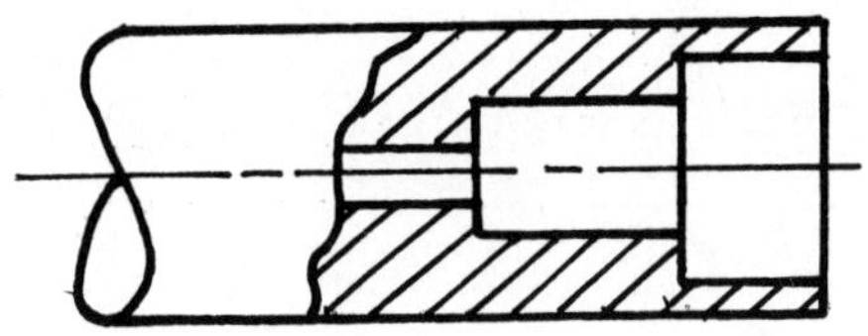

Multiple Diameters, Figure 10-18

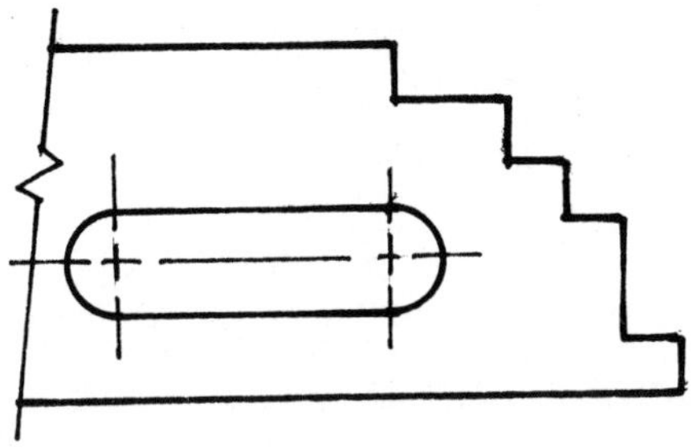

Slot and Steps, Figures 10-19, 10-66

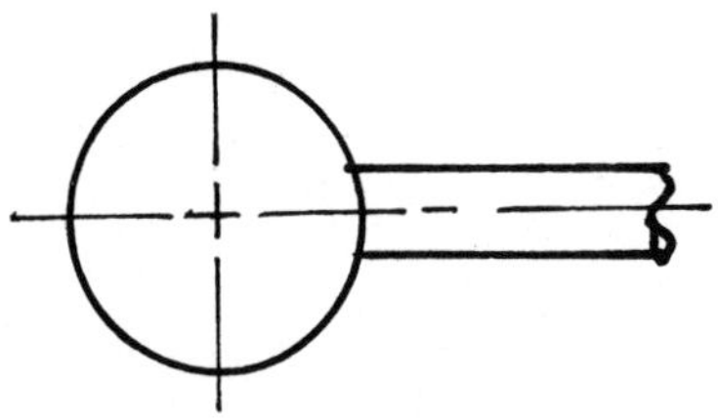

Sphere, Figure 10-47

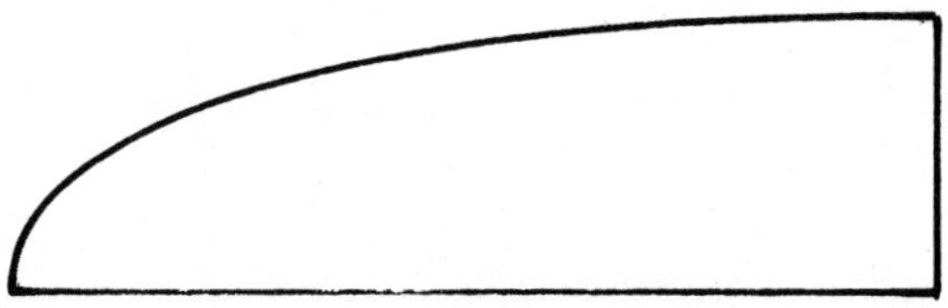

Irregular Curve, Figure 10-40

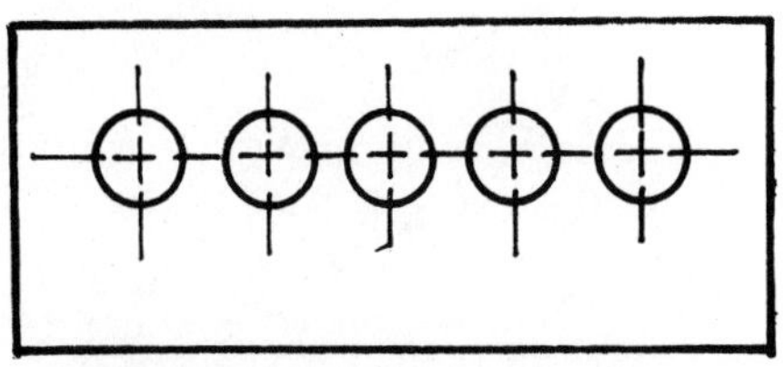

Repetitive Feature, Figure 10-54

Ch10 Dimensioning and Notation			Problem
Section-Chapter-Problem 2 10 1	**DIMENSIONING**	Course__________ Section_____ Student_______________ Date______	2-10-1

Use the grid to determine all dimensions in decimal inches to two places. Redraw the views shown, in the spaces indicated. Assume the grid has 1/2" spacing. Leave 1" actual spaces between views. Scale 1" = 3".

CHECK LIST FOR DIMENSIONING
- Size Dimensions
- Location Dimensions
- Placement
- Spacing
- Extension Lines
- Dimension Lines
- Leaders
- Arrow Heads
- Lettering
- Figures

A

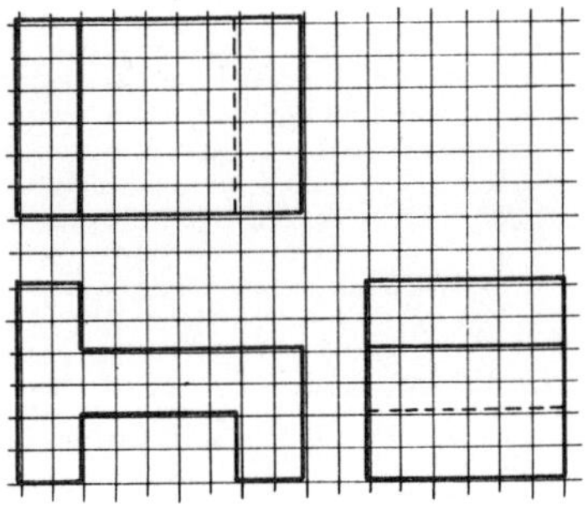

B

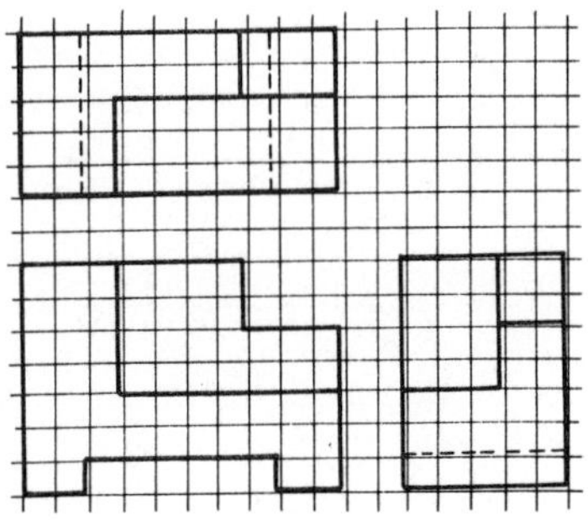

Ch10 Dimensioning and Notation			Problem 2-10-2
Section-Chapter-Problem 2 10 2	Dimensioning Figure 10-15	Course________ Section____ Student__________ Date____	

Use the grid to determine all dimensions in decimal inches to two places. Redraw the views shown, in the spaces indicated. Assume the grid has 1/2" spacing. Leave 1" actual spaces between views. Scale 1" = 2".

CHECK LIST FOR DIMENSIONING
- Size Dimensions
- Location Dimensions
- Placement
- Spacing
- Extension Lines
- Dimension Lines
- Leaders
- Arrow Heads
- Lettering
- Figures

A

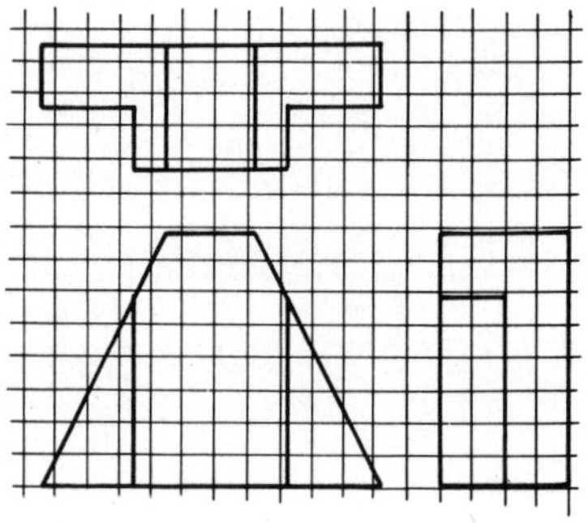

B

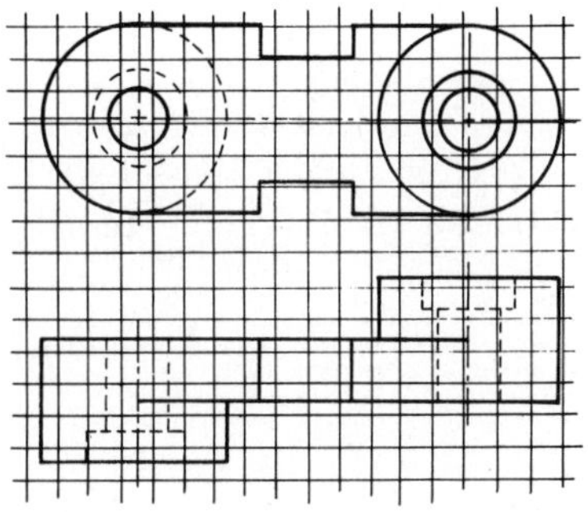

Ch10 Dimensioning and Notation

Section-Chapter-Problem			Geometric breakdown of objects Figure 10-30	Course______ Section______ Student______ Date______	Problem 2-10-3
2	10	3			

Use the grid to determine all dimensions in decimal inches to two places. Redraw the views shown, in the spaces indicated. Assume the grid has 1/2" spacing. Leave 1" actual spaces between views. Scale 1" = 2".

CHECK LIST FOR DIMENSIONING

- Size Dimensions
- Location Dimensions
- Placement
- Spacing
- Extension Lines
- Dimension Lines
- Leaders
- Arrow Heads
- Lettering
- Figures

A

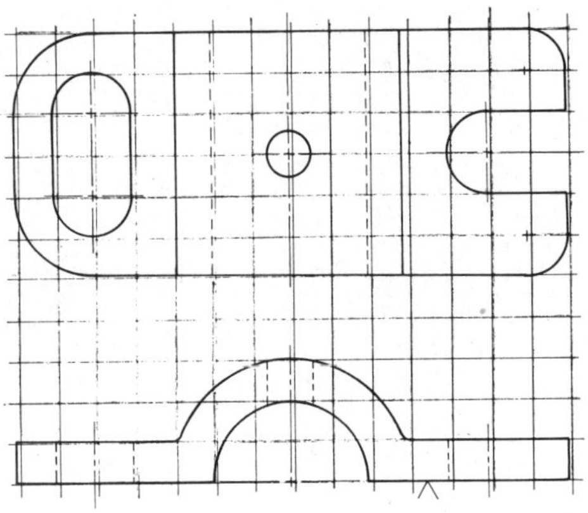

B

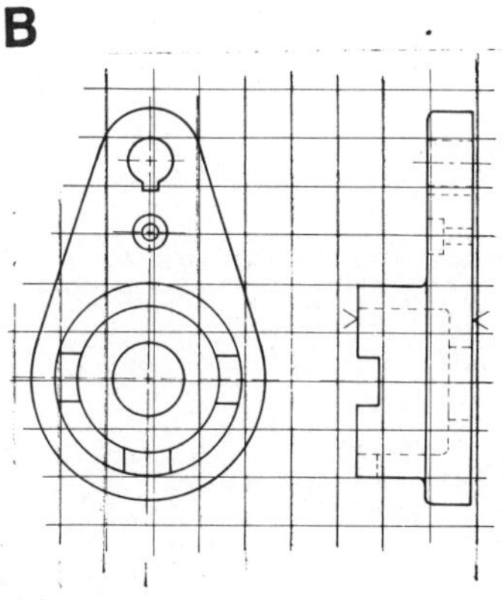

Ch10 Dimensioning and Notation			Problem
Section-Chapter-Problem 2 10 4	Geometric breakdown of objects Figure 10-30	Course______ Section______ Student______ Date______	2-10-4

Use the grid to determine all dimensions in decimal inches to two places. Redraw the views shown, in the spaces indicated. Assume the grid has 1/2" spacing. Leave 1" actual spaces between views. Scale 1" = 2".

CHECK LIST FOR DIMENSIONING
- Size Dimensions
- Location Dimensions
- Placement
- Spacing
- Extension Lines
- Dimension Lines
- Leaders
- Arrow Heads
- Lettering
- Figures

A

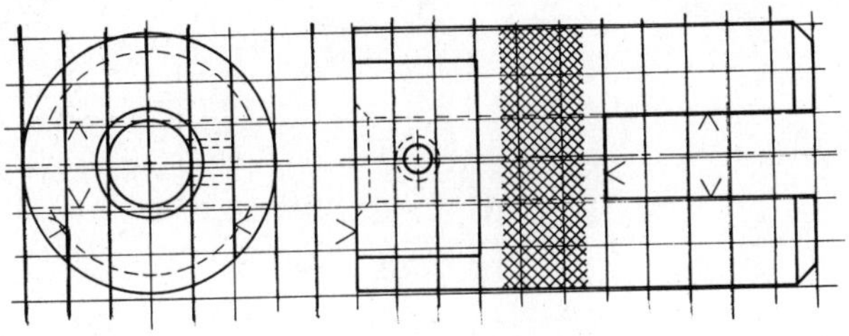

B

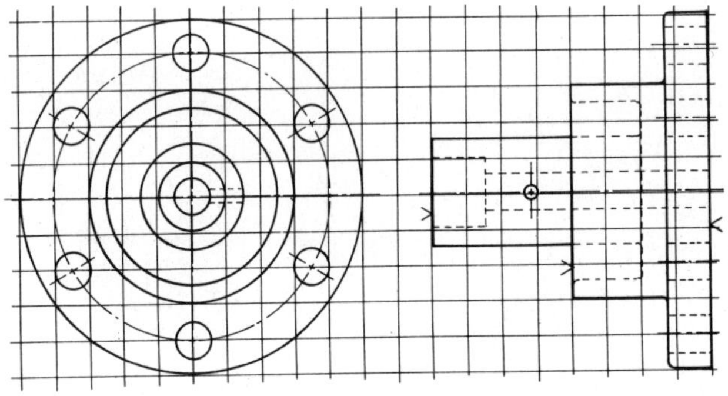

Ch10 Dimensioning and Notation			Problem
Section-Chapter-Problem 2 10 5	Geometric breakdown of objects Figure 10-30	Course__________ Section_____ Student________________ Date______	2-10-5

A

Heavy-in the phantom lines in the top and front views which outline the shape of a computer mouse. Measure and dimension the two views: The top view is composed of arcs and straight lines; the front view, an irregular curve. Decimal millimeters. Accuracy, one place. Scale 1mm = 2mm.

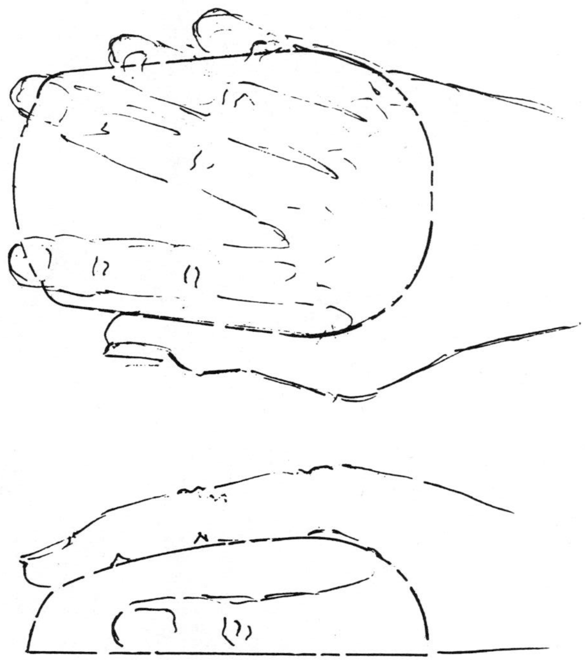

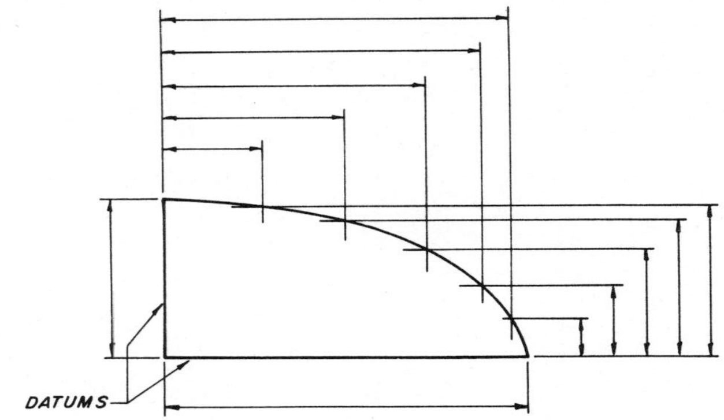

B

Measure and dimension the symmetrical gasket as irregular curves. Decimal millimeters. Accuracy, one place. Scale 1mm = 2mm.

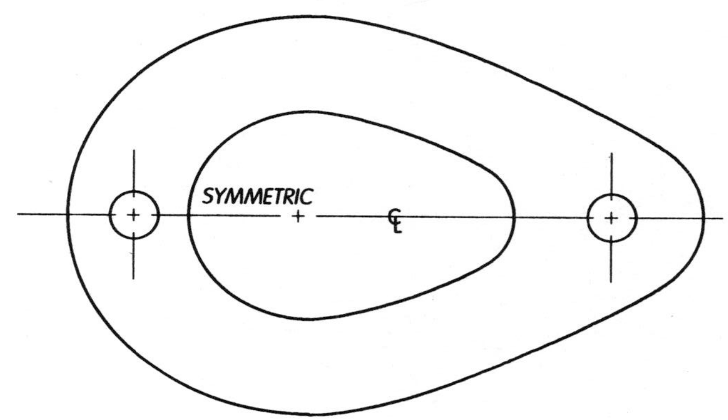

Ch10 Dimensioning and Notation			Problem
Section-Chapter-Problem 2 10 6	Radii dimensioning Figure 10-36 Irregular Curves Figure 10-40	Course________ Section_____ Student__________ Date_____	2-10-6

A Dimension the object shown using coordinate dimensioning for the hole locations. Use standard practices. Decimal millimeters. Accuracy, one place. Measure to centers of lines. Scale 1mm = 2mm.

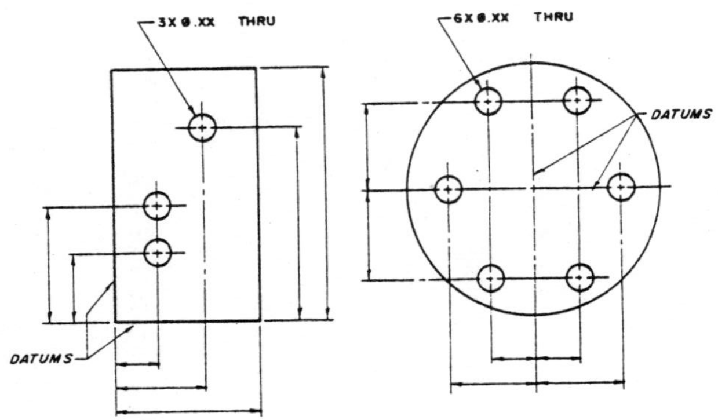

Rectangular coordinate dimensioning

CHECK LIST FOR DIMENSIONING
- Size Dimensions
- Location Dimensions
- Placement
- Spacing
- Extension Lines
- Dimension Lines
- Leaders
- Arrow Heads
- Lettering
- Figures

B Dimension the object shown using standard practices. Decimal inches. Accuracy to two places. Measure to centers of lines. Scale 1" = 2"

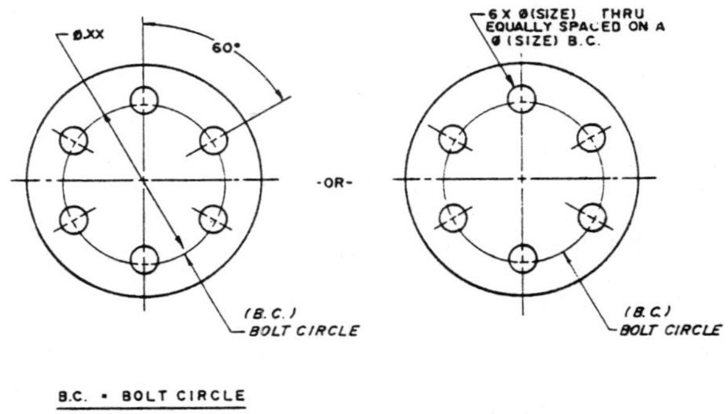

Dimensioning holes on a bolt center diameter

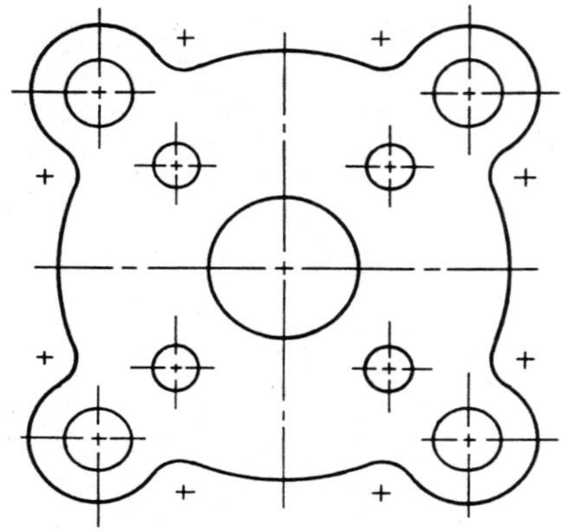

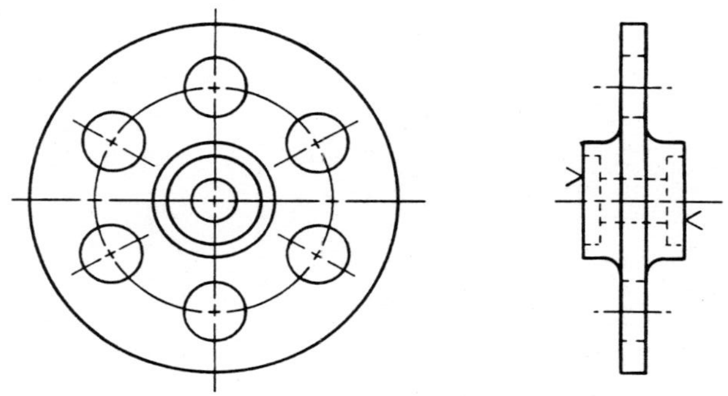

Ch10 Dimensioning and Notation

Section-Chapter-Problem		Course_______ Section_____	Problem
2 10 7	Radii dimensioning Figure 10-36 Rectangular coordinates and Bolt center dimensioning Figures 10-90, 91	Student_______ Date_____	2-10-7

A Dimension the object shown using standard practices. Decimal millimeters. Accuracy, one place. Measure to centers of lines. Scale 1mm = 3mm.

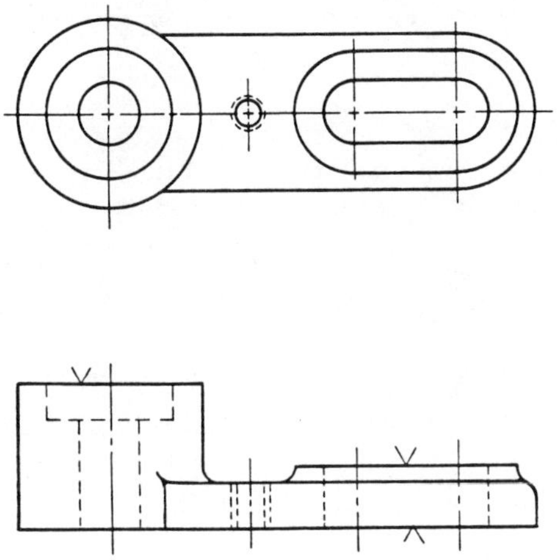

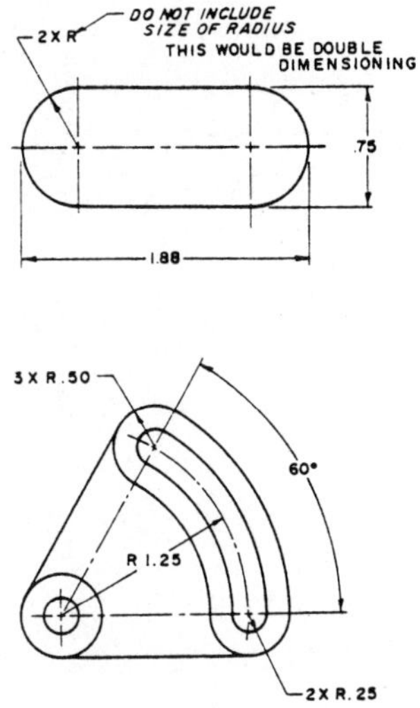

Dimensioning internal slots

CHECK LIST FOR DIMENSIONING

Size Dimensions
Location Dimensions
Placement
Spacing
Extension Lines
Dimension Lines
Leaders
Arrow Heads
Lettering
Figures

B Dimension the object shown using standard practices. Decimal inches. Accuracy to two places. Measure to centers of lines. Scale 1" = 4"

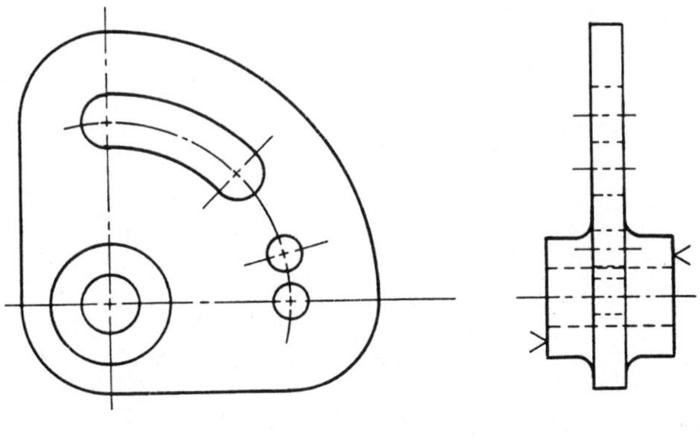

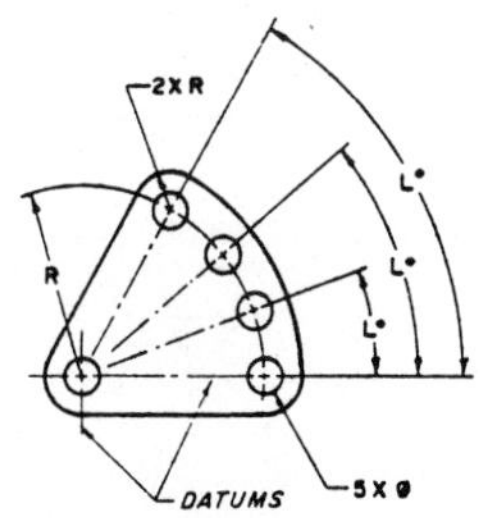

Locating holes on center lines and concentric arcs

Ch10 Dimensioning and Notation

Section-Chapter-Problem	Dimensioning practices	Course__________ Section_____	Problem
2 10 8	Figures 10-66,68,69 and 10-53,67	Student_______________ Date_____	2-10-8

A **Dimension the object shown using S= size and L = location dimensions. Use standard dimensioning practices.**

CHECK LIST FOR DIMENSIONING
Size Dimensions
Location Dimensions
Placement
Spacing
Extension Lines
Dimension Lines
Leaders
Arrow Heads
Lettering
Figures

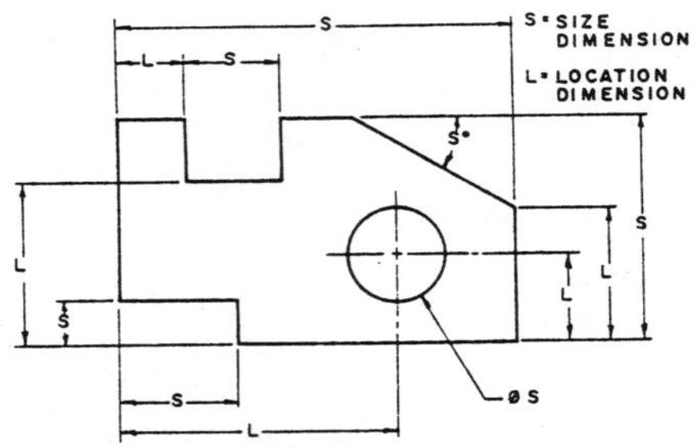

Size and location dimensions

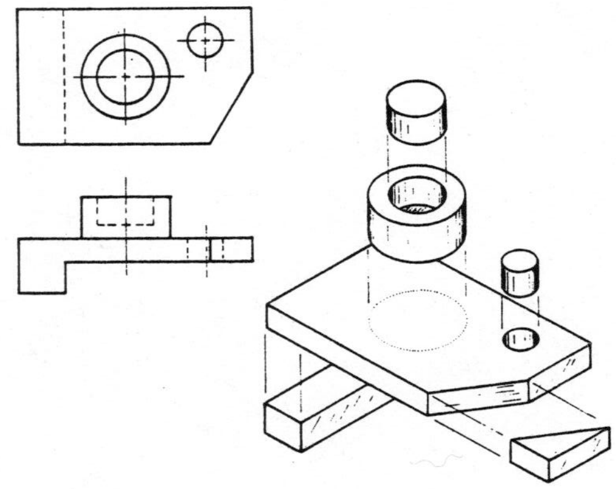

Geometric breakdown of an object

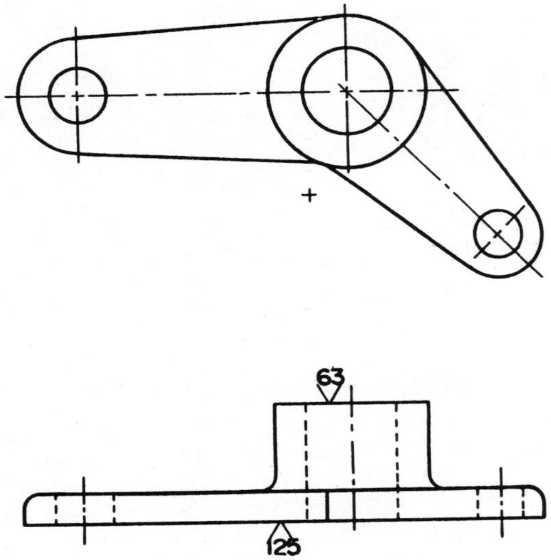

B **Dimension the object shown using S= size and L = location dimensions. Use standard dimensioning practices.**

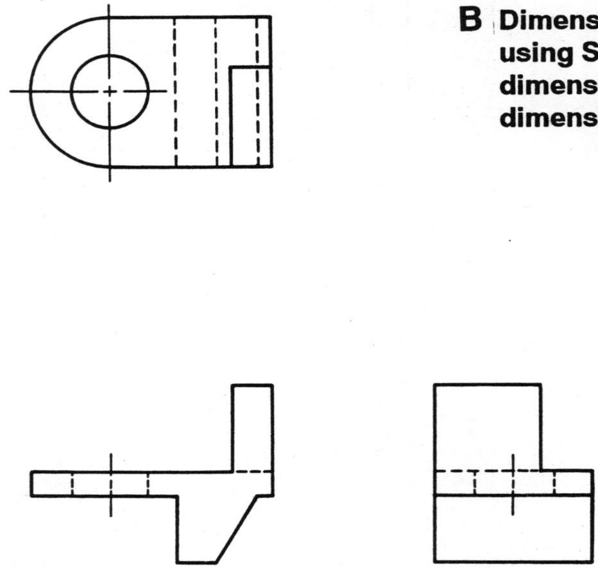

Ch10 Dimensioning and Notation			**Problem**
Section-Chapter-Problem 2 10 9	Size and location dimensions Figure 10-29 Geometric breakdown of an object Figure 10-30	Course______ Section____ Student________ Date____	**2-10-9**

Place the following NOTES and CALLOUTS, in appropriate locatons, on the drawing below.

1. **ALL DIMENSIONS IN INCHES**
2. **ALL UNMARKED RADII, R.06**
3. **Φ.625/.620 THRU**
4. **Φ.44-20 UNF-2 INLINE WITH Φ.625/.620 HOLE**
5. **Φ.38 THRU Φ.75 ↓.25**
6. **R.62**
7. **Φ1.25**

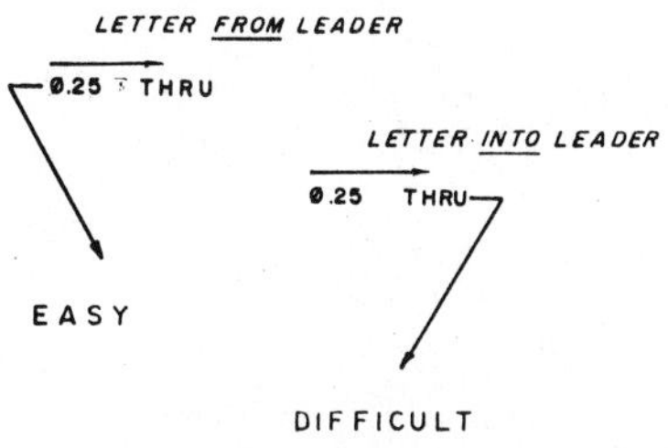

Lettering and leaders

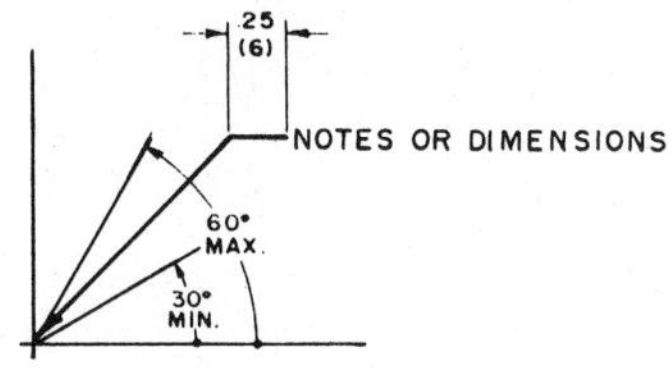

Angle of a leader line

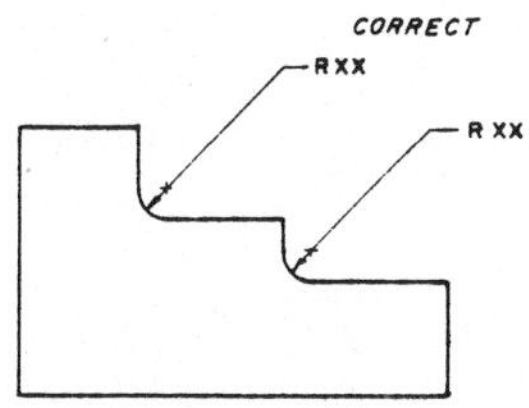

One dimension per leader line is preferred

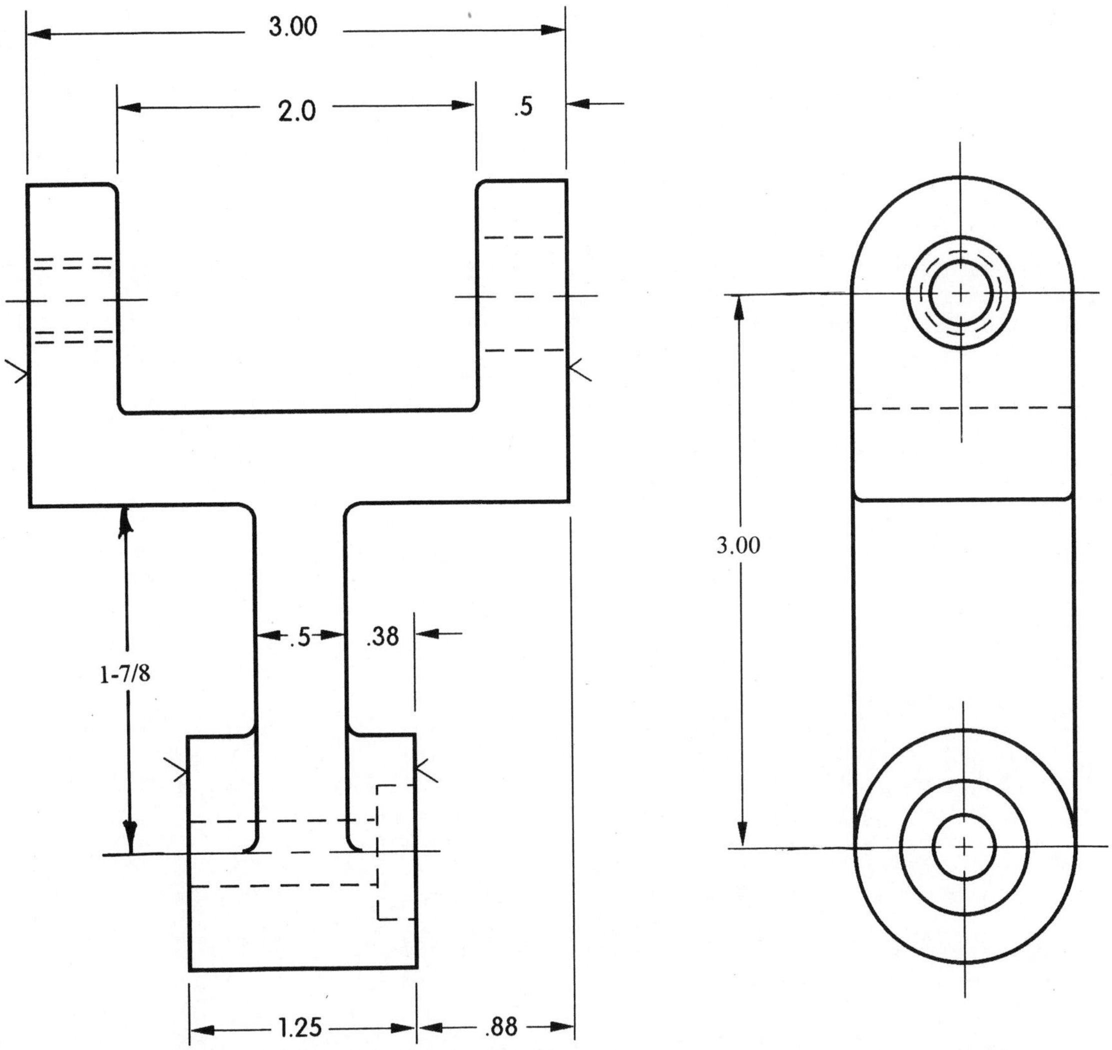

Ch10 Dimensioning and Notation			**Problem**
Section-Chapter-Problem 2 10 10	Notation Figures 10-122,125	Course________ Section____ **Student**____________ Date_____	**2-10-10**

A Dimension the object shown using standard practices and callouts. Decimal millimeters. Accuracy, one place. Measure to centers of lines. Scale 1mm = 2mm.

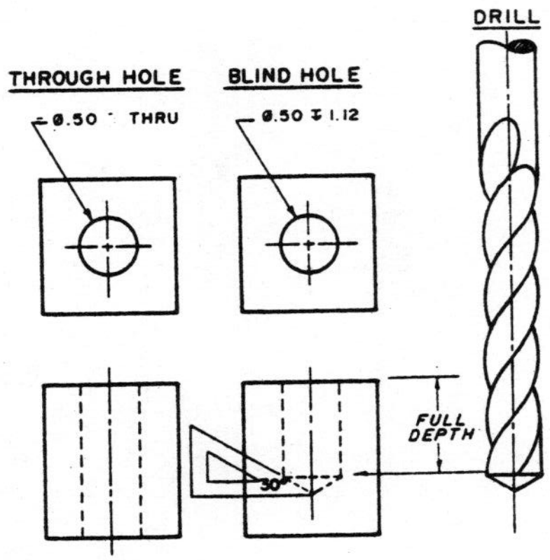

B Dimension the object shown using standard practices. Decimal inches. Accuracy to two places. Measure to centers of lines. Scale 1" = 4"

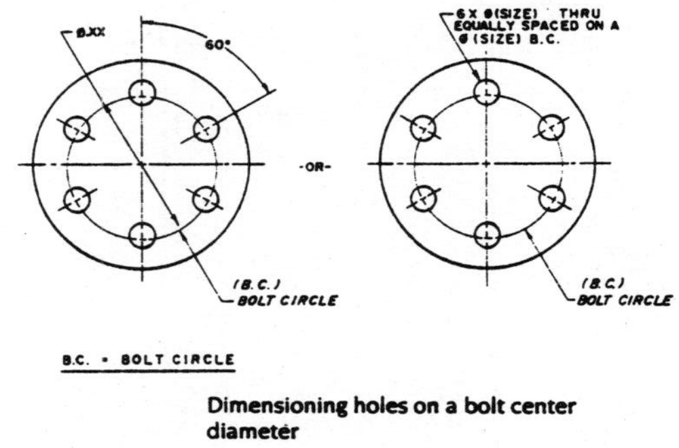

Dimensioning holes on a bolt center diameter

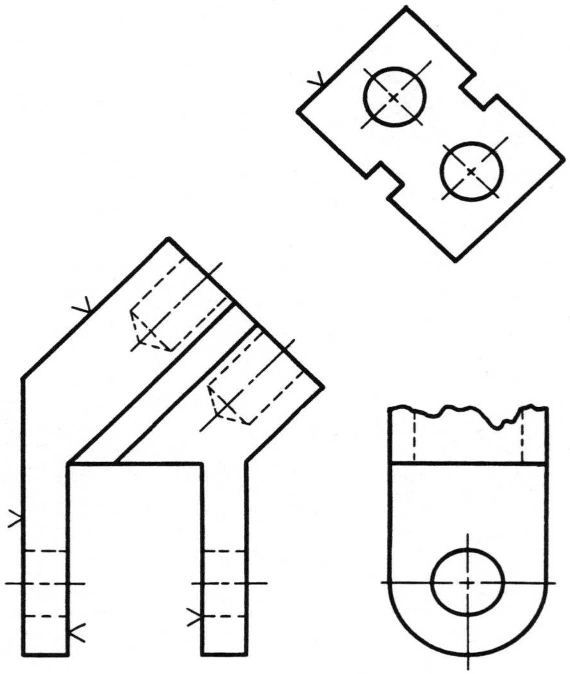

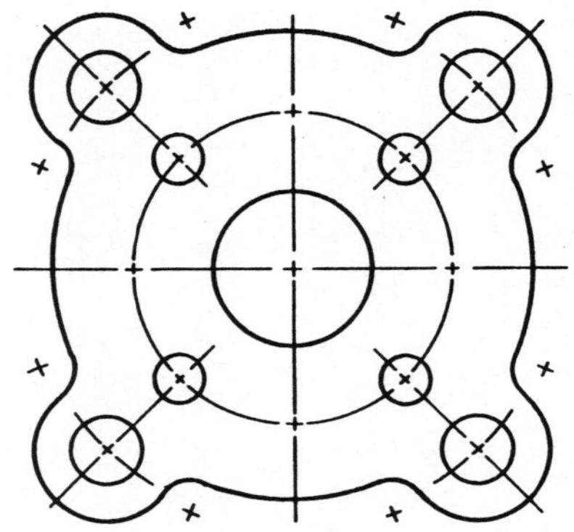

CHECK LIST FOR DIMENSIONING

Size Dimensions
Location Dimensions
Placement
Spacing
Extension Lines
Dimension Lines
Leaders
Arrow Heads
Lettering
Figures

Ch10 Dimensioning and Notation

Section-Chapter-Problem			Machined holes calouts Figure 10-49 Dimensioned holes on a bolt center diameter Figure 10-91	Course_____ Section_____ Student_____ Date_____	Problem 2-10-11
2	10	11			

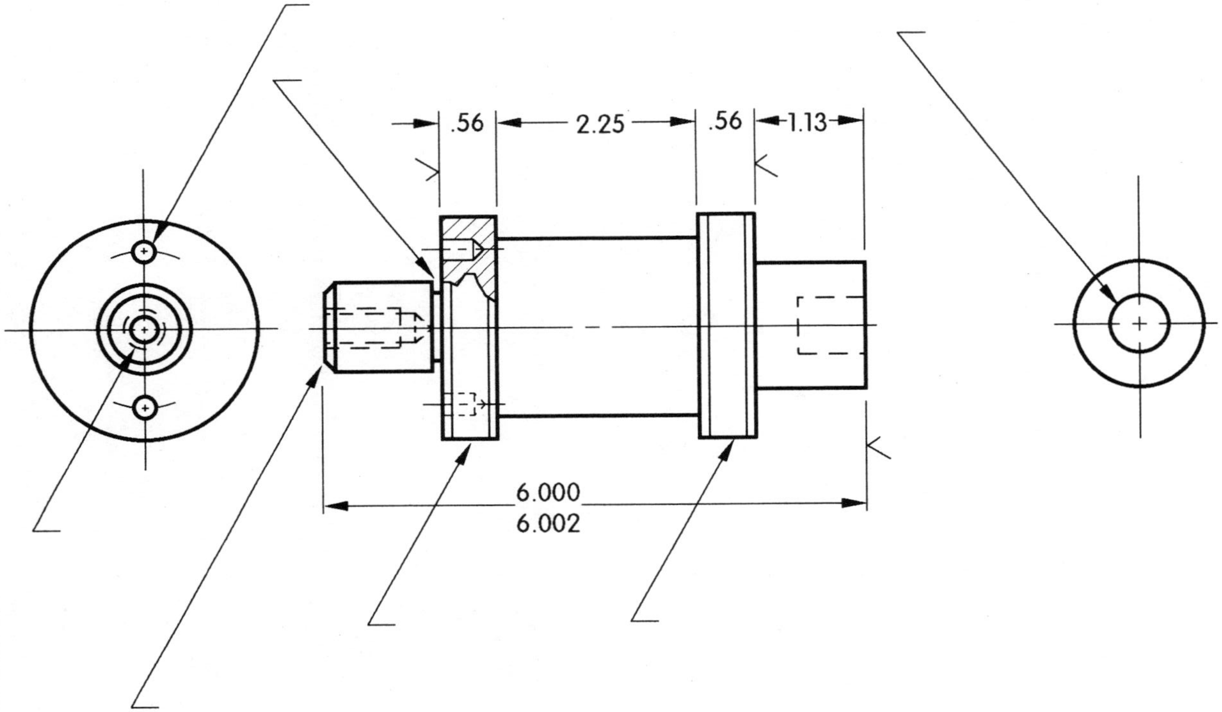

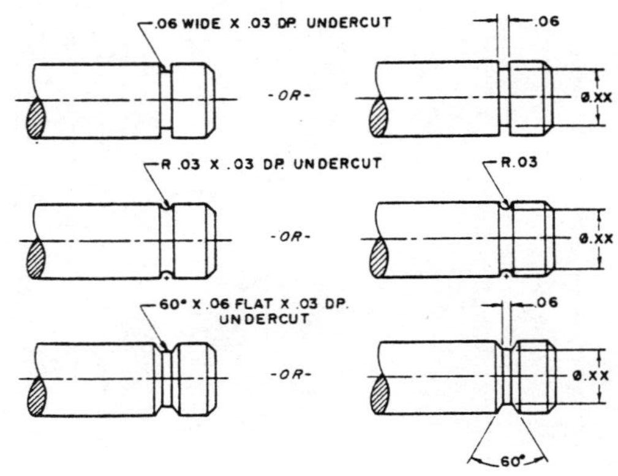

Dimensioning necks and undercuts

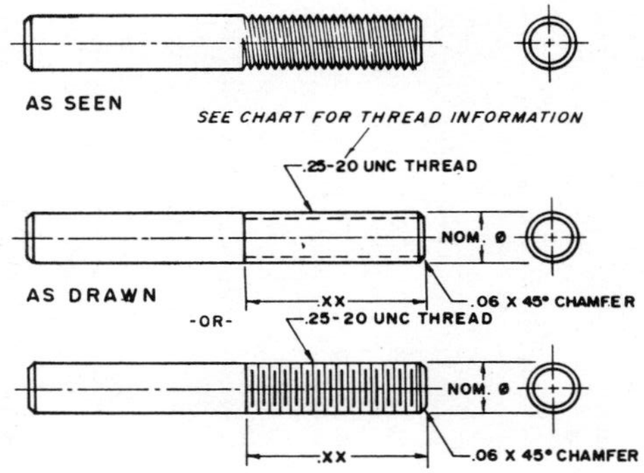

Dimensioning threads

Place the following NOTES and CALLOUTS, in appropriate locations, on the drawing above.

1. **ALL DIMENSIONS IN INCHES**
2. **KNURL 64 DP**
3. **Φ.38-24 UNF-2**
4. **.06 x 45° CHAMFER**
5. **UNDERCUT .09 X .O6 DEEP**
6. **2 X .252/.250 ↓.38**
7. **Φ.62↓.75**

(Note: Some dimensions of the object have been omitted for this problem.)

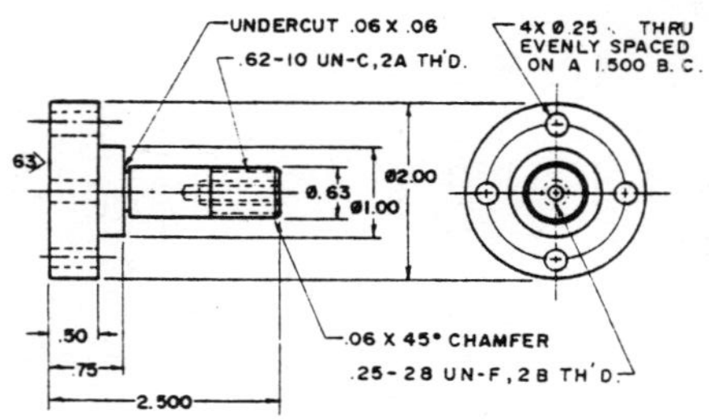

Notes are placed horizontally

Ch10 Dimensioning and Notation

Section-Chapter-Problem	Undercuts Figure 10-76, Threads Figure 10-83	Course_____ Section_____	Problem
2 10 12	Notations Figure 10-125	Student_____ Date_____	2-10-12

LIMIT DIMENSIONING AND FITS FOR MATING COMPONENTS

EXAMPLE:

GIVEN: American National Standard Runing and Sliding Fits
ANSI B4.1-1967 (R1994) (Courtesy ASME)
Exerpt from RC-3 shown beow.

TO FIND: Limit Dimensions for Φ1/2" shaft in Φ1/2" hole.

PROBLEM:
Determine limit dimensions for Φ5/8" hole/shaft, RC-3.
Show your calculations.

Nominal Size Range Inches	Class RC3		
	Clearance	Standard Tolerance Limits	
		Hole H7	Shaft f6
	Values shown are in thousandths of an inch		
0.40--0.71	0.6 1.7	+0.7 0	-0.6 -1.0

HOLE BASIS: (Note: For the HOLE BASIS one limit dimension of the hole is equal to the decimal equivalent of the nominal size, e.g., Φ1/2"=0.5000.)

0.5000	0.5000	0.5000	0.5000
+0.0007	0.0000	-0.0006	-0.0010
0.5007	0.5000	0.4994	0.4990

(Note: The dimension the machinist comes to first is on top.)

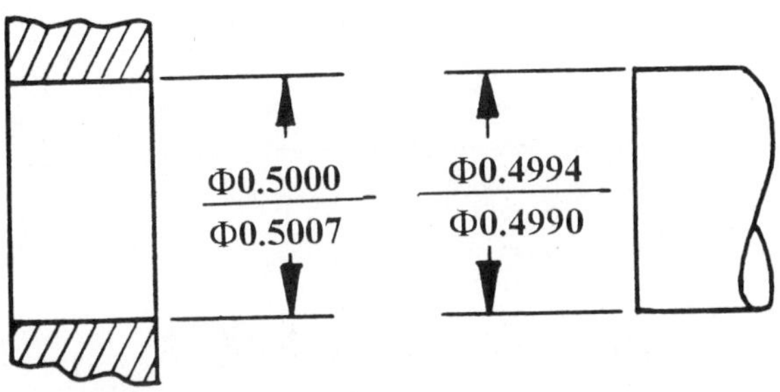

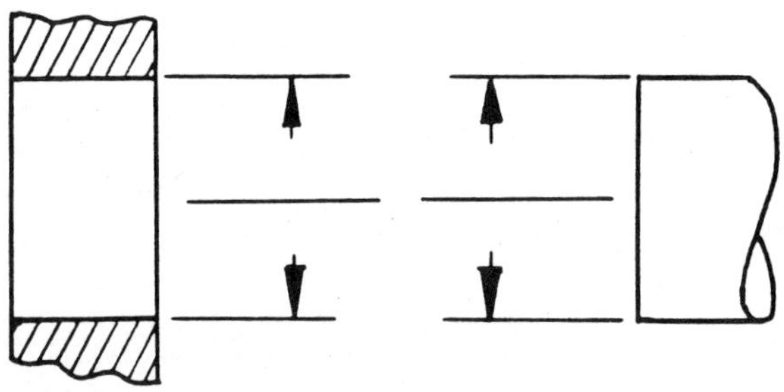

Next check ALLOWANCE and CLEARANCE using equations.

ALLOWANCE = tightest fit = smallest hole-largest shaft
= 0.5000-0.4994 = 0.0006 or .6

CLEARANCE = loosest fit = largest hole-smallest shaft
= 0.5007-0.4990 = 0.0017 or 1.7

Ch10 Dimensioning and Notation

Section-Chapter-Problem	RC fits Figure 10-116	Course____________ Section_____	Problem
2 10 13	HOLE BASIS	Student____________________ Date________	2-10-13

LIMIT DIMENSIONING AND FITS FOR MATING COMPONENTS

EXAMPLE SHAFT BASIS:

GIVEN: A shaft Φ3/4" nominal, .7500±.0002

TO FIND: Limit Dimensions for a Class RC-2 Fit
Use ALLOWANCE and CLEARANCE equations to determine the limit dimensions.

ALLOWANCE: tightest fit = smallest hole-largest shaft
.0003 = smallest hole-.7502
smallest hole = .7505
CLEARANCE: loosest fit = largest hole-smallest shaft
.0012 = largest hole-.7498
largest hole = .7510

(Note: The dimension the machinist comes to first is on top.)

PROBLEM:
Determine limit dimensions for Φ5/8" ..6250 ±.0002, SHAFT BASIS, RC-2 Fit.
Show your calculations.

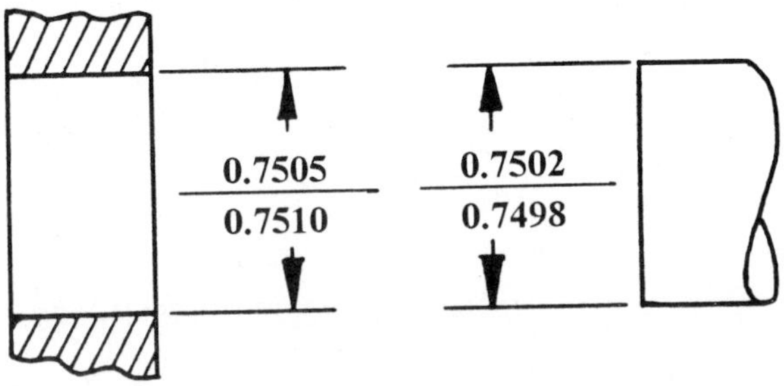

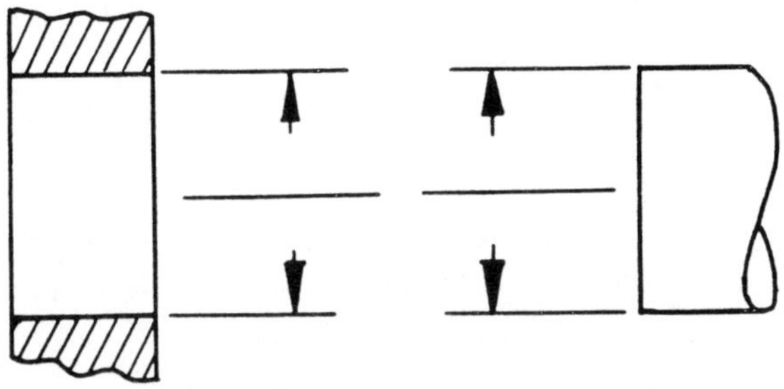

Next check ALLOWANCE and CLEARANCE using equations.

ALLOWANCE = .7505-.7502 = .0003 or .3*
CLEARANCE = .7510-.7498 = .0012 or 1.2*
* = thousandths of an inch

Ch10 Dimensioning and Notation			Problem
Section-Chapter-Problem 2 10 14	RC fits Figure 10-116 SHAFT BASIS	Course__________ Section_____ Student________________ Date______	2-10-14

LIMIT DIMENSIONING AND FITS FOR MATING COMPONENTS

EXAMPLE SHAFT BASIS:

COMMENT: One typical application for the SHAFT BASIS for establishing limit dimensions occurs in bearing installations.

GIVEN: An off-the-shelf ball bearing shown:

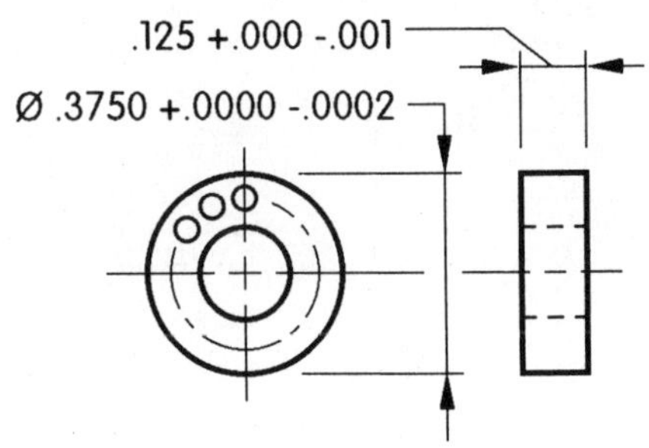

Next check ALLOWANCE and CLEARANCE using equations.

ALLOWANCE = .3736-.3750 = -.0014 or 1.4*interference
CLEARANCE = .3744-.3748 = -.0004 or .4* interference
* = thousandths of an inch

TO FIND: Limit Dimensions for the two holes in the linkage shown for an FN-2 interference fit for the bearings, SHAFT BASIS. Note that interference values, not clearance values, are listed in the FN-2 table.

Use ALLOWANCE and CLEARANCE equations to determine the limit dimensions.

ALLOWANCE: tightest fit (most interference)= smallest hole-largest shaft

-.0014 = smallest hole-.3750
smallest hole = .3736

CLEARANCE: loosest fit (least interference) = largest hole-smallest shaft

-.0004 = largest hole-.3748
largest hole = .3744

(minus signs indicate interference)

PROBLEM:
Assume that the bearings have an outside diameter of .6249 ±.0001, determine limit dimensions for the linkage holes, using the SHAFT BASIS, and an FN-2 Fit.
Show your calculations.

(Note: The dimension the machinist comes to first is on top.)

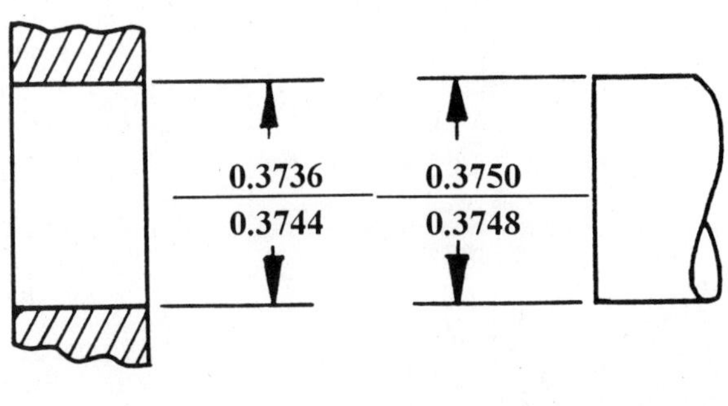

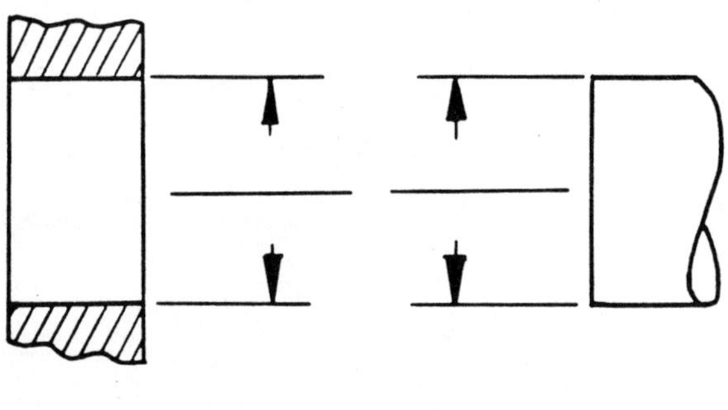

Ch10 Dimensioning and Notation			Problem
Section-Chapter-Problem 2 10 15	FN fits Figure 10-120 SHAFT BASIS: INTERFERENCE	Course__________ Section_____ Student_____________ Date______	2-10-15

A. Dimension the object shown using standard practices. Assume full scale in inches. Use accuracy to two places except for the large hole. Measure to centers of lines. The large hole is to have limit dimensions for an RC-4 Fit, HOLE BASIS. Show your calculations.

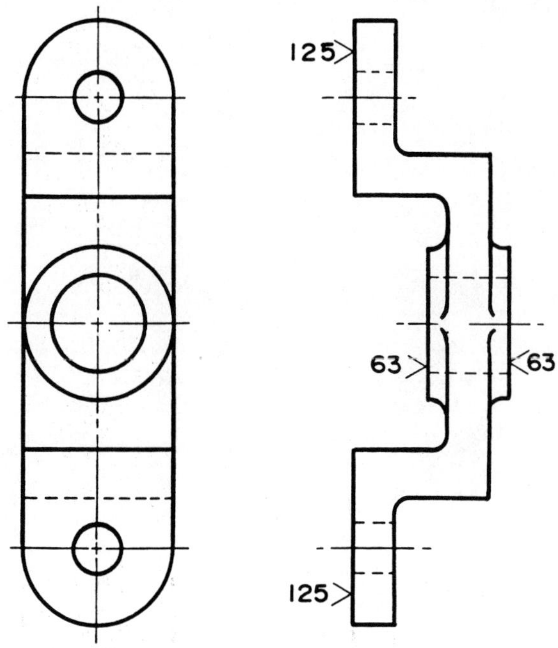

B. Assume that a ball bearing with an outside diameter of .5624 ±.0001" is to be installed in the large hole. Use the SHAFT BASIS method to determine limit dimensions for an FN-2 Fit. Show your calculations.

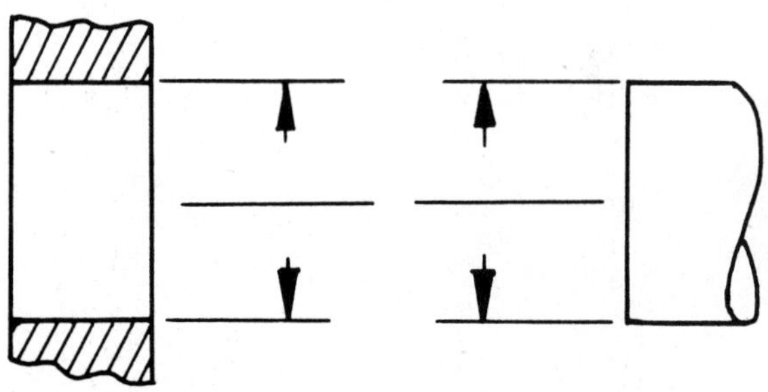

Ch10 Dimensioning and Notation			Problem
Section-Chapter-Problem 2 10 16	RC and FN fits Figures 10-116 and 10-120	Course__________ Section_____ Student________________ Date_____	2-10-16

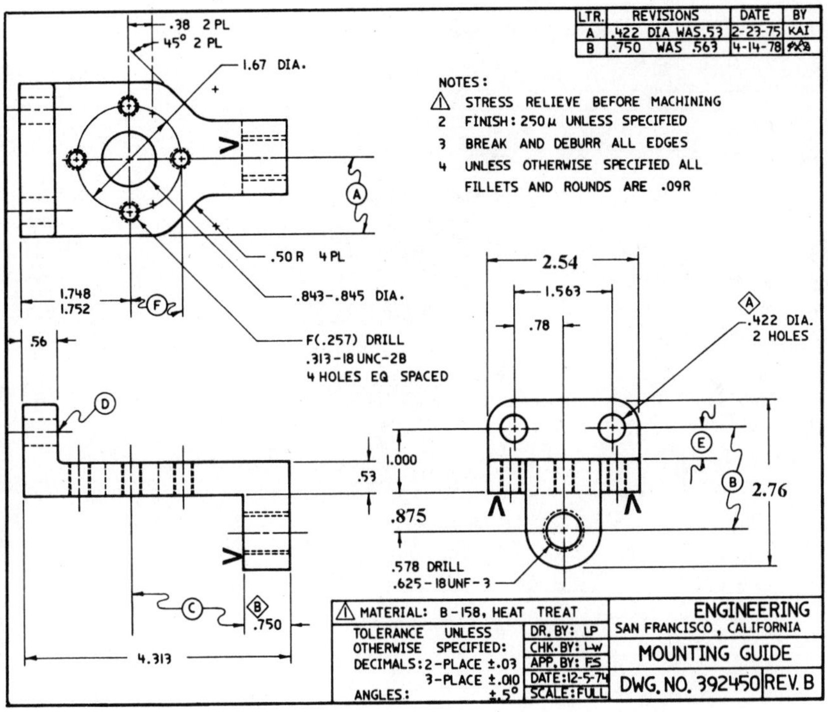

Study the drawing of the MOUNTING GUIDE and answer the following questions.

1. **How much would a MOUNTING GUIDE weigh if it were made of...**
   a. **steel at 0.28 lb/cu inch?**
   b. **aluminum at 0.10 lb/cu inch?**
   c. **brass at 0.30 lb/cu inch?**
   Show your calculations.
2. **What is the overall height of the guide**
3. **What is dimension (A)?**
4. **What tolerance is expected for the 2-place dimensions?**
5. **What tolerance is expected for the 3-place dimensions?**
6. **What is the maximum dimension of (F)?**
7. **What tolerance do angles have?**
8. **What does the V mark mean? What is the current practice for these marks?**

ANSWERS

1---

a---

b---

c---

2---

3---

4---

5---

6---

7---

8---

Ch10 Dimensioning and Notation			Problem
Section-Chapter-Problem 2 10 17	Dimensioning and Tolerancing	Course__________ Section_____ Student__________ Date_____	2-10-17

1. Add dimensions and notes to the drawing shown in file "10-1.dxf". Put all dimensions and notes on separate layers.

B The object in drawing file "10-2.dxf" is used to check the accuracy of depth and size of three holes in a piece of wood. The manufacturer has changed the object according to the specifications according to the engineering change orders (ECO).

<table>
<tr><td colspan="3">Ch10 Dimensioning and Notation</td><td rowspan="2">Problem

2-10-18</td></tr>
<tr><td>Section-Chapter-Problem
2 10 18</td><td>Dimensions
ECO</td><td>Course___________Section_____
Student___________________Date______</td></tr>
</table>

Section 3
Objectives:

After studying and doing the exercises and problems in Section 3 of the workbook you should be able to:

- recognize and read drawings with geometric dimensioning and tolerancing (GD&T), and prepare basic GD&T drawings.
- create the necessary callouts on drawings to prepare holes for fasteners, and to draw the fasteners if required.
- prepare drawings of compression springs.
- prepare a displacement drawing for a cam, and a dimensioned drawing of the cam.
- calculate basic geometric parameters for spur gears.
- develop a design idea with freehand sketches and design development layout drawings.
- communicate a design idea through detail and assembly drawings.
- prepare isometric, oblique, and two point perspective pictorial drawings.

Supplemental Comments: Section 3

Chapter 12 Geometric Dimensioning and Tolerancing

Problem 3-12-1: Geometric Dimensioning and Tolerancing (GD&T) provides a more accurate means of describing a component than traditional dimensioning practices can. This problem introduces you to one of the basic steps of GD&T—establishing and using datums.

Problem 3-12-2: You may have noticed that drawings prepared with GD&T can readily be identified because of the number of "boxes" used for BASIC dimensions, DATUMS, and FEATURE CONTROL SYMBOLS.

Problem 3-12-3: Note the difference between a datum plane (usually a machine tool) and a datum surface on the part.

Problems 3-12-4 and **3-12-5:** Parallelism, concentricity, perpendicularity, and runout are four of the building blocks of GD&T.

Problem 3-12-6: Primary, secondary, and tertiary datum surfaces may be established by three, two, and one point, respectively.

Problems 3-12-7 and **3-12-8:** A datum may be a plane or a cylinder.

Problems 3-12-9A and **3-12-9B:** These problems illustrate the need for more precise dimensioning when parts manufactured at different locations are to be assembled.

Problem 3-12-10: Use a CAD program to select datums and add tolerances to the drawing.

Chapter 13 Fasteners

Problem 3-13-1 requires standard dimensioning practices of holes to accommodate different fasteners.

Problem 3-13-2: Drawing fasteners is an excellent step toward becoming acquainted with the different types of fasteners.

Problems 3-13-3 and **3-13-4** Fasteners in section views are not sectioned unless you need to show a modification to the fastener.

Problem 3-13-5: An obvious advantage of fastening devices such as bolts, screws, pins, and retaining rings is that they can be removed. In Section 4, we will consider permanent fastening—welding.

Problem 3-13-6: Save time by using simplified methods of drawing threads.

Problem 3-13-7: Use a CAD program to draw bolts and holes as indicated.

Chapter 14 Springs

Problems 3-14-1 and **3-14-2:** Compression springs only are featured in these two problems. The longest overall length of a compression spring is its free-length dimension. Conversely, the shortest overall length of an extension spring is its free length.

Problem 3-14-2: Use a CAD program to draw a spring per the given specifications.

Chapter 15 Cams

Problems 3-15-1 and **3-15-2:** The cam follower for the cam design specified on these two pages is in line with the center of the cam. The displacement diagram utilizes two of four types of motion for cams—harmonic and uniform. Uniform velocity and modified uniform velocity are the other two types used.

Problem 3-15-3: Use a CAD program to draw a displacement diagram for the cam shown.

Chapter 16 Gears

Problems 3-16-1 and **3-16-2:** The same set of spur gears is used for these two pages that focus on gear geometry and nomenclature. Note that the diametral pitch **must** be identical for spur and bevel gear sets.

Problem 3-16-3: Use a CAD program to draw a second and third gear to increase the speed of the motor.

Chapter 17 Assembly and Detail Drawings for Design

Problems 3-17-1A through **3-17-1D:** A completed set of drawings is included on these four pages for your guidance to prepare the following types of drawings: **freehand sketches, design development layout** (generally called layout), **detail,** and **assembly.**

Problems 3-17-2A through **3-17-2E:** Note that the topic heading for these pages usually indicates which drawing is to be done on the page, instead of having a reference to a figure in our textbook. For example, page **3-17-2C** specifies "Details of Toggle Clamp". Also note that usually only one object is on a detail drawing (see example drawings in problem **3-17-1**); however, for the purposes of this workbook, and to shorten your drawing time, include more than one component on the detail drawings for this problem. Also, for convenience the parts list is on the same page as the assembly drawing.

Problems 3-17-3A through **3-17-3H:** This problem introduces you to a typical activity in a drafting department in industry—change orders. Also, one advantage of doing this problem is that the required drawings are very similar to those in problem **3-17-1.** Note that the topic heading for these pages usually indicates which drawing is to be done on the page, instead of having a reference to a figure in our textbook. For example, page **3-17-3B** specifies "ECO Design Layout" and **3-17 3D** specifies "ECO Spacer

Details." Note that only one object, or one weldment, is on a detail drawing for this problem.

Problems 3-17-4A through **3-17-4E:** The required drawings for the Paravane Stabilizer include: design layout, two detail drawings (one object or weldment on a drawing) and an assembly drawing with a parts list on the same page as the drawing.

Problems 3-17-5A through **3-17-5F:** The required drawings for the Paravane Stabilizer due to an Engineering Change Order (ECO) include: design layout, three detail drawings (one object or weldment on a drawing) and an assembly drawing with a parts list on the same page as the drawing.

Problem 3-17-6: Use a CAD program to create an assembly drawing from the multiview drawing of a palette.

Chapter 18 Pictorial Drawings

Problems 3-18-1 and **3-18-2:** We highly recommend drawing an "outline box" prior to filling in details, whether the drawing is to be pictorial or an orthographic multiview.

Problem 3-18-3: Oblique pictorial drawings are easy to draw because "circles remain circles;" however, isometric drawings are more popular because "circles" appear closer to what we normally see—ellipses.

Problems 3-18-4A and **3-18-4B:** We also recommend using an "outline box" for perspective drawings.

Problems 3-18-5A and **3-18-5B:** Analyzing the multiview drawings will help you to draw the perspectives.

Problem 3-18-6: Use a CAD program to draw an isometric and oblique drawing from the orthographic projection.

Add dimensions to the LASER LOCATOR drawing in accordance with the following:

1. Use unilateral tolerances for the height, to be machined to a range of .492 t o .500 inches.
2. Identify the bottom surface as Datum A.
3. The top must be parallel to the bottom within .003 inches.
4. The top must be flat within .ooo5.
5. The right surface must be perpendicular to the bottom within .002.
6. The back surface must be parallel to the front within .004, and perpendicular to the bottom within .001.

Scale the drawing for other dimensions. All dimensions in inches.

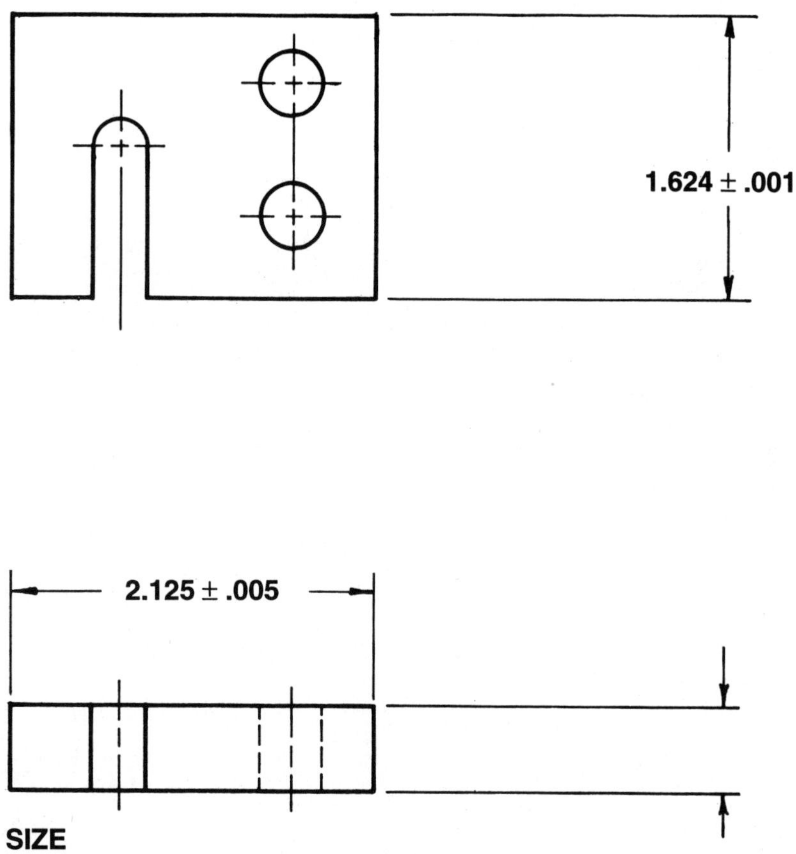

SCALE: FULL SIZE

Ch12 Geometric Dimensioning and Tolerancing			Problem
Section-Chapter-Problem	Unilateral Tolerancing Figure 12-2	Course____________Section_____	3-12-1
3 12 1	Orientation Tolerances Figure 12-6	**Student**________________Date______	

The more precise system of tolerancing called GEOMETRIC DIMENSIONING and POSITIONING TOLERANCING allows designers to ensure a component's function by establishing realistic tolerances. Some of the tolerances are tolerances of SIZE, tolerances of FORM, tolerances of ORIENTATION, and tolerances of LOCATION. This more precise system is recognizable on technical drawings because the information is usually presented in three different "BOXES". The three boxes shown are from ASME Y14.5-1994

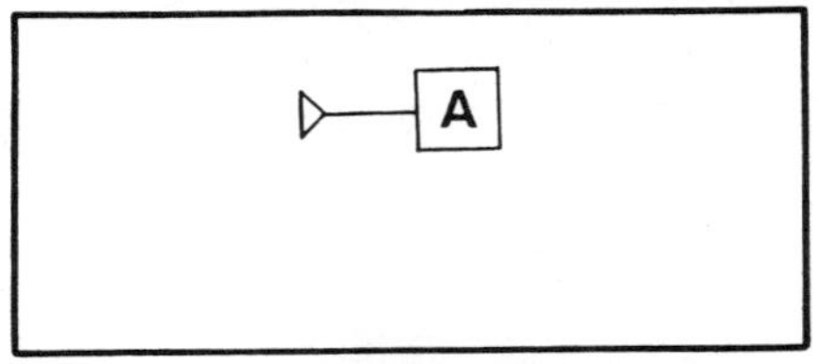

DATUM IDENTIFYING SYMBOL

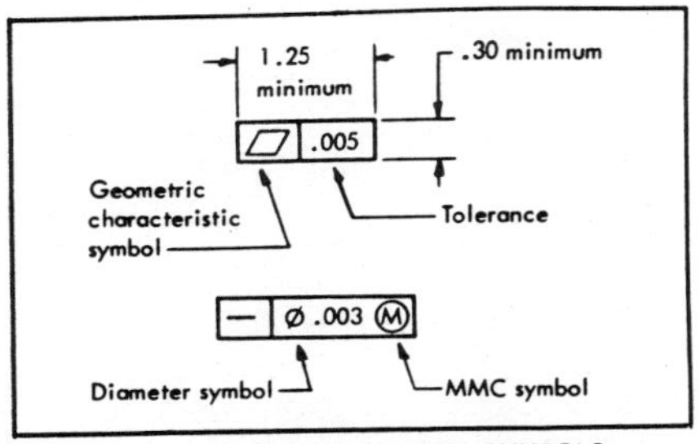

FEATURE CONTROL SYMBOLS

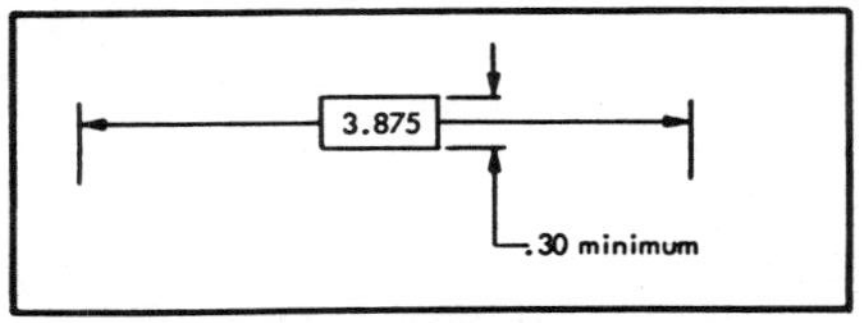

BASIC DIMENSION SYMBOL

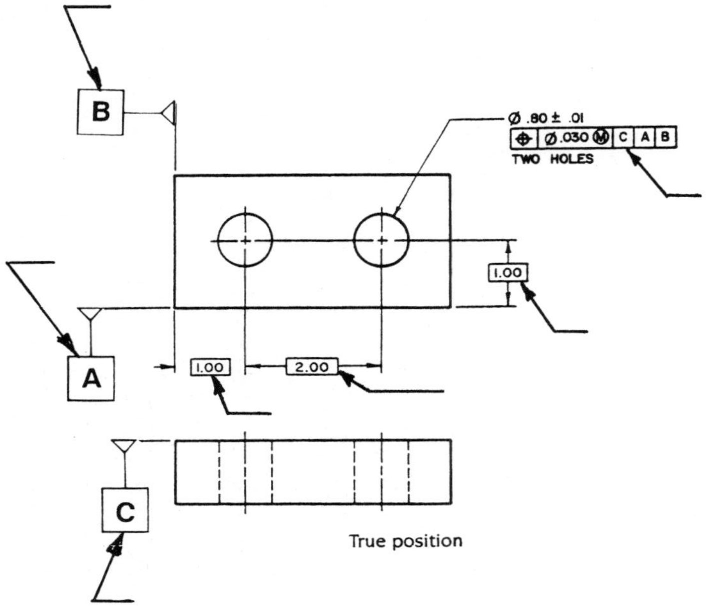

Do the following, related to the figure shown:

1. Print BASIC DIMENSION where basic dimensions occur.
2. Print DATUM where datums occur.
3. Print FEATURE CONTROL SYMBOL at the proper location.
4. Explain briefly, the following:

   a. ϕ.80 ±.01 __________

   b. ⌖ __________

   c. ϕ.030 Ⓜ __________

5. What is the tolerance zone for the holes if the diameters of the holes are ϕ.81? Explain.

6. What would be the tolerance zone for the ϕ.81 holes if the modifier was Ⓢ instead of Ⓜ? __________

Ch12 Geometric Dimensioning and Tolerancing			Problem
Section-Chapter-Problem 3 12 2	Geometric Dimensioning and Positioning Tolerancing Figure 12-5	Course____ Section____ Student____ Date____	3-12-2

GENERAL QUESTIONS: DATUMS

Fill in the blanks.

1. **The P, Q and R planes shown are all ______________ to each other, and in drawings, usually all dimensions are either ______________ or ______________ to the three planes.**

   (Note: Actual datum planes are usually surfaces of machine tool tables, machine tool surfaces, or inspection equipment surfaces.)

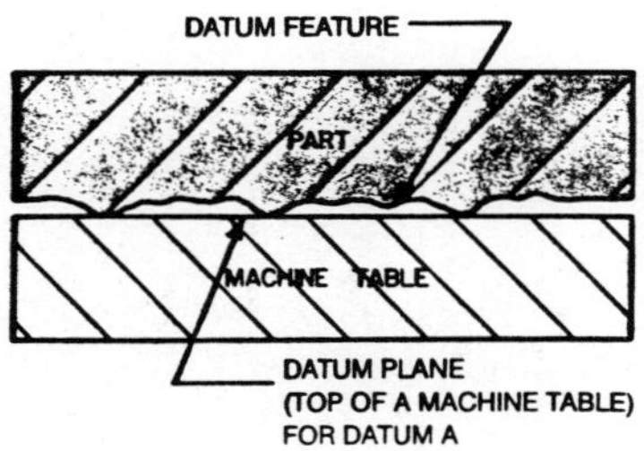

Establishing datums

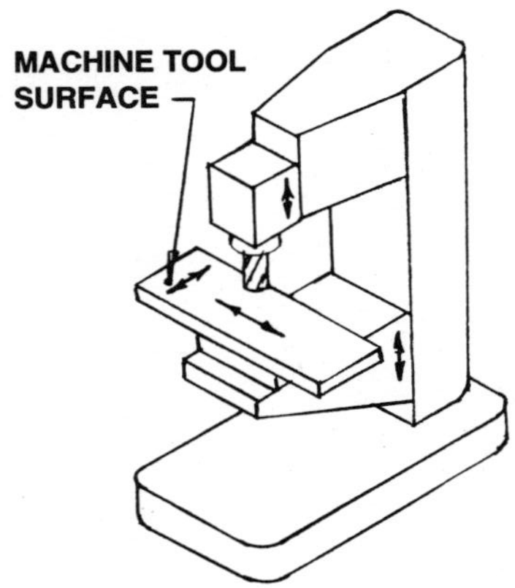

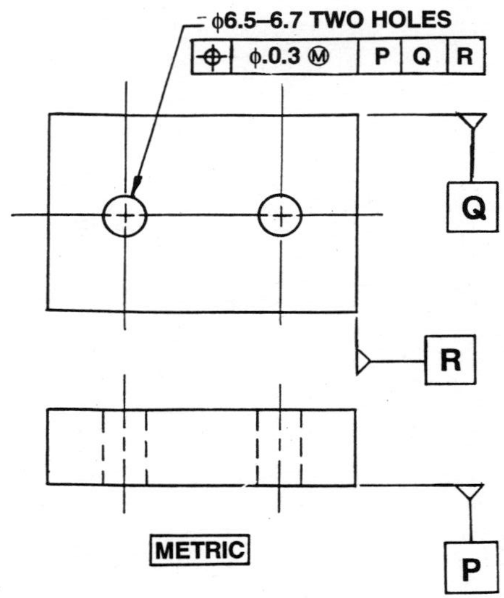

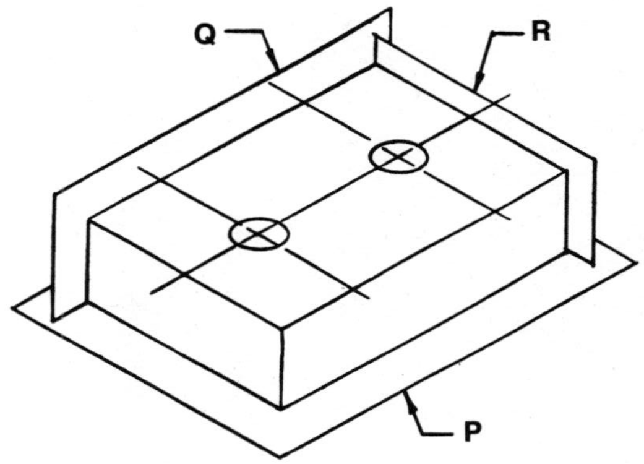

2. **What would be the tolerance-zone-diameter for the hole at the left if the hole diameter was:**

   a. **6.5mm ______________ ?**

   b. **6.7mm ______________ ?**

Ch12 Geometric Dimensioning and Tolerancing			Problem
Section-Chapter-Problem 3 12 3	Geometric Dimensioning and Positioning Tolerancing Figure 12-5 Datum Surfaces and MMC Figure 12-8, 12-9	Course ________ Section _____ Student ____________ Date _____	3-12-3

A Add a feature control symbol with specifications that the top surface of this object is to be parallel to datum A within .003″.

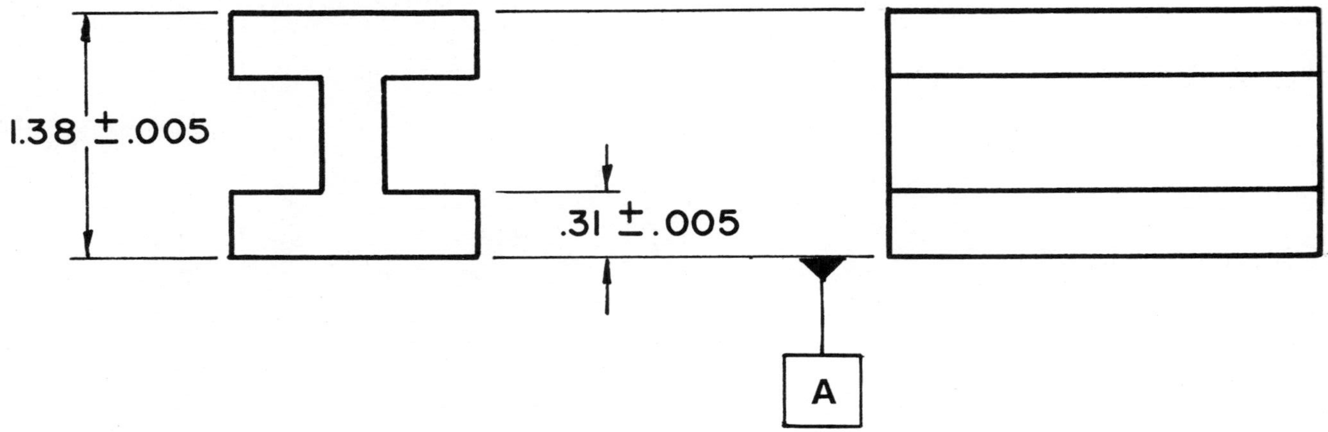

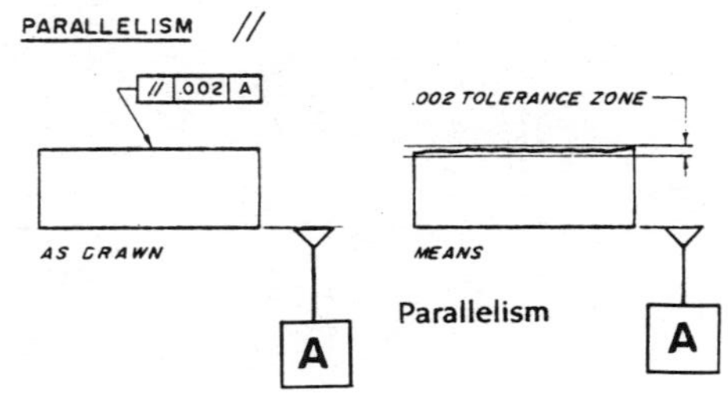

Parallelism

B Add a feature control symbol with specifications that the smaller diameter of this object is to be concentric to datum A within .003″.

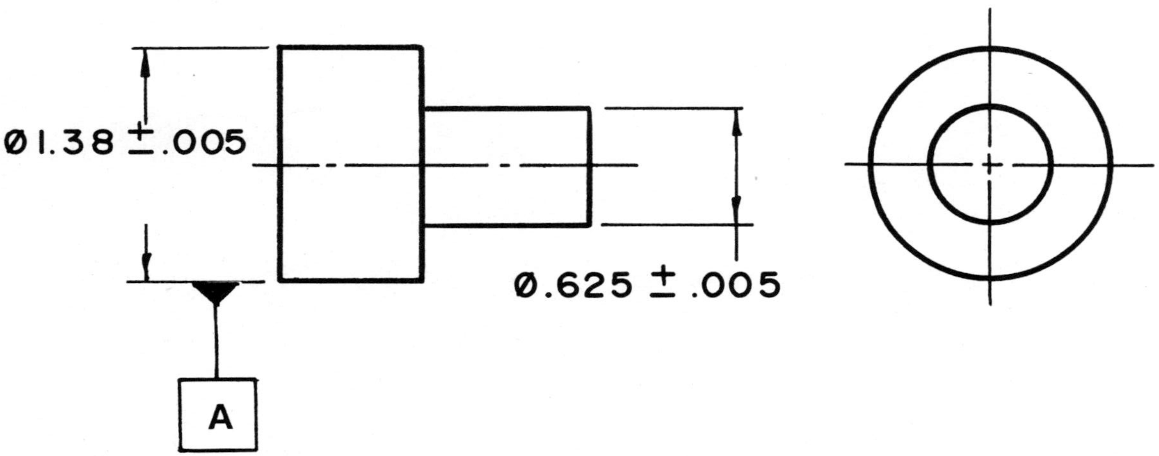

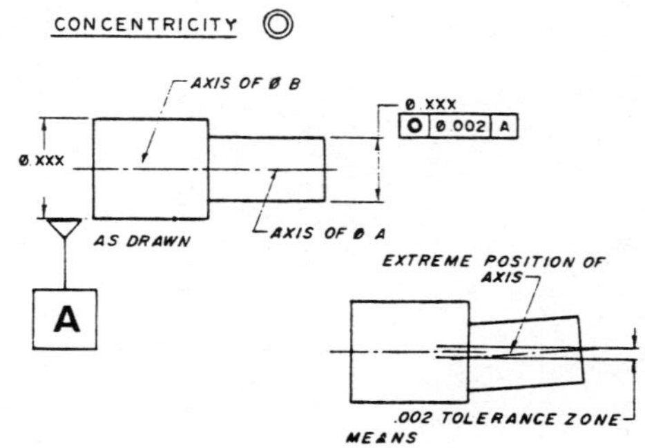

Concentricity

Ch12 Geometric Dimensioning and Tolerancing			Problem 3-12-4
Section-Chapter-Problem 3 12 4	Parallelism and Concentricity Figures 12-59, 12-75	Course_____ Section_____ Student_____ Date_____	

A **Add a feature control symbol with specifications that the left-side surface of this object is to be perpendicular to datum A within .003″.**

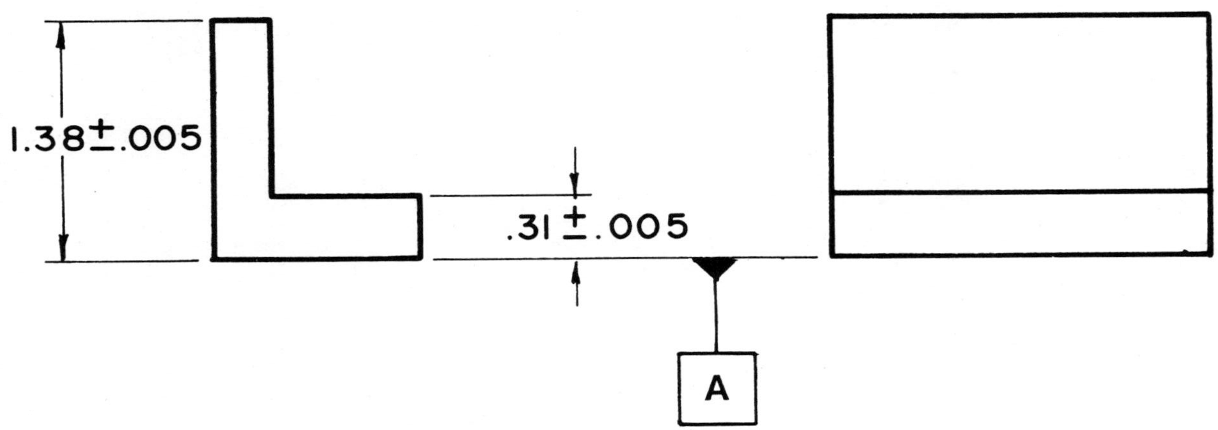

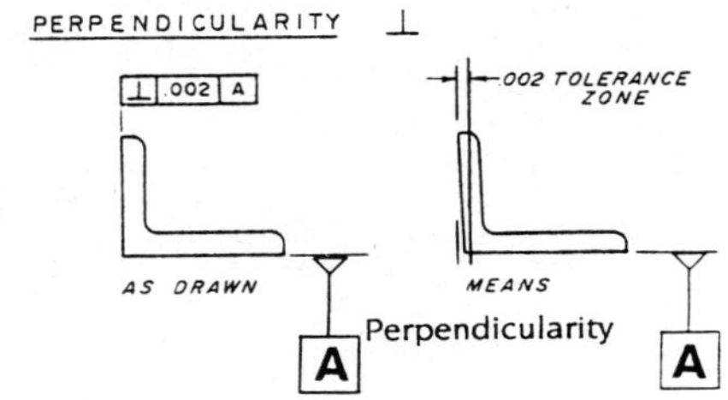

B **Add a feature control symbol with specifications that surface X is to have a total runout of .003″ with respect to datum A.**

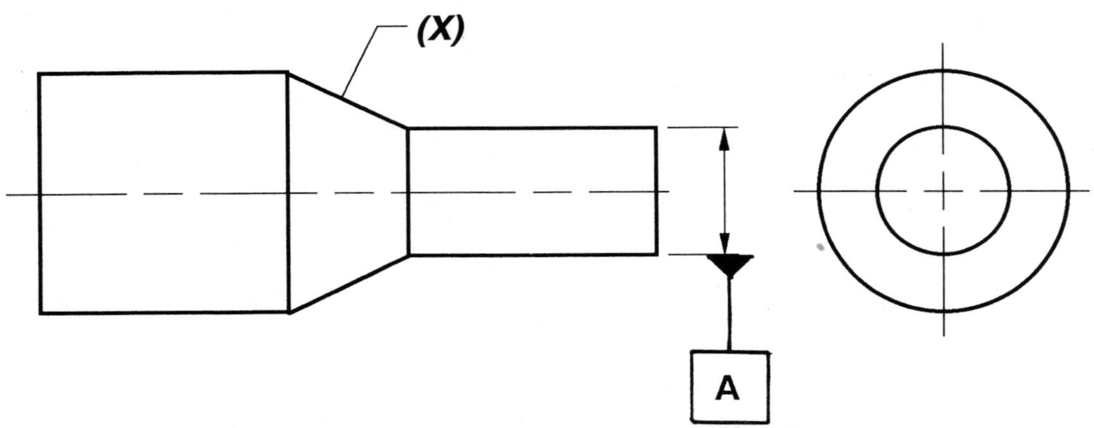

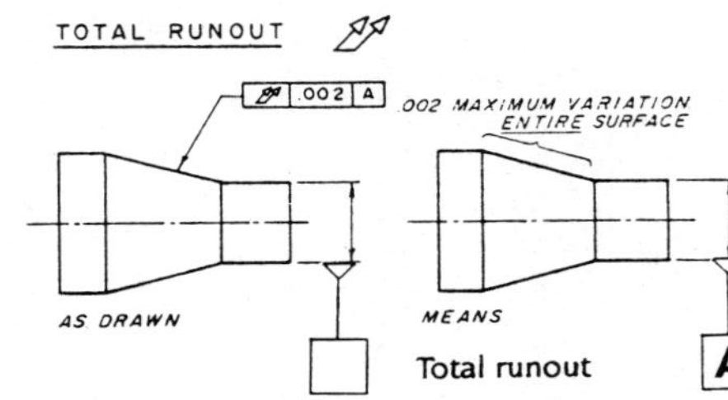

Ch12 Geometric Dimensioning and Tolerancing			Problem 3-12-5
Section-Chapter-Problem 3 12 5	Perpendicularity and Total Runout Figures 12-63, 12-74	Course______ Section____ **Student**________ Date____	

Use **DATUM TARGET** symbols and arbitrarily locate, using **BASIC DIMENSION** symbols, three specific points on surface A, two specific points on surface B, and one specific point on surface C.

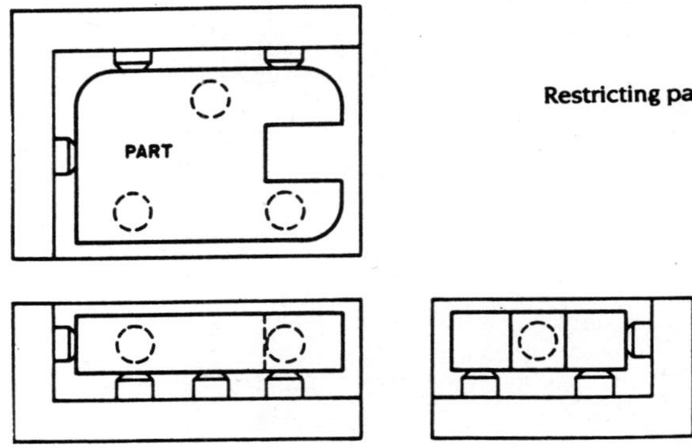

Restricting part movement

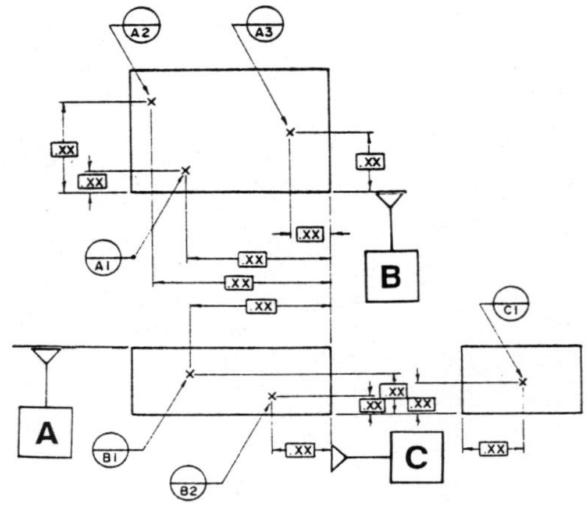

Dimensioning datum points

Ch12 Geometric Dimensioning and Tolerancing

Section-Chapter-Problem	Datum Targets Figure 12-88	Course______ Section______	Problem
3 12 6		Student______ Date______	3-12-6

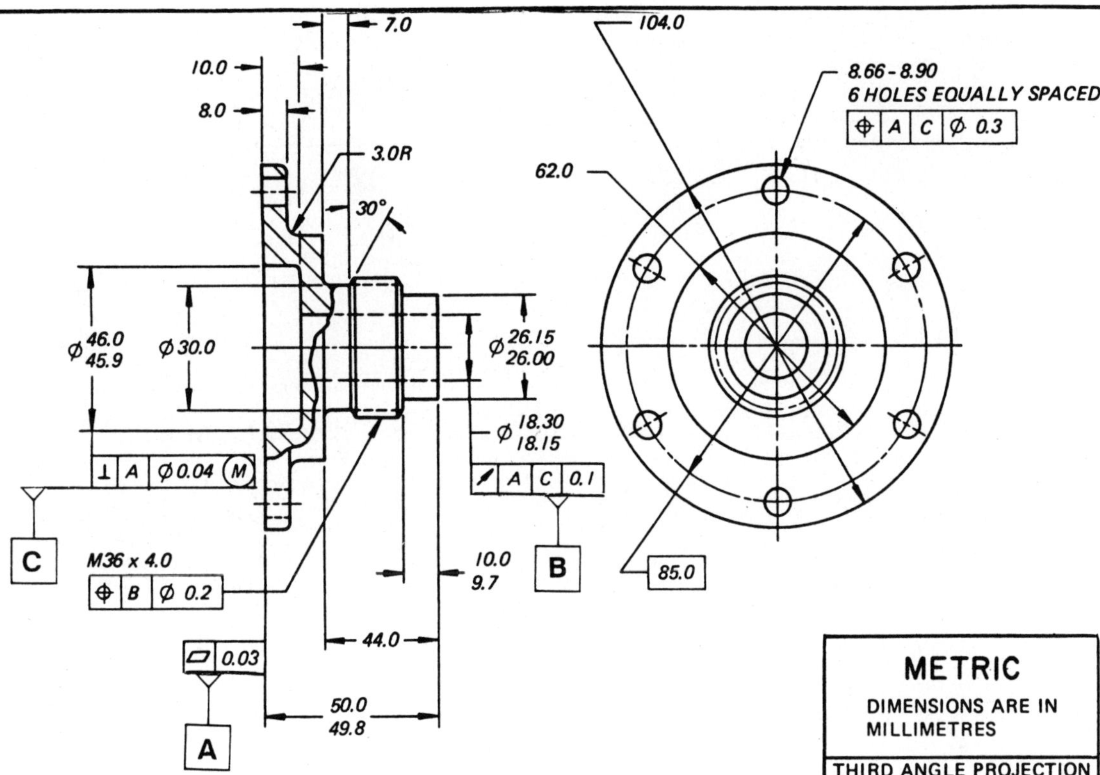

METRIC DIMENSIONS ARE IN MILLIMETRES
THIRD ANGLE PROJECTION

UNLESS OTHER WISE SPECIFIED:
±0.5 TOL ON MACH DIM.
±1.0 TOL ON CSTG DIM.

Refer to the above drawing and answer the following questions:

1. **What general tolerances are specified for the part?** ____________________
2. **What feature of the part is datum A? datum B? datum C?** ____________________
3. **What relationship must exist between the six equally spaced holes and datums A and C?** __________
4. **What specification is given for surface A?** ____________________
5. **Interpret the feature control symbol for the 18.30 – 18.15 Dia. hole.** ____________________
6. **What relationship must exist between the features identified as datum A and datum C?** __________
7. **What does the rectangular frame around the 85.0 dimension designate?** ____________________
8. **Interpret the dimension M36 × 4.0.** ____________________
9. **What relationship must exist between the threaded feature and datum B?** ____________________
10. **What does the symbol Ⓜ designate?** ____________________

Ch12 Geometric Dimensioning and Tolerancing			Problem
Section-Chapter-Problem 3 12 7	GD & T Reading Drawings Figure 12-5	Course________ Section____ **Student**__________ Date_____	3-12-7

Complete the dimensioning of this INDEX STOP to comply with the following design requirements. Use proper symbols and lettering.

1. The small cylinder is cylindrical within .05, and is concentric with the large cylinder within .10.
2. The large cylinder is cylindrical within .05.
3. The left end is perpendicular to the large cylinder within .10, and parallel to the other end of the large cylinder within .15.
4. The flat surface on the large cylinder is parallel to the center line of the large cylinder within .05, and flat within .02.

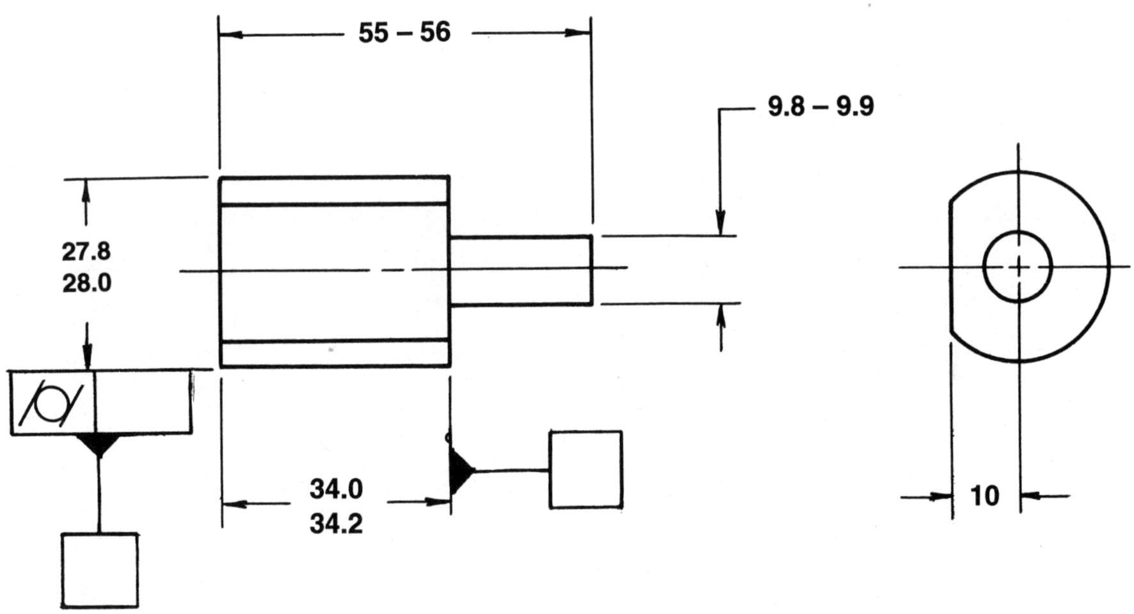

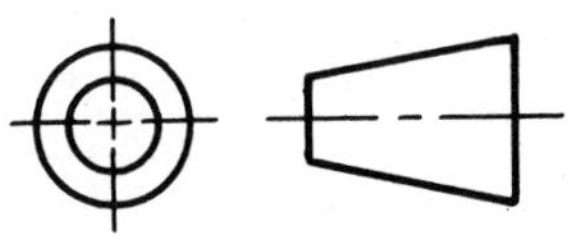

METRIC

Third Angle Projection

Ch12 Geometric Dimensioning and Tolerancing			Problem 3-12-8
Section-Chapter-Problem 3 12 8	GD & T Cylindricity Perpendicularity & Parallelism	Course__________ Section_____ **Student**________________ Date______	

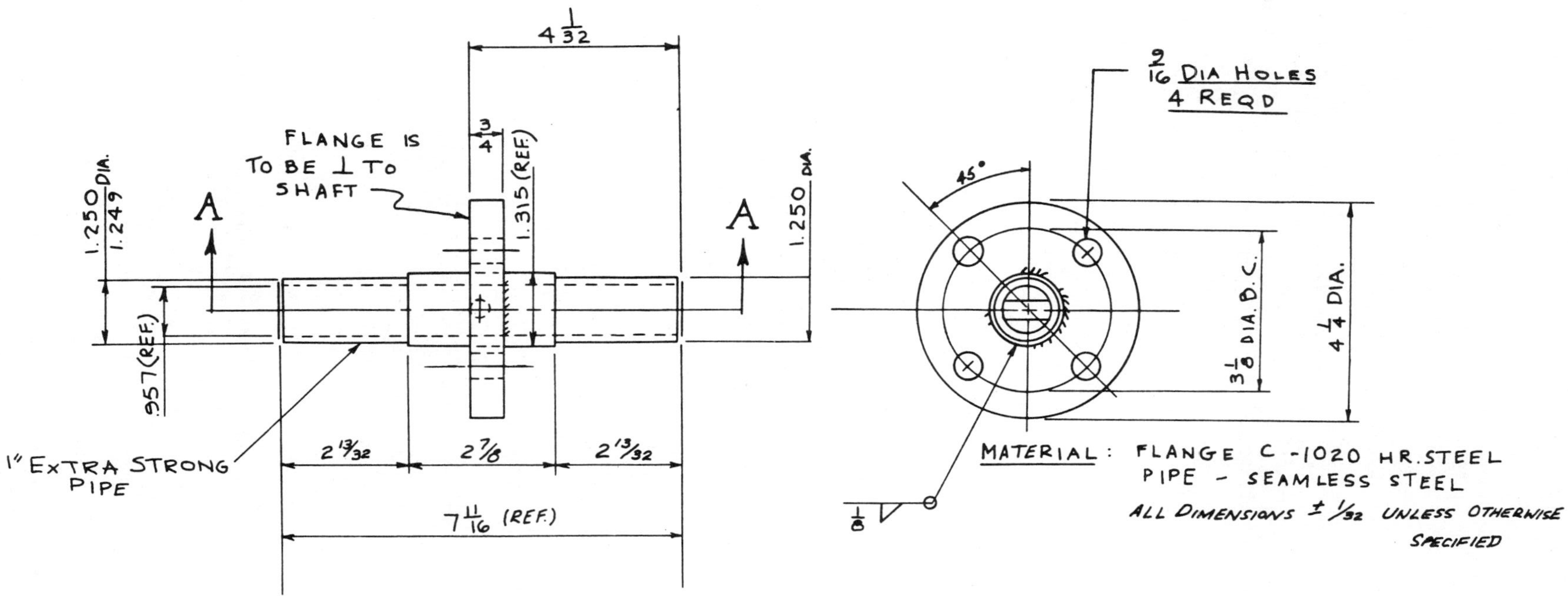

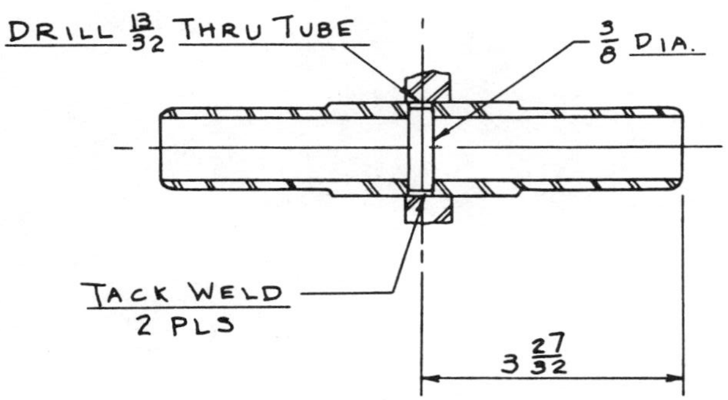

The two views and section drawing of the control-blade-shaft were used in a machine shop which was in the same building as the designer who prepared the drawings. Any questions that the shop had were resolved readily because the designer visited the shop frequently. For example, the designer informed the machinist that the left side of the 4-1/4″ diameter flange had to be perpendicular to the centerline of the tubular shaft, and that the four holes in the flange were to accurately match a blade (not shown) with the same four hole pattern.

In the future, the control-blade-shaft is to be made by a shop in another city, and it must assemble with the blades made by the designer's company, so more precise dimensioning is required. Therefore, redo the drawing using the 1.250″ diameter on the left side as Datum A, and use geometric dimensioning and tolerancing to ensure that the flange will meet the requirements.

Ch12 Geometric Dimensioning and Tolerancing			Problem
Section-Chapter-Problem 3 12 9A	Traditional Dimensioning & Problem Statement Control-Blade Shaft	Course________ Section_____ Student____________ Date______	3-12-9A

Ch12 Geometric Dimensioning and Tolerancing

Section-Chapter-Problem	Revision with GD & T Dimensioning	Course____________Section_____	Problem
3 12 9B	Control-Blade Shaft	**Student**____________________Date______	**3-12-9B**

A A positional stop block is shown in drawing file "12-1.dxf". Select datums and add positional tolerance of .002 at MMC to the holes, a flatness tolerance of .001 to the vertical stop, and a perpendicularity tolerance of .004 to the vertical leg of the angle.

B Shown in drawing file "12-2.dxf" is a specialized chuck for a toy maker's lathe. It is not dimensioned and does not contain any tolerance information. Enlarge the center hole to 1.75", make the vertical hole threaded, and add dimensions and appropriate tolerances. Keep in mind that the machining process for this company requires extreme precision and care.

Given: Top view and side view of a steel plate. On the centerlines shown, draw the kinds of holes indicated. Add the correct callouts. All dimensions in inches.

A. Through hole, .06 clearance for a 1/2-13 UNC-2A hex head cap screw.

B. Blind hole for a .18 dia pin (FN-2 Fit), .50 deep.

C. Counterbored hole for a 1/2-13 UNC-2A Allen head, socket cap screw, .05 allowance for the screw body and head.

D. Flat head cap screw .38 dia. Allowance .o5 for the screw body.

E. Spot face for a 5/8-11 UNC-2A hex head cap screw. Allow .06 clearance for the hex head corners, .02 for the screw body.

F. Threaded through hole, 1/2 coarse thread.

G. Threaded blind hole, 1/2 coarse thread, .44 full threads.

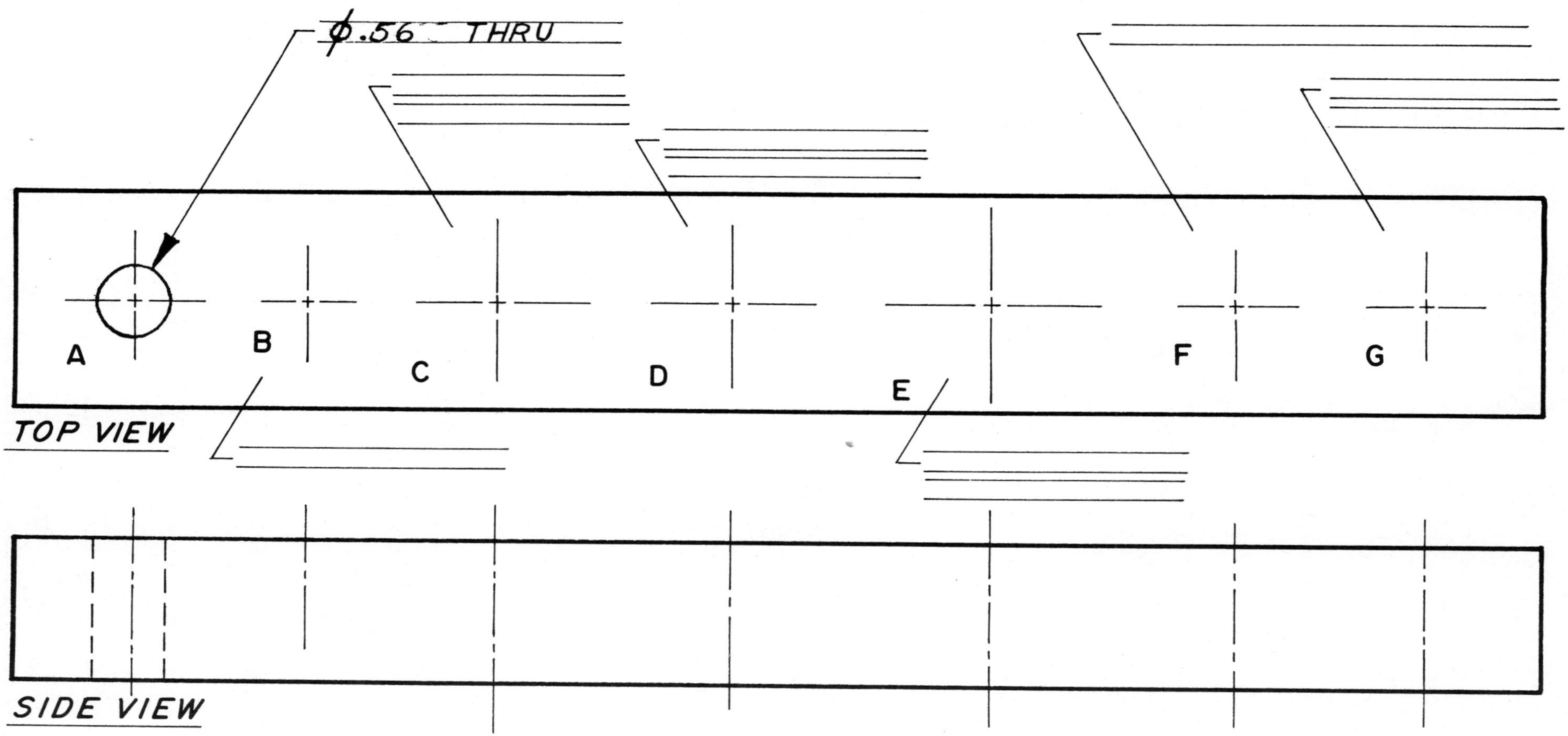

Ch13 Fasteners			Problem 3-13-1
Section-Chapter-Problem 3 13 1	Holes for Fasteners Figures 13-15, & 13-22	Course______ Section_____ Student______ Date_____	

On the centerlines provided below, draw two views (end view/ side view) of the fasteners:

A

Flat head machine screw #10–24 UNC–2A x 1.25 long

B

Round head cap screw .31–24 UNF–2A x 2.00 long

C

Oval head cap screw .75–10 UNC–2A x 1.75 long

D

Socket head cap screw (Allen head) .50–28 UNF–2A x 3.00 long

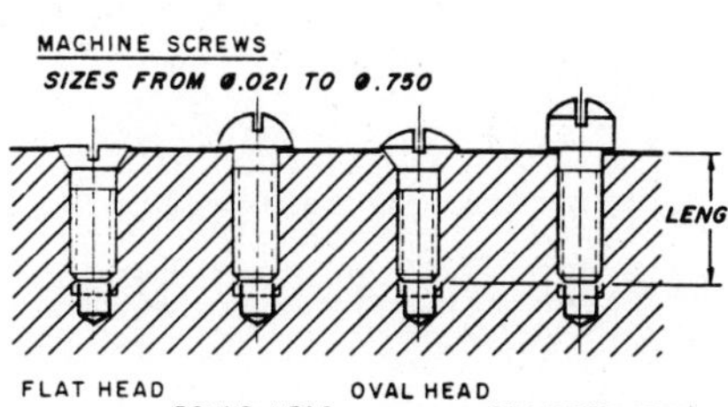

Machine screws

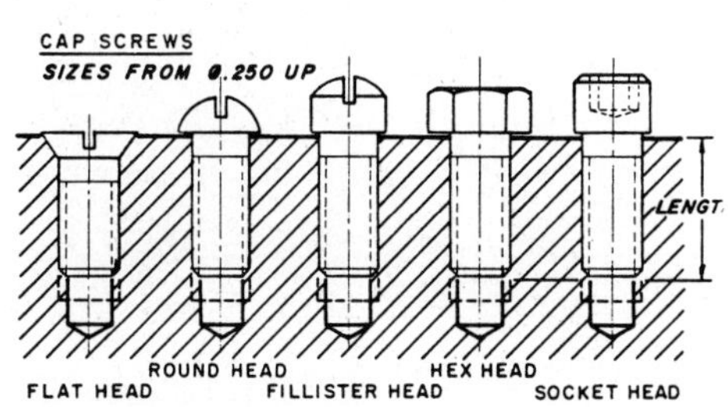

Cap screws

Ch13 Fasteners			Problem 3-13-2
Section-Chapter-Problem 3 13 2	Drawings of Fasteners Figures 13-25--30	Course________ Section____ Student__________ Date____	

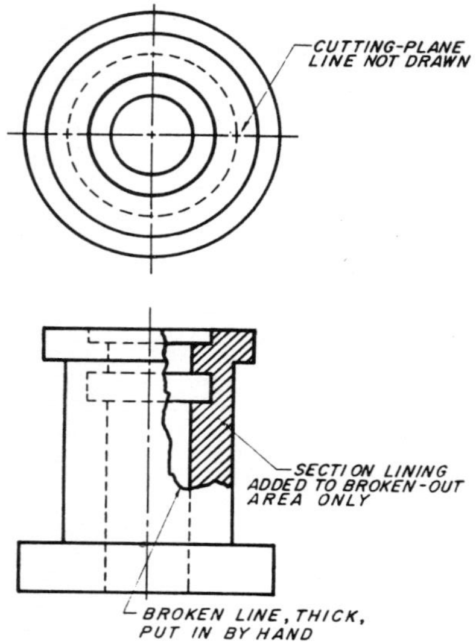

Broken-out section

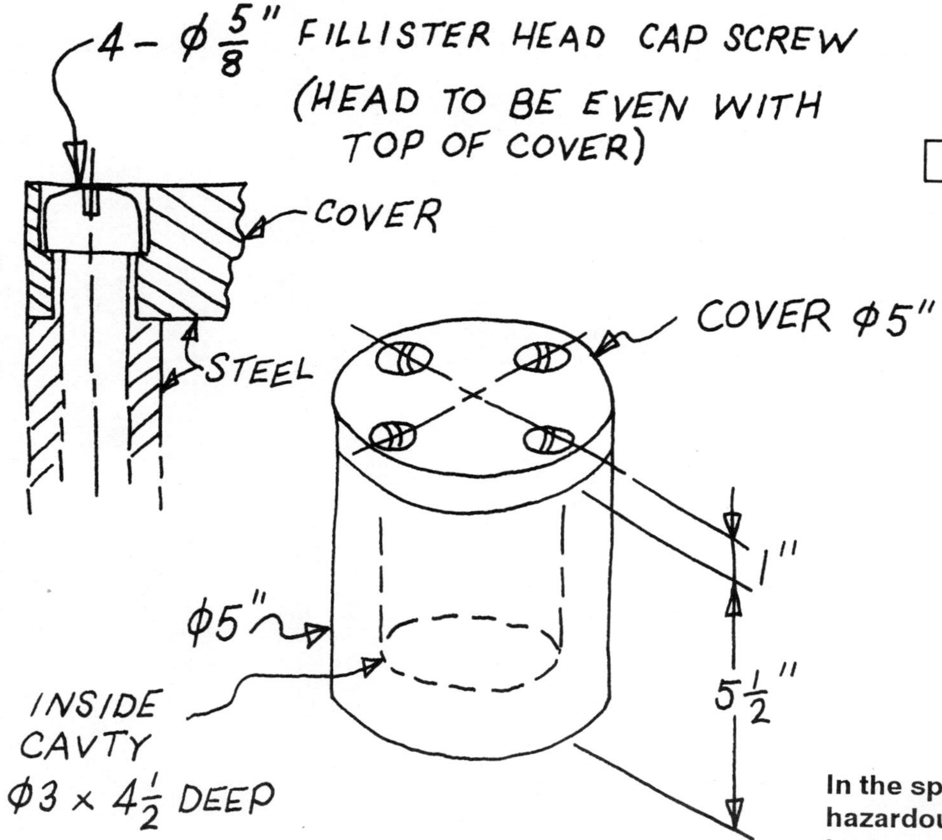

In the space provided prepare an assembly drawing, front and top views, of the hazardous-radiaton container, sketched below.Draw to half scale and use a broken-out section to show one fastener in place. Include callouts for the holes.

Ch13 Fasteners			Problem
Section-Chapter-Problem 3 13 3	Drawings of Fasteners Figures 10-58 & 13-32	Course__________ Section_____ Student________________ Date______	3-13-3

Complete this two view drawing and fill-in the blank spaces.

Clearance hole tolerances for the ϕ.25 fasteners are to be +.010 and +.015.

Key size = approximately 1/4 of shaft diameter

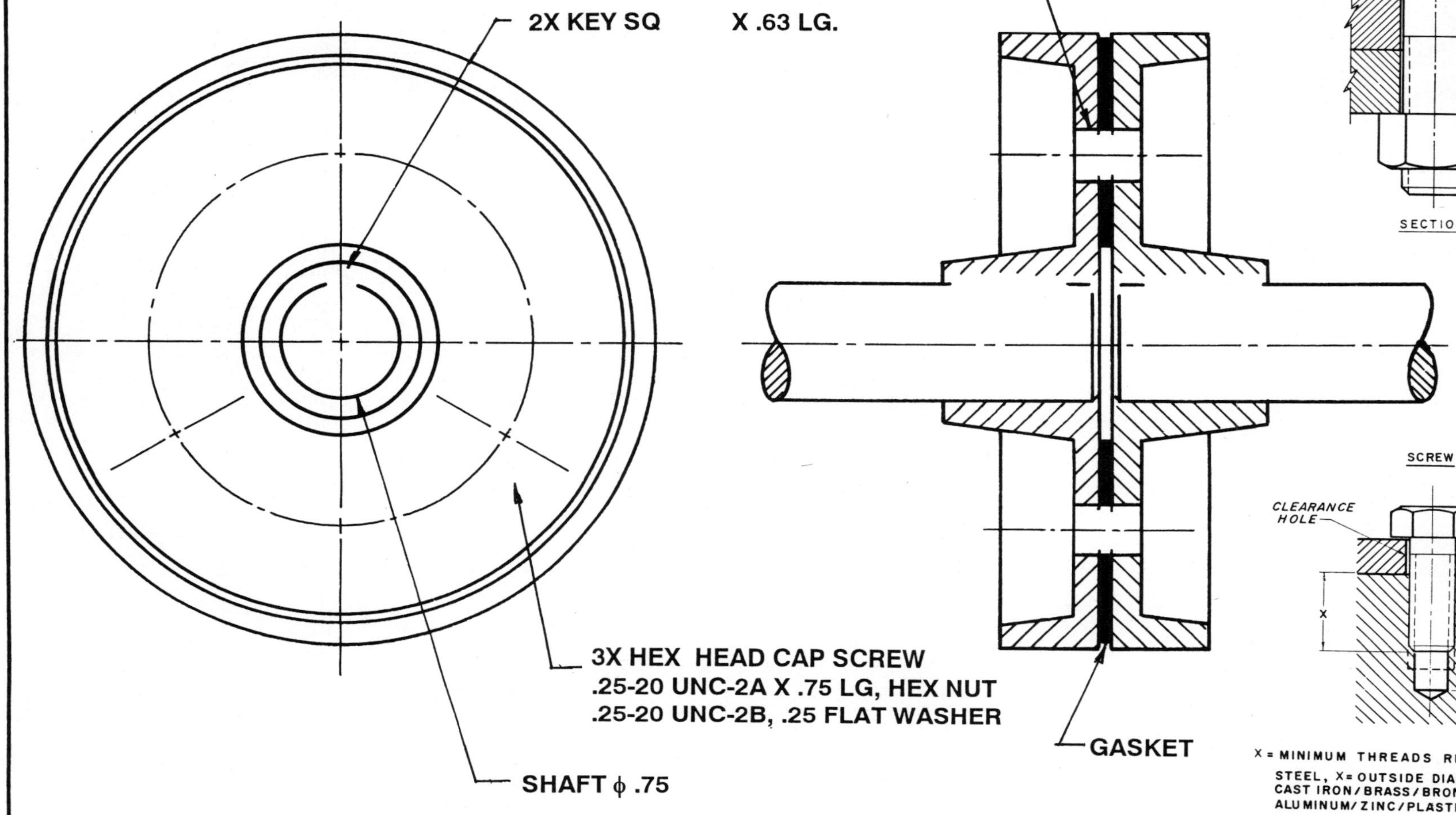

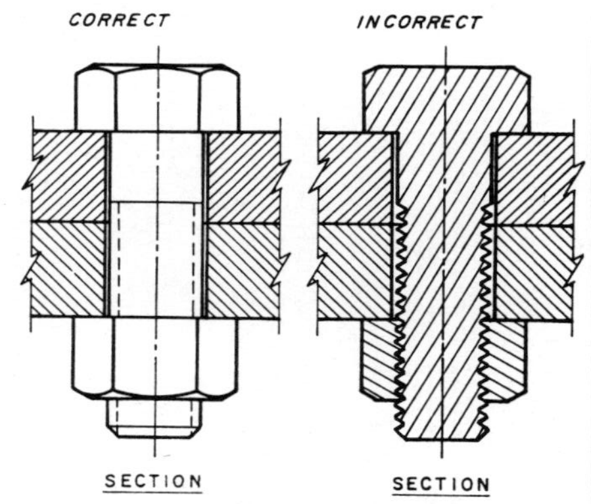

Fasteners in section

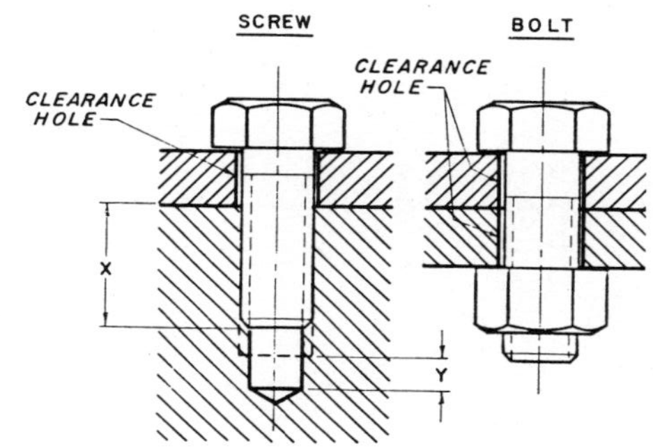

X = MINIMUM THREADS REQUIRED:
STEEL, X = OUTSIDE DIAMETER
CAST IRON/BRASS/BRONZE, X = 1.5 X OUTSIDE DIAMETER
ALUMINUM/ZINC/PLASTIC, X = 2 X OUTSIDE DIAMETER

CLEARANCE HOLE:

0 TO .375 (9) = .03 (1) LARGER THAN OUTSIDE DIAMETER
.375(9) UP = .06 (2) LARGER THAN OUTSIDE DIAMETER

Ch13 Fasteners			
Section-Chapter-Problem 3 13 4	Fasteners in section Figure 6-51	Course______ Section______ Student______ Date______	Problem 3-13-4

Dimension the ϕ 2.5 collar and the ϕ 1.5 shaft end for a ϕ .25 x 2.5" long spring pin. Half scale. Accuracy to two places except for the spring-pin hole callout. The spring-pin manufacturer recommends a drilled hole ϕ .250-.256 for the spring-pin. Specify on the drawing that the hole is to be drilled with the collar assembled on the shaft.

Spring pin

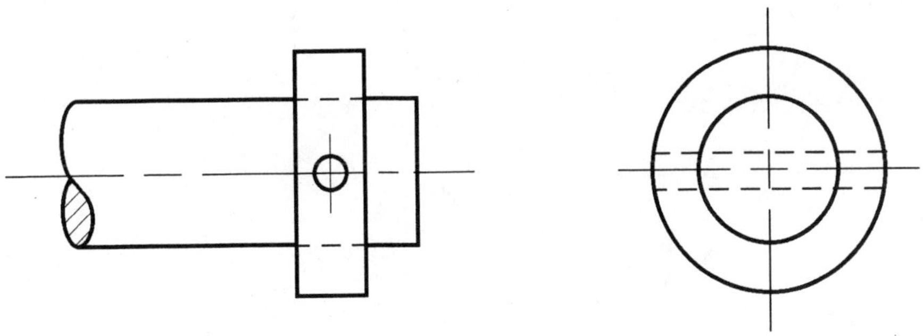

Dimension the ϕ 3.50 shaft end to accomodate a 5100-350 Retaining Ring, at the location indicated.

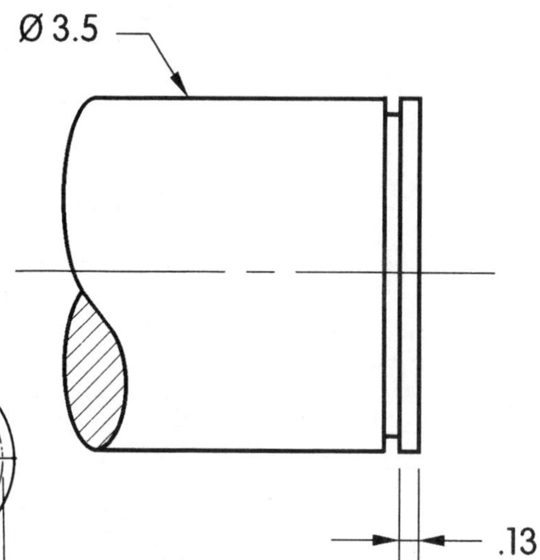

• Sizes -206 thru -1000 are available in banded or tape-wrapped Rol-Pak cartridges.

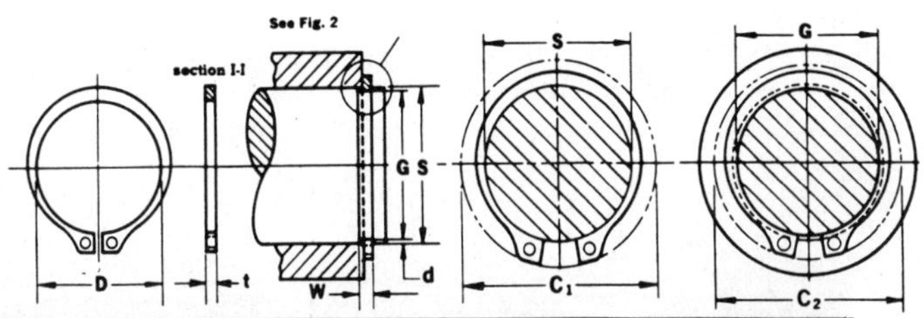

SHAFT DIAMETER			MIL-R-21248 MS 16624 EXTERNAL SERIES 5100	TRUARC RING DIMENSIONS (Thickness t applies only to unplated rings. For plated and stainless steel (Type H) rings, add .002" to the listed maximum thickness. Maximum ring thickness will be at least .0002" less than the listed minimum groove width (W).)					GROOVE DIMENSIONS (T.I.R. (total indicator reading) is the maximum allowable deviation of concentricity between groove and shaft.)					APPLICATION DATA			
				FREE DIA.		THICKNESS		Approx. weight per 1000 pieces	DIAMETER		WIDTH		Nominal groove depth	CLEARANCE DIAMETER		ALLOW. THRUST LOAD (lbs.) Sharp Corner Abutment	
Dec. equiv. inch	Approx fract. equiv. inch	Approx mm												When sprung over shaft S	When sprung into groove G	RINGS (standard material) Safety factor = 4	GROOVES (Cold Rolled steel shafts) Safety factor = 2
S	S	S	size — no.	D	tol.	t	tol.	lbs.	G	tol.	W	tol.	d	C_1	C_2	P_r	P_g
3.438	3 7/16	87.3	5100-343	3.179		.093		66.0	3.257		.103		.090	4.14	3.96	37700	21900
3.500	3 1/2	88.9	5100-350	3.237		.109		72.0	3.316		.120		.092	4.25	4.07	44900	22800
3.543	--	90.0	5100-354	3.277	+.020 −.030	.109		73.0	3.357	±.006	.120		.093	4.29	4.11	45500	23300
3.625	3 5/8	92.1	5100-362	3.352		.109	±.003	76.0	3.435	.006 T.I.R.	.120	+.005 −.000	.095	4.37	4.18	46600	24300
3.688	3 11/16	93.7	5100-368	3.410		.109		80.0	3.493		.120		.097	4.43	4.24	47300	25300

External series 5100 (*Courtesy Waldes Kohinoor, Inc.*)

Ch13 Fasteners			Problem 3-13-5
Section-Chapter-Problem 3 13 5	Spring Pin Figure 13-58 Retaining Ring Figure 13-74	Course________ Section____ Student__________ Date_____	

Complete the three views. Dimension them and use standard calloffs for the threads and chamfers.

A. Thread at hole A: ϕ1.00 coarse thread, average fit, 1.00″ deep.
B. Thread at B: ϕ1.5 fine, from left end to shoulder, average fit. Add a thread relief next to the shoulder.
C. Thread at C: ϕ2.00 coarse thread, average fit. Chamfer at right end to depth of thread.
D. Thread at hole D: ϕ.50 fine thread, average fit, 1.25″ deep, full thread. Add a counterbore to accommodate a ϕ.50 Fillister head cap screw, 1.0″ long (head to be flush with right-end surface).

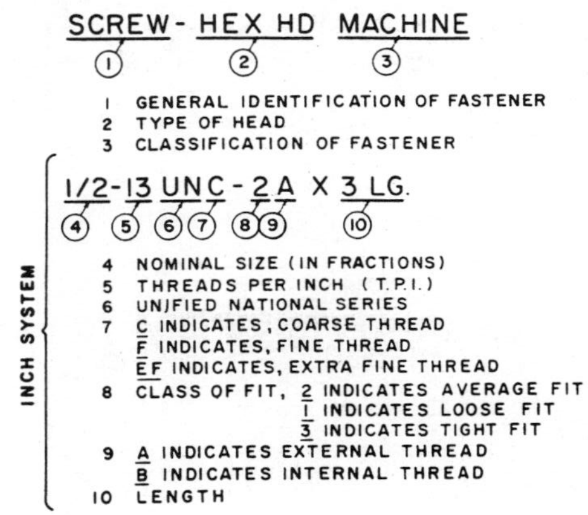

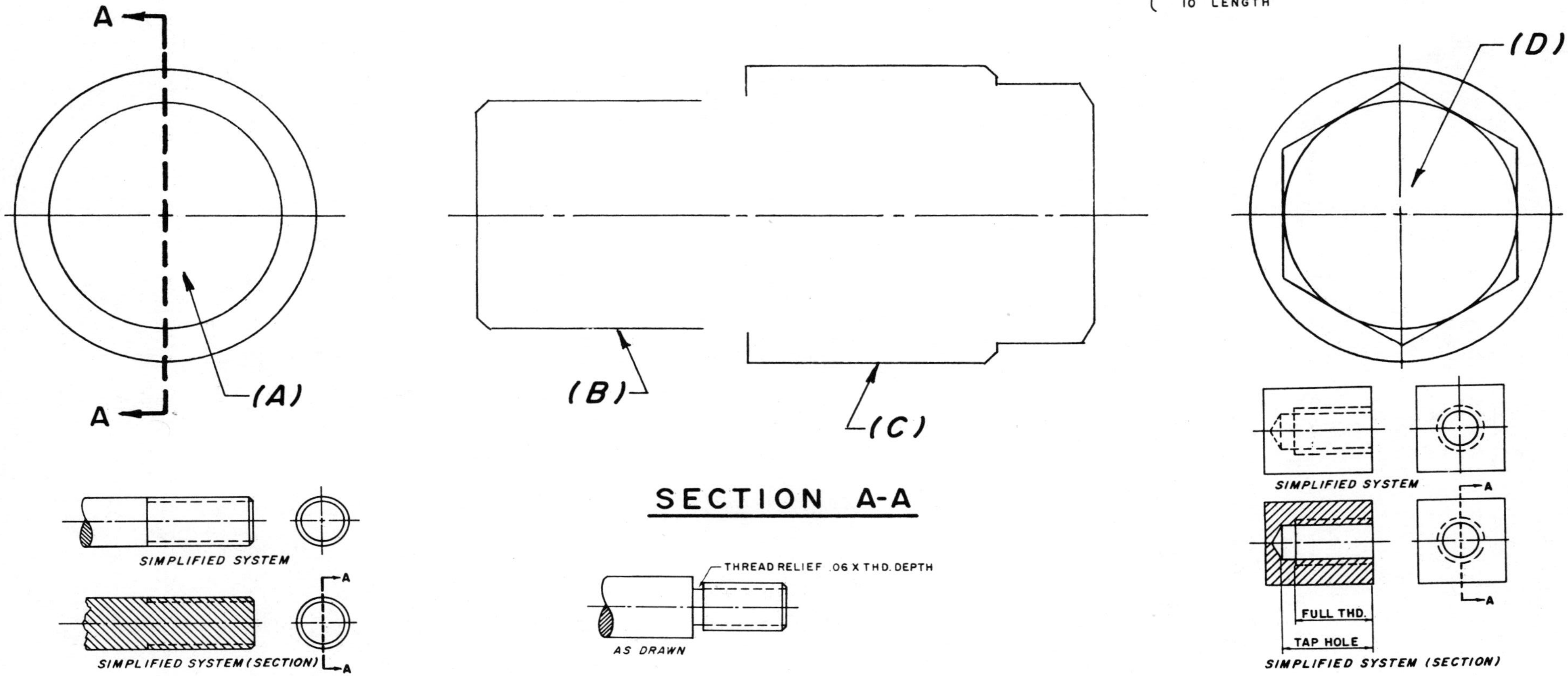

Ch13 Fasteners			Problem
Section-Chapter-Problem 3 13 6	Thread Specifications Figure 13-32 Thread Standards f or drawings Figure 13-13, 13-14 & 13-20	Course____________Section_____ Student____________________Date______	3-13-6

A In each of the four quadrants of drawing file "13-1.dxf", draw the bolts and holes as indicated.

B The spacer and hole locator block shown in drawing file "13-2.dxf" now requires a bolt that is a tight fit, 5/16" in diameter, and has 20 threads/inch, and can extend only 3/8" past the surface of the plate. Each hole that requires these bolts is indicated on the drawing. Be sure to correctly label the holes and add a parts list for the required bolts and nuts. Finally, add the remaining dimensions that pertain to the threaded holes.

<table>
<tr><td colspan="3">Ch13 Fasteners</td><td rowspan="2">Problem

3-13-7</td></tr>
<tr><td>Section-Chapter-Problem
3 13 7</td><td>Bolts
Dimensioning</td><td>Course________ Section____
Student______________ Date______</td></tr>
</table>

Draw the specified spring, to be used with an automobile valve mechanism, twice size, with the following specifications:

Free length 1.9" Outside diameter ϕ1.15" Wire size ϕ.2" Active coils 4 Ground closed ends Right hand coils Zinc plate finish Heat treat to relieve coiling stresses Hard drawn spring wire

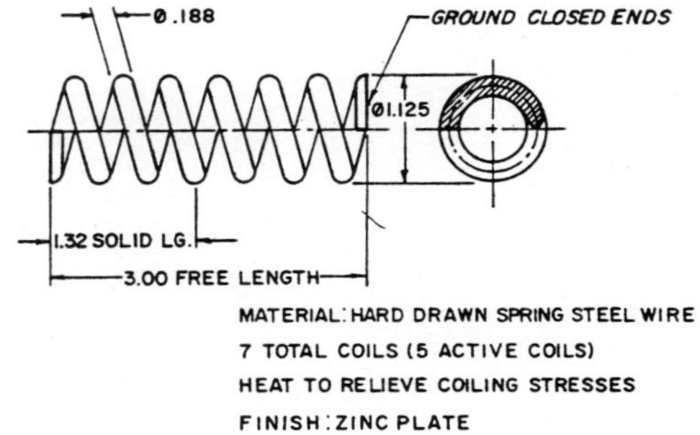

SPECIFICATIONS

Ch14 Springs			
Section-Chapter-Problem 3 14 1	Ground, Closed Ends, Spring Figure 14-13	Course________ Section____ Student__________ Date____	Problem 3-14-1

Draw the specified spring, to be used with a child's POGO STICK, half size, with the following specifications:

Free length 8.5" Outside diameter ϕ2.5" Wire size ϕ.25 Active coils 12 Plain closed ends Right hand coils Black oxide finish Heat treat to relieve coiling stresses Hard drawn spring wire

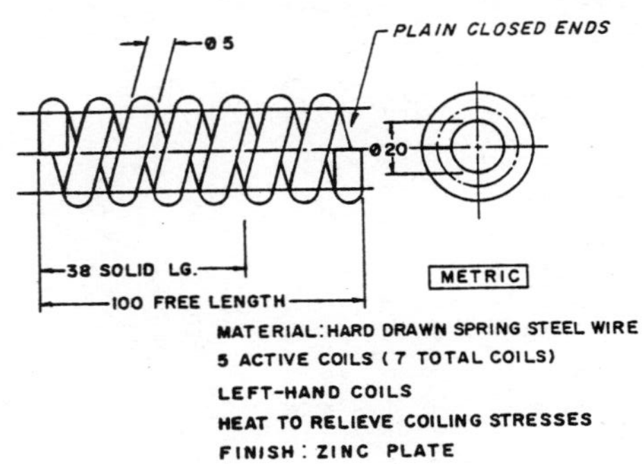

SPECIFICATIONS

____
____
____
____
____
____

Ch14 Springs			Problem 3-14-2
Section-Chapter-Problem 3 14 2	Plain closed Ends, Spring Figure 14-12	Course______ Section______ Student______ Date______	

A Draw a spring that would be used to return a pump handle to its original state which is up. Follow the specifications given:

a. Free length = 7 ½"
b. OD = 3 ¼"
c. Wire size = ¼"
d. Active coils = 10
e. Plain closed end
f. Right hand coils
g. Zinc plated
h. Heat to relieve cooling stresses

B Shown in drawing file "14-2.dxf" are two hooks that are connected to a gate for a swimming pool. The gate should close automatically so children can not enter through an open door. Design an appropriate spring for this gate. You may want to consider visiting a local hardware store for some ideas.

<table>
<tr><td colspan="3">Ch14 Springs</td><td rowspan="2">Problem

3-14-3</td></tr>
<tr><td>Section-Chapter-Problem
3 14 3</td><td>Spring Design</td><td>Course____________ Section_____
Student____________________ Date_______</td></tr>
</table>

Given: displacement diagram grid. Layout a diagram with the following specifications:
Rise 1.00" 90° harmonic motion, dwell 30°, Rise .50" 60° harmonic motion to highest point on cam, dwell 30°, Fall to starting level 135°uniform acceleration, Dwell 15°. Clockwise rotation.

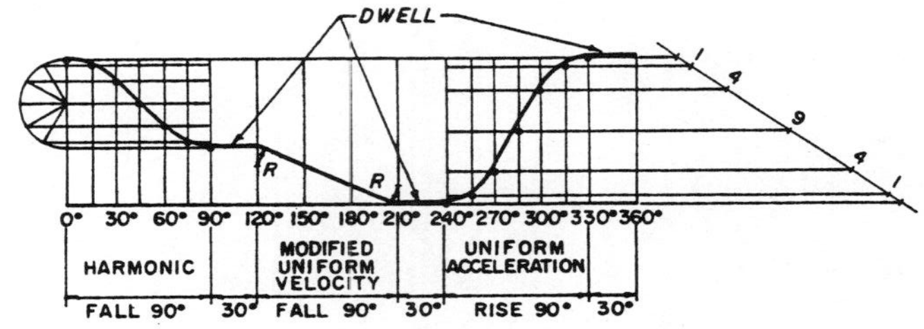

Combination of motions

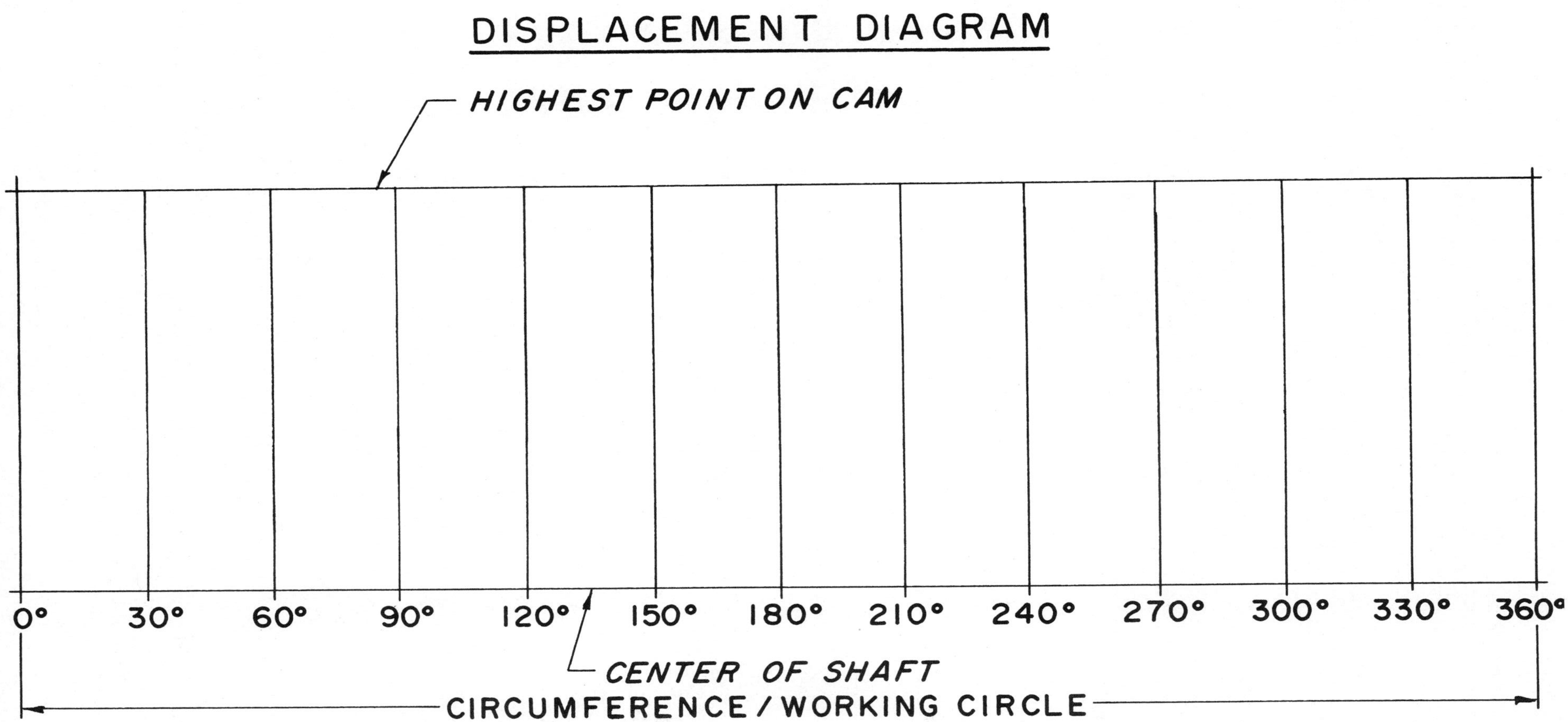

Ch15 Cams			Problem 3-15-1
Section-Chapter-Problem 3 15 1	Cam displacement diagram Figure 15-14	Course______ Section______ **Student**______ Date______	

Construct and fully dimension the cam described in the diagram on page 3-15-1.

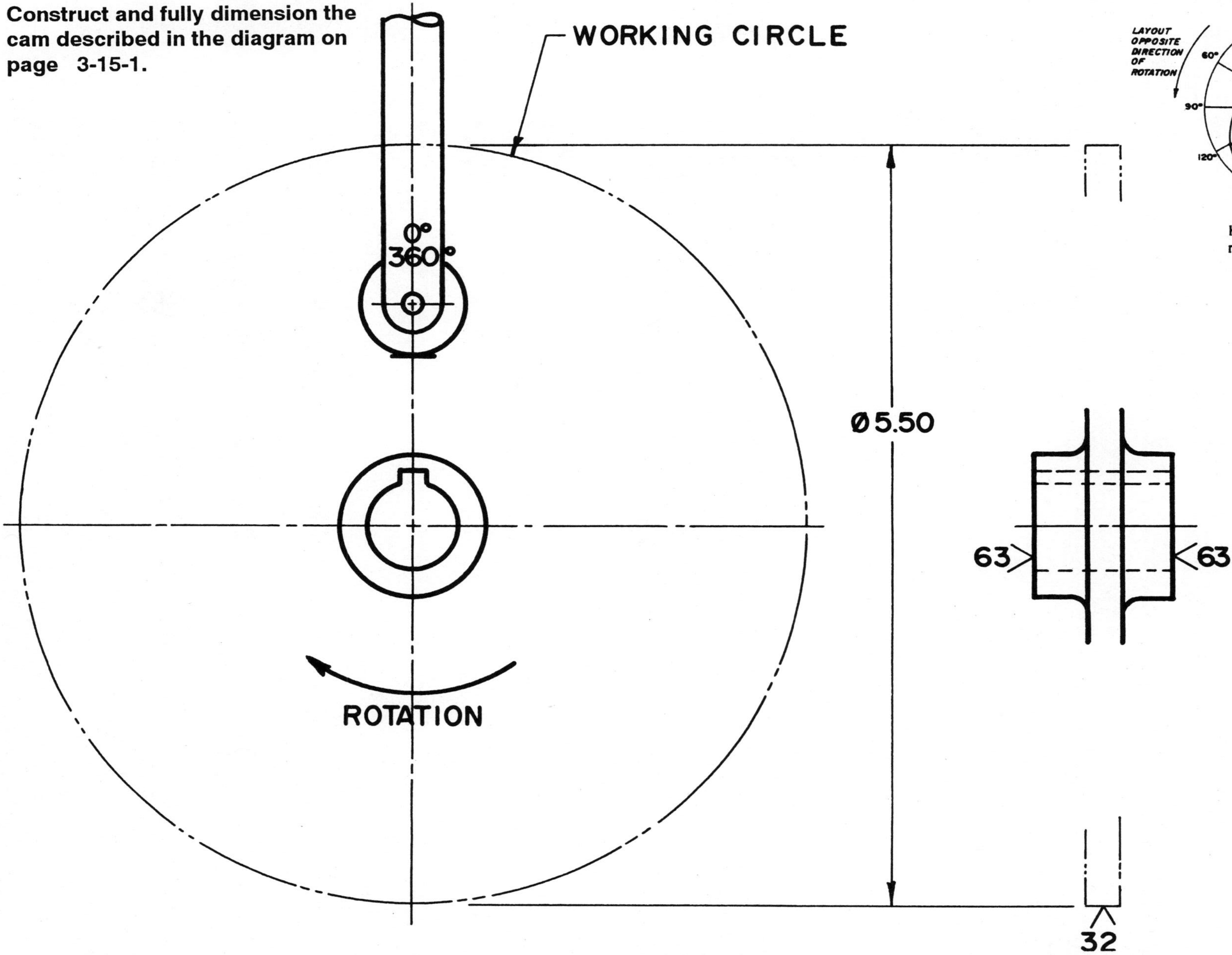

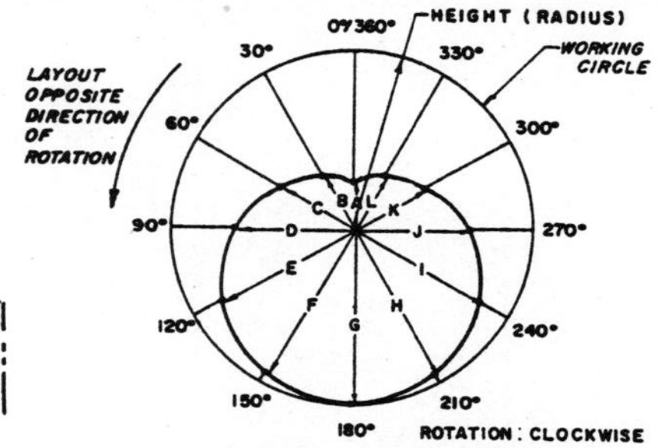

Heights transferred from displacement diagram

Ch15 Cams			Problem
Section-Chapter-Problem 3 15 2	Drawing a cam Figure 15-15 Dimensioning a cam Figure 15-20	Course________ Section____ **Student**__________ Date____	3-15-2

A A cam is shown in file "15-1.dxf". Draw the displacement diagram for it in the space provided.

B Design two cams on the same shaft that will raise and lower the rods shown in drawing file "15- 1.dxf" a distance of .25" and remain at their top position for 15 degrees. The rod on the right should raise 60 degrees after the one on the left.

Ch15 Cams			**Problem**
Section-Chapter-Problem 3 15 3	Displacement Cam Design	Course__________ Section_____ **Student**________________ Date______	**3-15-3**

Fill-in the chart for the gear set shown, pinion gear at left, and gear at right. Diametral Pitch P = 10 Also answer the two questions. (Note: A point on a rotating gear, for one revolution, travels a distance = π 2r (1). Where r = the radius to the point.

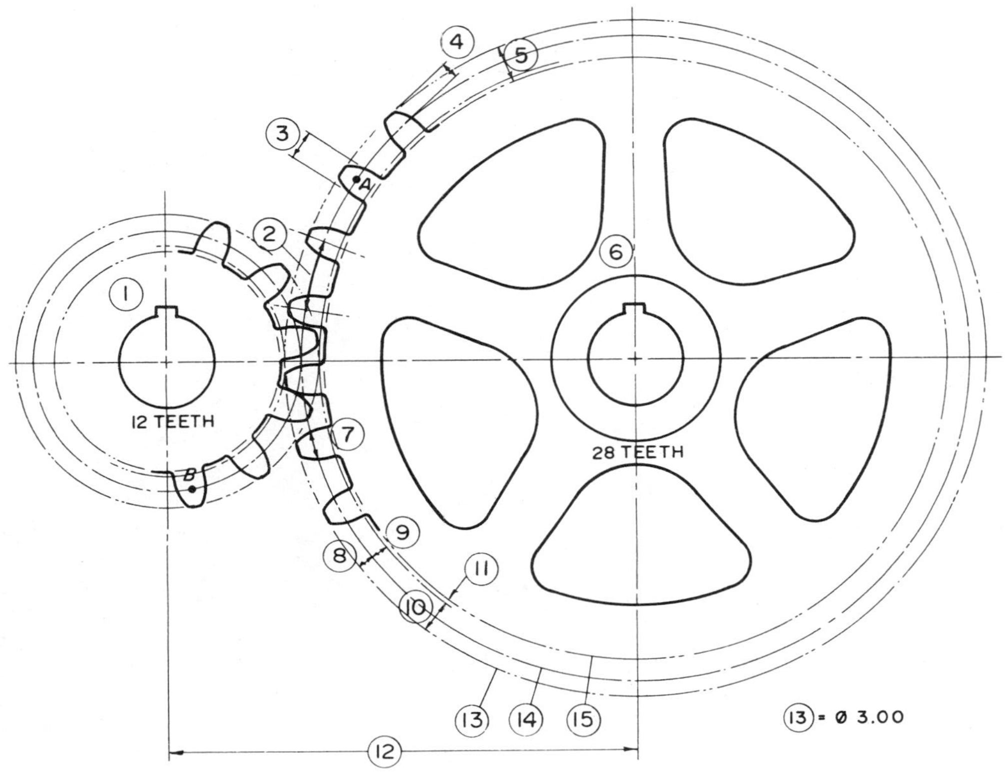

NUMBER	TERM
1	______
2	______
3	______
4	______
5	______
6	______
7	______
8	______
9	______
10	______
11	______
12	______
13	______
14	______
15	______

FOR EACH REVOLUTION OF GEAR (6), GEAR (1) MAKES ______ REV

HOW MANY FEET DOES POINT *A* TRAVEL IN 5 MINUTES AT 850 R.P.M. ? ______

Ch16 Gears

Section-Chapter-Problem			
3 16 1	Gear nomenclature Figure 16-16 Gear nomenclature Problem 3-16-2	Course______ Section_____ **Student**______ Date_____	**Problem** **3-16-1**

ITEM	PINION GEAR	SPUR GEAR
NUMBER OF TEETH (N)	12	28
DIAMETRAL PITCH (P)	10	10
PRESSURE ANGLE (Ø)	20°	20°
PITCH DIAMETER (D)		
WHOLE DEPTH (ht)		
OUTSIDE DIAMETER (DO)		
ADDENDUM (a)		
WORKING DEPTH (hk)		
CIRCULAR THICKNESS (t)		
CHORDAL THICKNESS		
CHORDAL ADDENDUM		
DEDENDUM (b)		
CENTER TO CENTER DISTANCE		

DO MATH IN THE SPACE BELOW:

Several helpful spur gear relationships:

P= N/D where P= diametral pitch
N= number of teeth
D= pitch diameter

Outside Diameter (OD)=D+(2/P)

Addendum= 1/P

Additional relationships may be found in Goetsch, Nelson and Chalk, TECHNICAL DRAWING, DELMAR PUBLISHERS, or *Machinery Handbook*

Ch16 Gears			Problem
Section-Chapter-Problem 3 16 2	Gear tooth data Figure 16-24	Course______ Section_____ Student______ Date_____	3-16-2

A A spur gear with 48 teeth and 1 ½" in diameter (Gear #1) is connected to a motor. These are shown in file "16-1.dxf". The motor is traveling at 1,500 RPM and the speed needs to be increased to 4,500 RPM and turn the same direction as Gear #1. Draw the second and third gear. How many teeth are needed for the third gear? What is the diameter of the third gear?

B The hand drill gear configuration is shown in drawing file "16-2.dxf". Currently, the handle must be rotated very fast for the drill bit to have enough speed to drill into wood. Redesign the gear train so the handle can be turned and the drill bit will move faster.

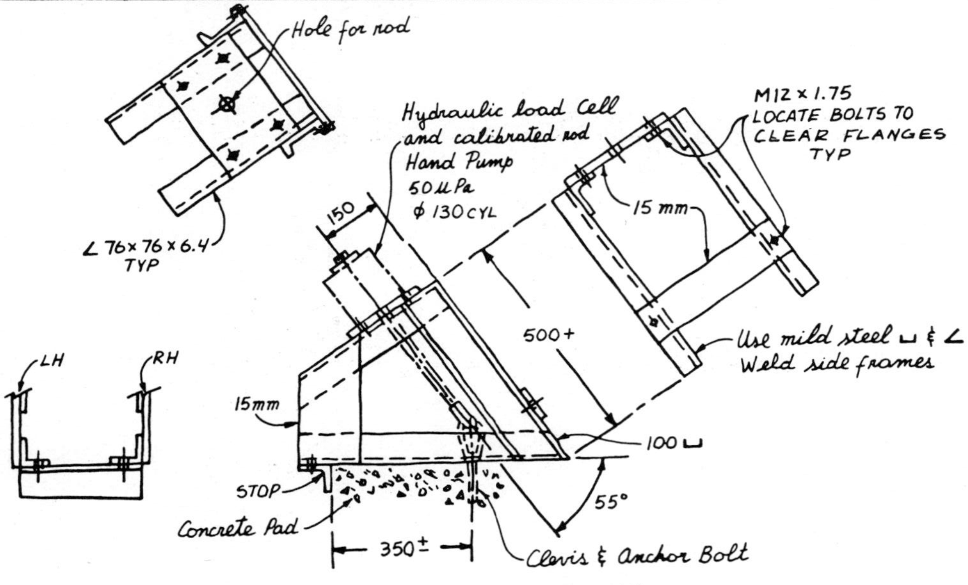

ANCHOR BOLT PULLER FRAME

Freehand sketch of a design idea

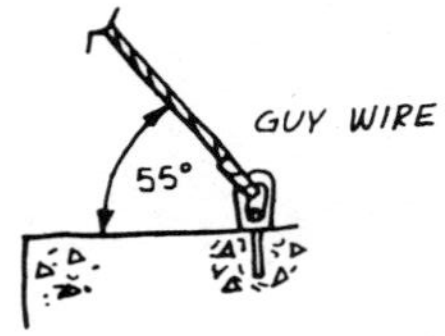

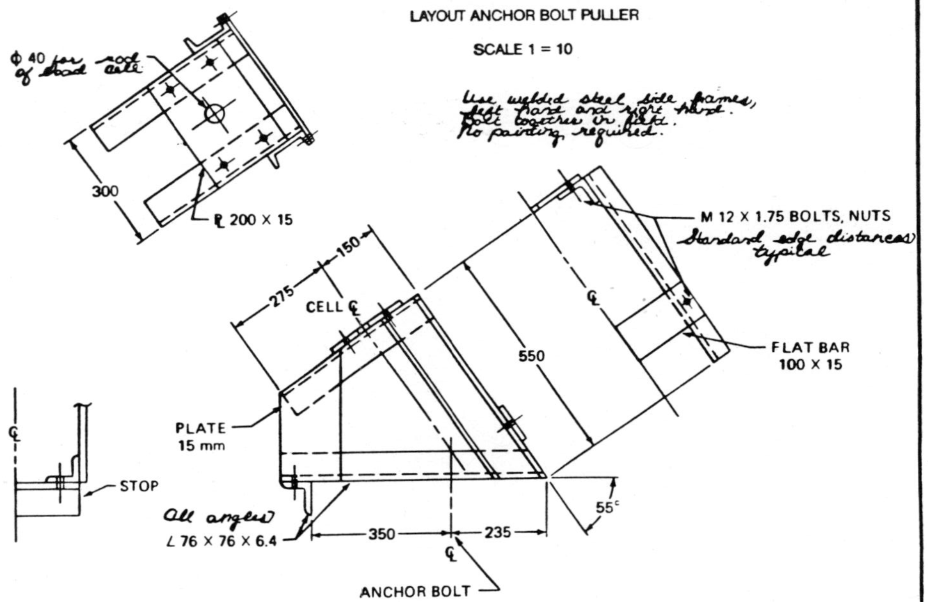

Design development layout

The next drawing is a design development layout drawn to scale, includes multivieews, auxiliaries, and sections. Has notes in writing and printing. A drafter or detailer uses the design development layout as a guide for preparing formal detail and assembly drawings, for appropriate shops.

The **ANCHOR BOLT PULLER** was used to measure the pull-out strength of anchor bolts. The Hydraulic Load Cell had a pounds force readout gage which indicated the force to dislodge the anchor bolt. Guy wires (for steadying a columnar structure) to be attached to the anchor bolts were to be at 55° to the horizontal.

Design development begins with a freehand sketch, which shows the approximate shape , size and features of a design idea. Materials and functional requirements are noted in writing and printing.

Ch17 Assembly and Detail Drawings for Design			Problem
Section-Chapter-Problem 3 17 1A	Freehand sketch Figure Ex 1-1 Design development layout Ex 1-2	Course_____ Section_____ Student_____ Date_____	3-17-1A

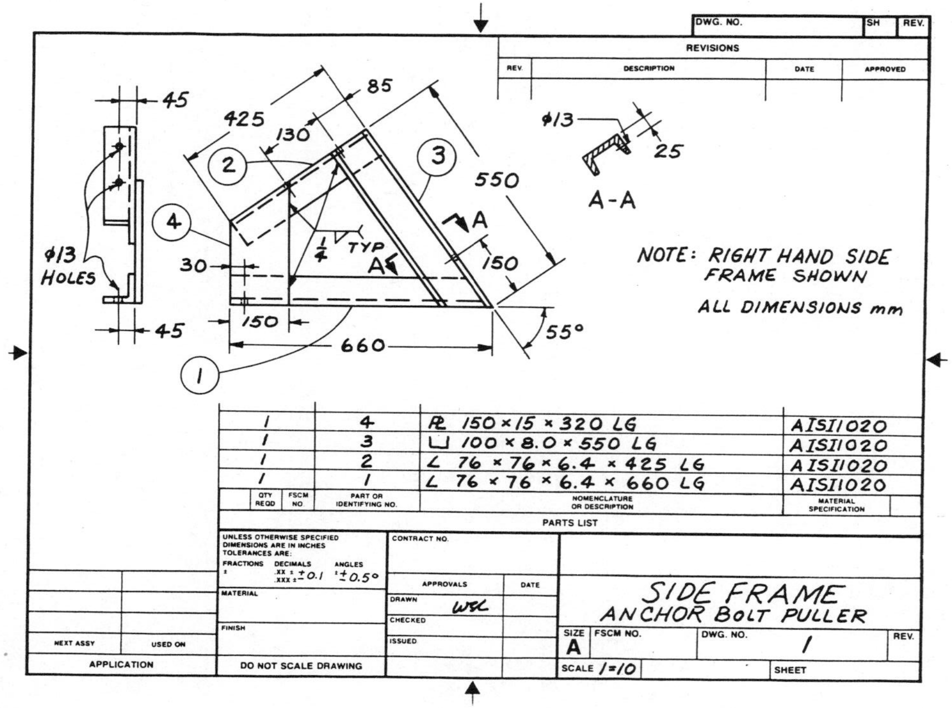

Technical detail drawing of sub-assembly

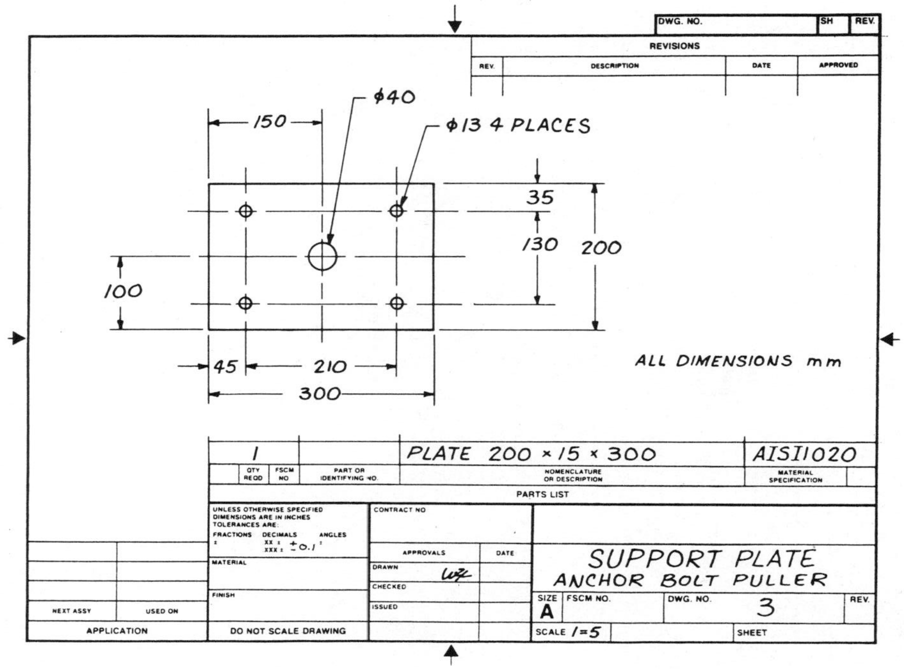

Technical detail drawing of a support plate

This detail drawing shows a welded frame made up of four pieces. But on the assembly drawing of the ANCHOR BOLT PULLER the welded frame is listed as only one item.

Detail drawings usually contain only one item for efficient routing in manufacturing. Different items could go to different shop operations.

Ch17 Assembly and Detail Drawings for Design

Section-Chapter-Problem	Detail drawing Figure Ex 1-3 & 1-4	Course____________ Section_____	Problem
3 17 1B		**Student**________________________ Date______	3-17-1B

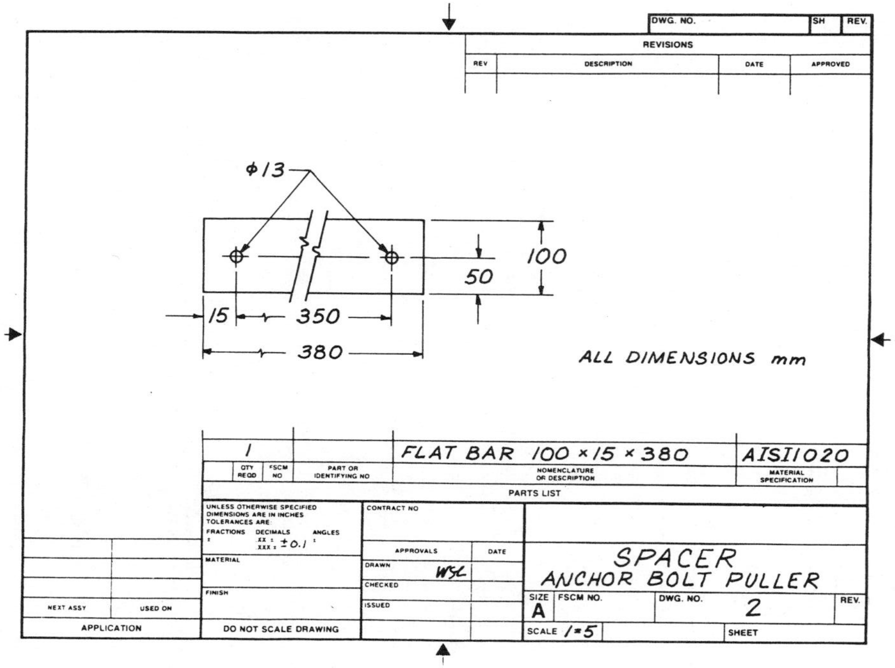

Technical detail drawing of a spacer

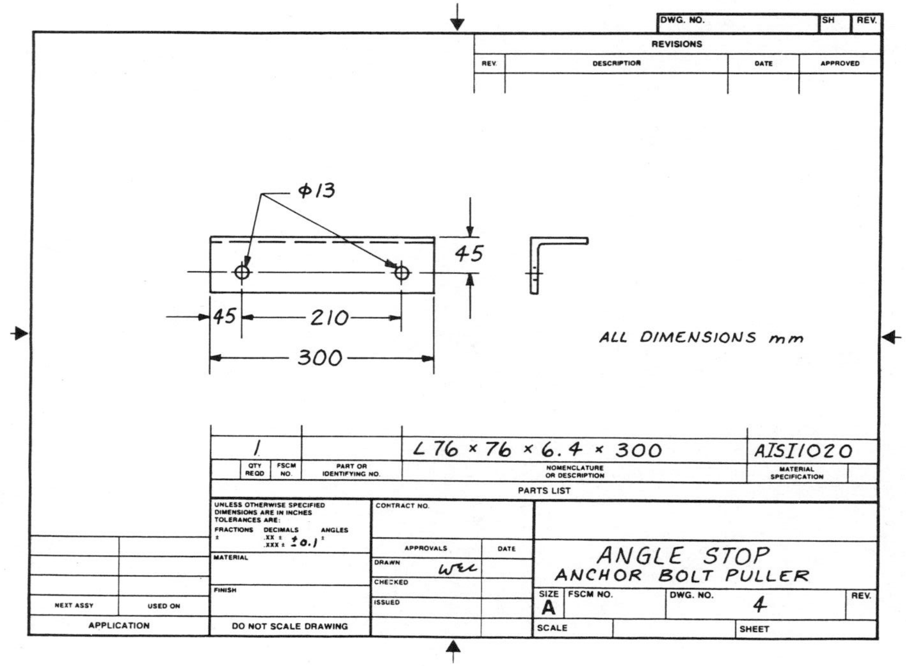

Technical detail drawing of an angle stop

Detail drawings usually contain only one item for efficient routing in manufacturing. Different items could go to different shop operations.

Ch17 Assembly and Detail Drawings for Design

Section-Chapter-Problem	Detail drawing Figure Ex 1-3 & 1-4	Course______ Section_____	Problem
3 17 1C		Student______ Date_____	3-17-1C

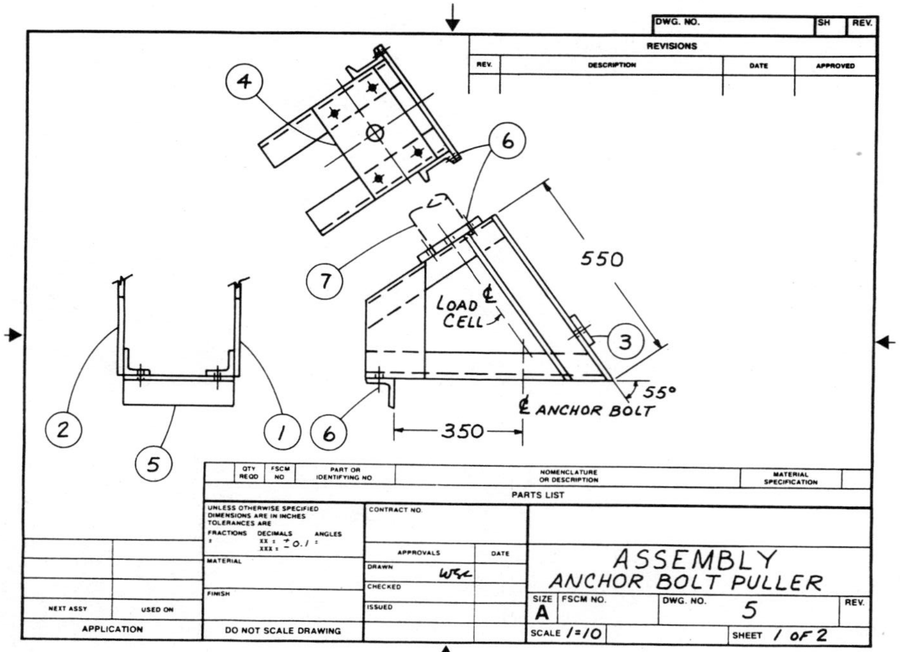

Assembly drawing of anchor bolt puller frame

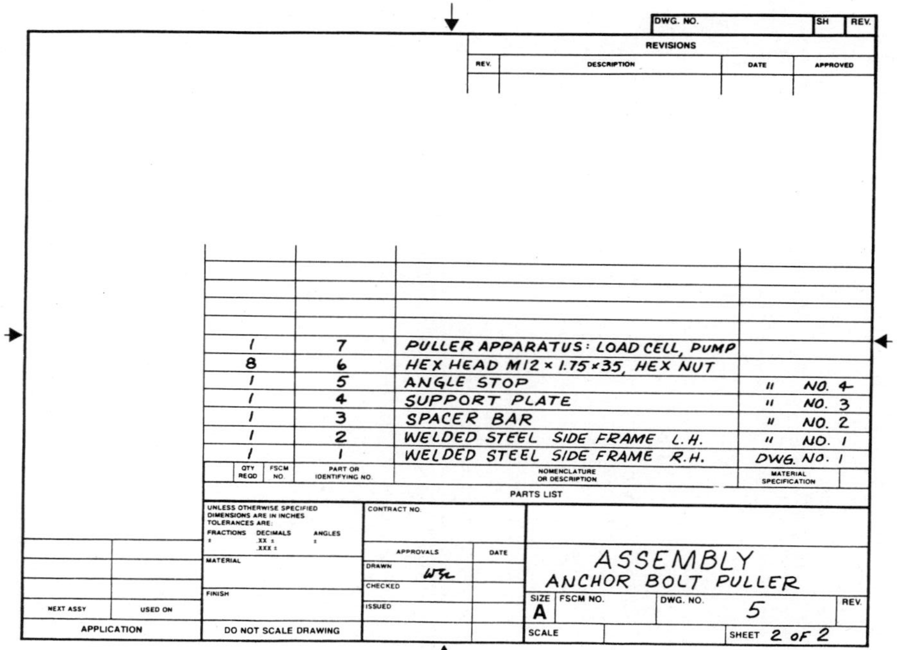

QTY REQD	FSCM NO	PART OR IDENTIFYING NO	NOMENCLATURE OR DESCRIPTION	MATERIAL SPECIFICATION
1		7	PULLER APPARATUS: LOAD CELL, PUMP	
8		6	HEX HEAD M12 × 1.75×35, HEX NUT	
1		5	ANGLE STOP	" NO. 4
1		4	SUPPORT PLATE	" NO. 3
1		3	SPACER BAR	" NO. 2
1		2	WELDED STEEL SIDE FRAME L.H.	" NO. 1
1		1	WELDED STEEL SIDE FRAME R.H.	DWG. NO. 1

Parts list for anchor bolt puller

Assembly drawings with parts lists (integral or separate) account for all the items which make up the design. Note that items 6 and 7 in the parts list were not on a detail drawing.

Ch17 Assembly and Detail Drawings for Design

Section-Chapter-Problem	Assembly drawing Figure Ex 1-5 & 1-6	Course_____ Section_____	Problem
3 17 1D		Student_____ Date_____	3-17-1D

A conceptual drawing of a Toggle Clamp is shown below. The drawing is essentially to scale, but still needs to be redrawn in a development layout to establish dimensions and some details. Also, detail and assembly drawings are needed.

The "toggle" occurs when B on the arm ABT passes below the line of centers between A and C, and pressure at P causes the elastic head to compress on an object to be clamped. Thus B is "locked" in position. A typical compressing distance is 1.5mm. Adjustment at E is provided to accomodate different objects.

Do the followin to develop the Toggle Clamp:

1. On Page 3-17-2B prepare a layout drawing from the information given. Scale the drawing (mm) and develop dimensions of the components, and locations of pivot points and bolt holes.
2. On Pages 3-17-2C and 3-17-2D prepare details of the parts of the toggle clamp. For this problem have more than one detail on a page.
3. On Page 3-17-2E prepare an assembly drawing and a parts list.

Suggestions:

1. Member DC could be "T" shaped with four holes in the base.
2. Have two BC members, one on each side of DC.
3. Member ADE could be slotted at the right end so ABT and DC would be inside the slot.
4. ABT would be a single thickness.
5. Use machined pins with retaining rings at piviot points A,B,C, & D.

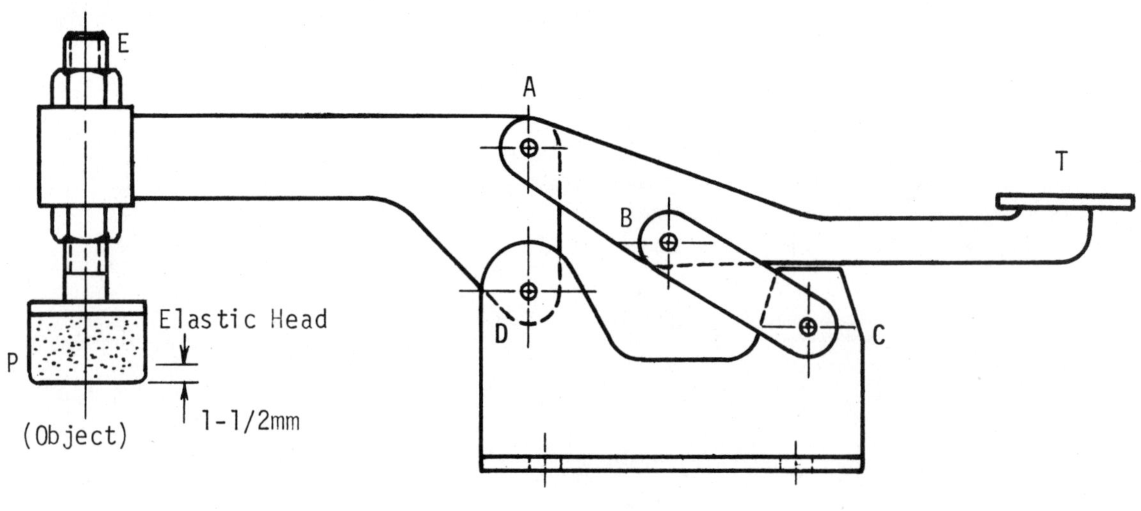

Ch17 Assembly and Detail Drawings for Design			Problem
Section-Chapter-Problem 3 17 2A	Conceptual Drawing Toggle Clamp	Course________ Section____ Student____________ Date_____	3-17-2A

Ch17 Assembly and Detail Drawings for Design			Problem
Section-Chapter-Problem 3 17 2B	Design Layout Drawing Toggle Clamp	Course____________Section_____ **Student**____________________Date______	**3-17-2B**

Details of the Toggle Clamp are to be drawn here. For this problem more than one detail may be on the page.

DWG. NO.	SH	REV.

REVISIONS

REV.	DESCRIPTION	DATE	APPROVED

	QTY REQD	FSCM NO.	PART OR IDENTIFYING NO.	NOMENCLATURE OR DESCRIPTION	MATERIAL SPECIFICATION	

PARTS LIST

UNLESS OTHERWISE SPECIFIED DIMENSIONS ARE IN INCHES TOLERANCES ARE:
FRACTIONS ± DECIMALS .XX ± .XXX ± ANGLES

MATERIAL

FINISH

NEXT ASSY	USED ON

APPLICATION

DO NOT SCALE DRAWING

CONTRACT NO.

APPROVALS	DATE
DRAWN	
CHECKED	
ISSUED	

SIZE	FSCM NO.	DWG. NO.	REV.
A			

SCALE SHEET

Ch17 Assembly and Detail Drawings for Design

Section-Chapter-Problem	Details of Toggle Clamp	Course____ Section____	Problem
3 17 2C		Student____ Date____	3-17-2C

Details of the Toggle Clamp are to be drawn here. For this problem more than one detail may be on the page.

DWG. NO.	SH	REV.

REVISIONS

REV.	DESCRIPTION	DATE	APPROVED

	QTY REQD	FSCM NO.	PART OR IDENTIFYING NO.	NOMENCLATURE OR DESCRIPTION	MATERIAL SPECIFICATION	

PARTS LIST

UNLESS OTHERWISE SPECIFIED DIMENSIONS ARE IN INCHES TOLERANCES ARE:
FRACTIONS : DECIMALS .XX : .XXX : ANGLES

MATERIAL

FINISH

DO NOT SCALE DRAWING

CONTRACT NO.

APPROVALS	DATE
DRAWN	
CHECKED	
ISSUED	

NEXT ASSY	USED ON

APPLICATION

SIZE A	FSCM NO.	DWG. NO.	REV.

SCALE		SHEET

Ch17 Assembly and Detail Drawings for Design

Section-Chapter-Problem	Details of Toggle Clamp	Course__________ Section____	Problem
3 17 2D		Student__________ Date_____	3-17-2D

An assembly drawing for the Toggle Clamp is to be prepared on this page.
The drawing and parts list are to be on the same page.

DWG. NO.	SH	REV.

REVISIONS

REV.	DESCRIPTION	DATE	APPROVED

	QTY REQD	FSCM NO.	PART OR IDENTIFYING NO.	NOMENCLATURE OR DESCRIPTION	MATERIAL SPECIFICATION	

PARTS LIST

UNLESS OTHERWISE SPECIFIED DIMENSIONS ARE IN INCHES TOLERANCES ARE:
FRACTIONS ± DECIMALS .XX ± .XXX ± ANGLES

MATERIAL

FINISH

DO NOT SCALE DRAWING

CONTRACT NO.

APPROVALS	DATE
DRAWN	
CHECKED	
ISSUED	

NEXT ASSY	USED ON

APPLICATION

SIZE A	FSCM NO.	DWG. NO.	REV.

SCALE | SHEET

Ch17 Assembly and Detail Drawings for Design

Section-Chapter-Problem
3 17 2E

Assembly Drawing
Toggle Clamp

Course__________ Section_____
Student________________ Date______

Problem
3-17-2E

ENGINEERING CHANGE ORDER (ECO) PROBLEM

Assume that you are a newly hired drafter in an engineering construction company and that your first assignment is to process an ECO. (Note that ECO's are used frequently to introduce new technical personnel to the company's paperwork and manufacturing procedures.) The ECO refers to the ANCHOR BOLT PULLER shown on pages 3-17-1A through 3-17-1D, and specifies thet the angle of pull has changed from 55° to 75°, and that the width of the puller is to be 25% larger. The 550mm dimension does not change.

FREEHAND SKETCH

On this page make a freehand sketch of the ANCHOR BOLT PULLER with the 75° and 25% revisions.

On the next seven pages do the following:

1. On 3-17-3B prepare a design layout drawing with the 75° and 25% changes incorporated.
2. Do detail drawings on pages 3-17-3C through 3-17-3F for the side-frame, spacer, support plate, and angle stop, respectively.
3. On 3-17-3G prepare an assembly drawing and a parts list on page 3-17-3H.

Ch17 Assembly and Detail Drawings for Design			Problem
Section-Chapter-Problem 3 17 3B	ECO Design Layout Anchor Bolt Puller	Course____________Section_____ **Student**____________________Date_______	3-17-3B

DWG. NO.	SH	REV.

REVISIONS

REV.	DESCRIPTION	DATE	APPROVED

QTY REQD	FSCM NO.	PART OR IDENTIFYING NO.	NOMENCLATURE OR DESCRIPTION	MATERIAL SPECIFICATION

PARTS LIST

UNLESS OTHERWISE SPECIFIED DIMENSIONS ARE IN INCHES TOLERANCES ARE:
FRACTIONS ± DECIMALS .XX ± .XXX ± ANGLES

MATERIAL

FINISH

CONTRACT NO.

APPROVALS	DATE
DRAWN	
CHECKED	
ISSUED	

NEXT ASSY	USED ON

APPLICATION

DO NOT SCALE DRAWING

SIZE A	FSCM NO.	DWG. NO.	REV.

SCALE SHEET

Ch17 Assembly and Detail Drawings for Design

Section-Chapter-Problem
3 17 3C

ECO Side Frame Details
Anchor Bolt Puller

Course____________Section_____
Student____________________Date______

Problem
3-17-3C

DWG. NO.	SH	REV.

REVISIONS

REV.	DESCRIPTION	DATE	APPROVED

	QTY REQD	FSCM NO.	PART OR IDENTIFYING NO.	NOMENCLATURE OR DESCRIPTION	MATERIAL SPECIFICATION	

PARTS LIST

UNLESS OTHERWISE SPECIFIED DIMENSIONS ARE IN INCHES TOLERANCES ARE:
FRACTIONS ± DECIMALS .XX ± .XXX ± ANGLES

MATERIAL

FINISH

NEXT ASSY | USED ON

APPLICATION

DO NOT SCALE DRAWING

CONTRACT NO.

APPROVALS	DATE
DRAWN	
CHECKED	
ISSUED	

SIZE A	FSCM NO.	DWG. NO.	REV.
SCALE		SHEET	

Ch17 Assembly and Detail Drawings for Design

Section-Chapter-Problem
3 17 3D

ECO Spacer Details
Anchor Bolt Puller

Course____________ Section_____
Student____________________ Date______

Problem
3-17-3D

DWG. NO.	SH	REV.

REVISIONS

REV.	DESCRIPTION	DATE	APPROVED

	QTY REQD	FSCM NO.	PART OR IDENTIFYING NO.	NOMENCLATURE OR DESCRIPTION	MATERIAL SPECIFICATION	

PARTS LIST

UNLESS OTHERWISE SPECIFIED DIMENSIONS ARE IN INCHES TOLERANCES ARE:
FRACTIONS DECIMALS ANGLES
.XX :
.XXX :

MATERIAL

FINISH

DO NOT SCALE DRAWING

CONTRACT NO.

APPROVALS	DATE
DRAWN	
CHECKED	
ISSUED	

NEXT ASSY	USED ON

APPLICATION

SIZE A	FSCM NO.	DWG. NO.	REV.
SCALE		SHEET	

Ch17 Assembly and Detail Drawings for Design

Section-Chapter-Problem
3 17 3E

ECO Support Plate Details
Anchor Bolt Puller

Course__________ Section_____
Student________________ Date______

Problem
3-17-3E

DWG. NO.	SH	REV.

REVISIONS

REV.	DESCRIPTION	DATE	APPROVED

	QTY REQD	FSCM NO.	PART OR IDENTIFYING NO.	NOMENCLATURE OR DESCRIPTION	MATERIAL SPECIFICATION	

PARTS LIST

UNLESS OTHERWISE SPECIFIED
DIMENSIONS ARE IN INCHES
TOLERANCES ARE:
FRACTIONS ± DECIMALS .XX ± .XXX ± ANGLES

MATERIAL

FINISH

DO NOT SCALE DRAWING

CONTRACT NO.

APPROVALS	DATE
DRAWN	
CHECKED	
ISSUED	

NEXT ASSY	USED ON

APPLICATION

SIZE A	FSCM NO.	DWG. NO.	REV.
SCALE		SHEET	

Ch17 Assembly and Detail Drawings for Design

Section-Chapter-Problem		
3	17	3F

ECO Angle Stop Details
Anchor Bolt Puller

Course__________ Section_____

Student________________ Date______

Problem 3-17-3F

DWG. NO.	SH	REV.

REVISIONS

REV.	DESCRIPTION	DATE	APPROVED

	QTY REQD	FSCM NO.	PART OR IDENTIFYING NO.	NOMENCLATURE OR DESCRIPTION	MATERIAL SPECIFICATION	

PARTS LIST

UNLESS OTHERWISE SPECIFIED DIMENSIONS ARE IN INCHES TOLERANCES ARE:
FRACTIONS ± DECIMALS .XX ± .XXX ± ANGLES

MATERIAL

FINISH

DO NOT SCALE DRAWING

NEXT ASSY	USED ON

APPLICATION

CONTRACT NO.

APPROVALS	DATE
DRAWN	
CHECKED	
ISSUED	

SIZE A	FSCM NO.	DWG. NO.	REV.
SCALE		SHEET	

Ch17 Assembly and Detail Drawings for Design

Section-Chapter-Problem		
3	17	3G

ECO Assembly Drawing
Anchor Bolt Puller

Course____________Section_____
Student____________________Date______

Problem 3-17-3G

DWG. NO.	SH	REV.

REVISIONS

REV.	DESCRIPTION	DATE	APPROVED

	QTY REQD	FSCM NO.	PART OR IDENTIFYING NO.	NOMENCLATURE OR DESCRIPTION	MATERIAL SPECIFICATION	

PARTS LIST

UNLESS OTHERWISE SPECIFIED
DIMENSIONS ARE IN INCHES
TOLERANCES ARE:
FRACTIONS DECIMALS ANGLES
.XX :
.XXX :

MATERIAL

FINISH

DO NOT SCALE DRAWING

CONTRACT NO.

APPROVALS	DATE
DRAWN	
CHECKED	
ISSUED	

NEXT ASSY	USED ON

APPLICATION

SIZE	FSCM NO.	DWG. NO.	REV.
A			

SCALE SHEET

Ch17 Assembly and Detail Drawings for Design

Section-Chapter-Problem		
3	17	3H

ECO Parts List

Anchor Bolt Puller

Course__________ Section_____

Student______________ Date_____

Problem 3-17-3H

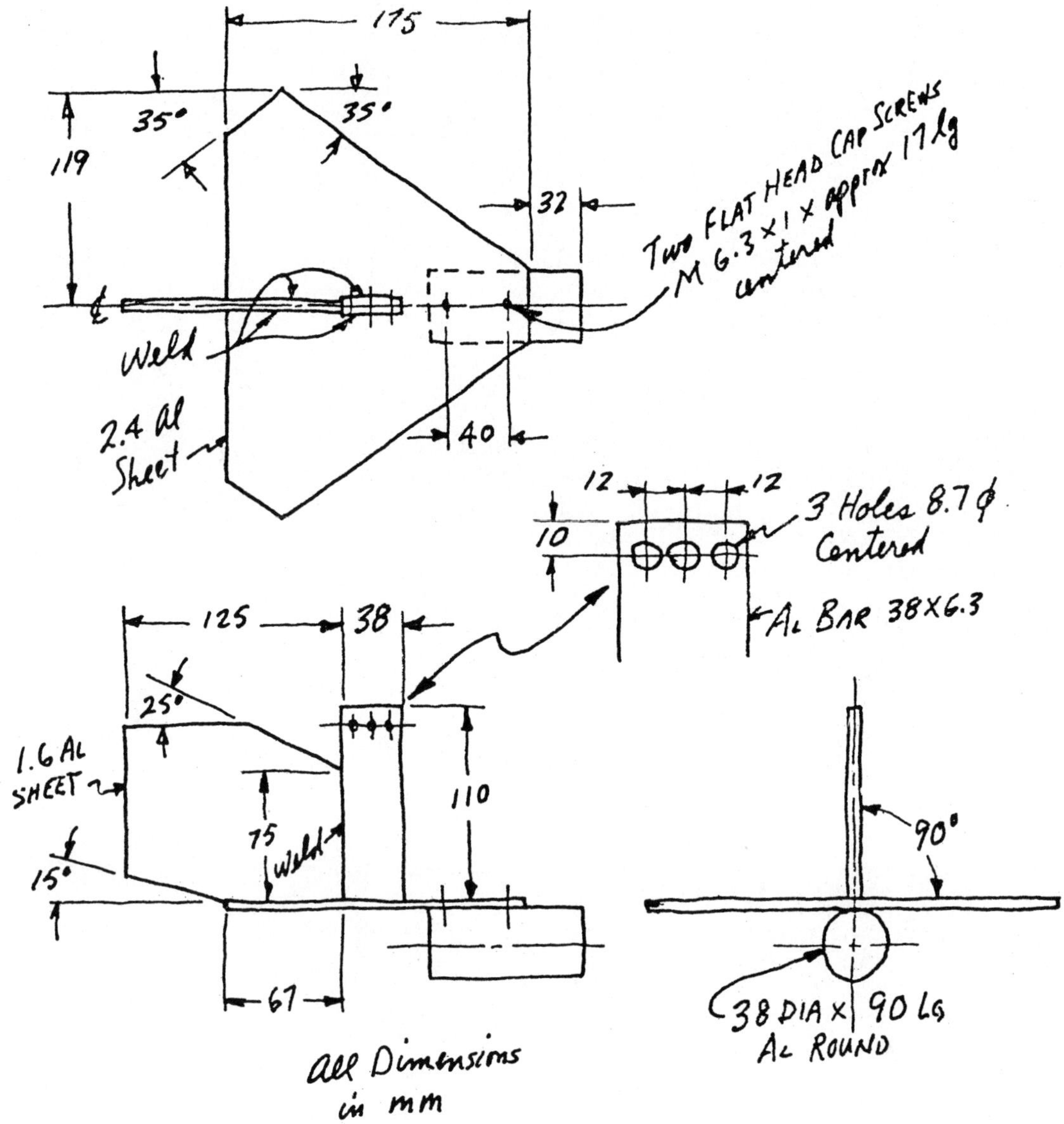

The freehand sketch shown is a Paravane Stabilizer used in conjunction with ship-model research.

1. On page 3-17-4B prepare a design development layout drawing of the Paravane Stabilizer.
2. On 3-17-4C prepare details of the vane (three piece weldment).
3. On 3-17-4D prepare details of the weight.
4. On 3-17-4E prepare an assembly drawing with parts list.

(Note:
for example drawings refer to Anchor Bolt Puller freehand sketch, design development layout, details and assembly drawings on pages 3-17-1A through 3-17-1D.)

Ch17 Assembly and Detail Drawings for Design			Problem
Section-Chapter-Problem 3 17 4A	Freehand sketch Paravane Stabilizer	Course________ Section_____ Student____________ Date_____	3-17-4A

Ch17 Assembly and Detail Drawings for Design			Problem 3-17-4B
Section-Chapter-Problem 3 17 4B	Design Development Layout Paravane Stabilizer	Course____________Section_____ **Student**_____________________Date______	

DWG. NO.	SH	REV.

REVISIONS

REV.	DESCRIPTION	DATE	APPROVED

	QTY REQD	FSCM NO.	PART OR IDENTIFYING NO.	NOMENCLATURE OR DESCRIPTION	MATERIAL SPECIFICATION	

PARTS LIST

UNLESS OTHERWISE SPECIFIED DIMENSIONS ARE IN INCHES TOLERANCES ARE:
FRACTIONS ± DECIMALS .XX ± .XXX ± ANGLES

MATERIAL

FINISH

DO NOT SCALE DRAWING

CONTRACT NO.

APPROVALS	DATE
DRAWN	
CHECKED	
ISSUED	

NEXT ASSY	USED ON

APPLICATION

SIZE A	FSCM NO.	DWG. NO.	REV.
SCALE		SHEET	

Ch17 Assembly and Detail Drawings for Design

Section-Chapter-Problem
3 17 4C

Vane Details
Paravane Stabilizer

Course__________ Section_____
Student________________ Date______

Problem
3-17-4C

DWG. NO.	SH	REV.

REVISIONS

REV.	DESCRIPTION	DATE	APPROVED

	QTY REQD	FSCM NO.	PART OR IDENTIFYING NO.	NOMENCLATURE OR DESCRIPTION	MATERIAL SPECIFICATION	

PARTS LIST

UNLESS OTHERWISE SPECIFIED DIMENSIONS ARE IN INCHES TOLERANCES ARE:
FRACTIONS ± DECIMALS .XX ± .XXX ± ANGLES

MATERIAL

FINISH

DO NOT SCALE DRAWING

CONTRACT NO.

APPROVALS	DATE
DRAWN	
CHECKED	
ISSUED	

NEXT ASSY	USED ON

APPLICATION

SIZE A	FSCM NO.	DWG. NO.	REV.
SCALE		SHEET	

Ch17 Assembly and Detail Drawings for Design

Section-Chapter-Problem
3 17 4D

Weight Detail
Paravane Stabilizer

Course__________ Section_____
Student_______________ Date______

Problem
3-17-4D

DWG. NO.	SH	REV.

REVISIONS

REV.	DESCRIPTION	DATE	APPROVED

	QTY REQD	FSCM NO.	PART OR IDENTIFYING NO.	NOMENCLATURE OR DESCRIPTION	MATERIAL SPECIFICATION	

PARTS LIST

UNLESS OTHERWISE SPECIFIED DIMENSIONS ARE IN INCHES TOLERANCES ARE:
FRACTIONS ± DECIMALS .XX ± .XXX ± ANGLES

MATERIAL

FINISH

DO NOT SCALE DRAWING

NEXT ASSY	USED ON
APPLICATION	

CONTRACT NO.

APPROVALS	DATE
DRAWN	
CHECKED	
ISSUED	

SIZE	FSCM NO.	DWG. NO.	REV.
A			
SCALE		SHEET	

Ch17 Assembly and Detail Drawings for Design

Section-Chapter-Problem		Course_______Section_____	Problem
3 17 4E	Assembly Drawing and Parts List Paravane Stabilizer	Student_______Date_____	3-17-4E

The freehand sketch shown below shows an ECO (Enginerring Change Order) revision to the Paravane Stabilizer depicted on page 3-17-4A. The proposed change will allow users to try out different vertical stabilizer fins.

To initiate this change do the following:

1. On page 3-17-5B prepare a design development layout drawing of the revised Paravane Stabilizer.

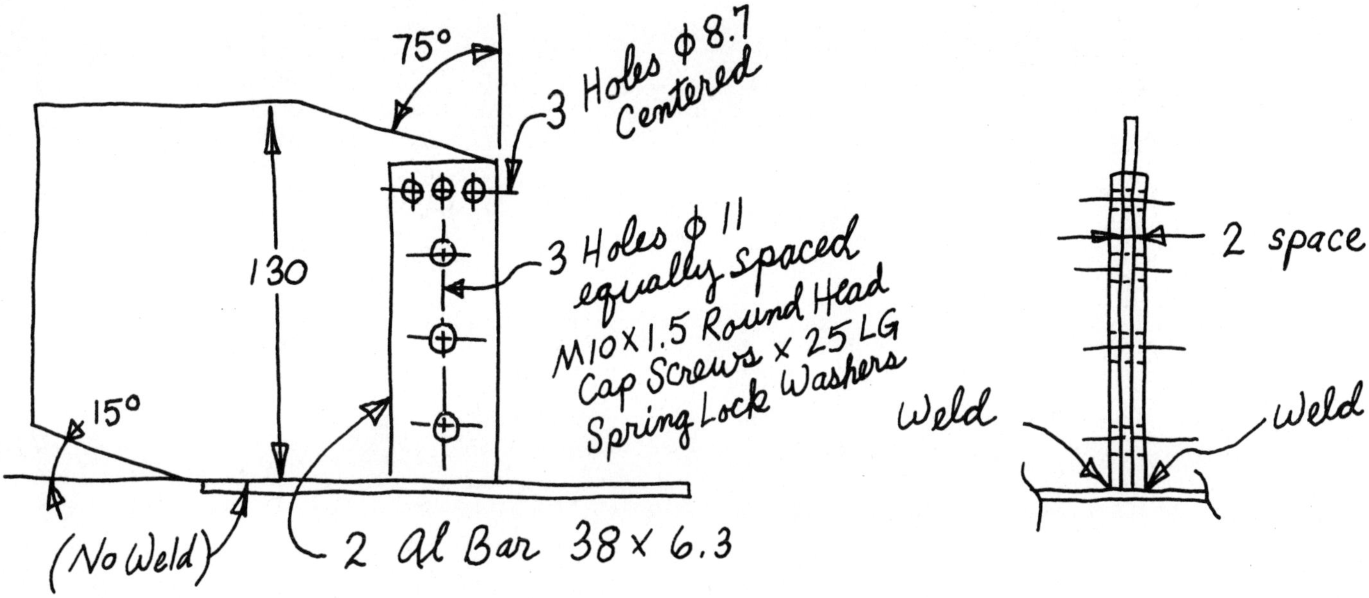

2. On 3-17-5C prepare details of the horizontal vane (three piece weldment).
3. On 3-17-5D prepare details of the vertical vane.
4. On 3-17-5D prepare details of the weight.
5. On 3-17-5E prepare an assembly drawing with parts list.

Ch17 Assembly and Detail Drawings for Design			Problem
Section-Chapter-Problem 3 17 5A	Freehand Sketch ECO Paravane Stabilizer	Course__________ Section____ Student______________ Date_____	3-17-5A

Ch17 Assembly and Detail Drawings for Design			Problem 3-17-5B
Section-Chapter-Problem 3 17 5B	Design development layout ECO Paravane Stabilizer	Course____________Section_____ **Student**______________________Date_______	

DWG. NO.	SH	REV.

REVISIONS

REV.	DESCRIPTION	DATE	APPROVED

	QTY REQD	FSCM NO.	PART OR IDENTIFYING NO.	NOMENCLATURE OR DESCRIPTION	MATERIAL SPECIFICATION	

PARTS LIST

UNLESS OTHERWISE SPECIFIED
DIMENSIONS ARE IN INCHES
TOLERANCES ARE:
FRACTIONS ± DECIMALS .XX ± .XXX ± ANGLES

MATERIAL

FINISH

DO NOT SCALE DRAWING

CONTRACT NO.

APPROVALS	DATE
DRAWN	
CHECKED	
ISSUED	

NEXT ASSY	USED ON

APPLICATION

SIZE	FSCM NO.	DWG. NO.	REV.
A			

SCALE SHEET

Ch17 Assembly and Detail Drawings for Design

Section-Chapter-Problem		Course______ Section______	Problem
3 17 5C	Details of Horizontal Vane (three-piece weldment) ECO Paravane Stabilizer	**Student**______ Date______	3-17-5C

DWG. NO.	SH	REV.

REVISIONS

REV.	DESCRIPTION	DATE	APPROVED

	QTY REQD	FSCM NO.	PART OR IDENTIFYING NO.	NOMENCLATURE OR DESCRIPTION	MATERIAL SPECIFICATION	

PARTS LIST

NEXT ASSY	USED ON
APPLICATION	

UNLESS OTHERWISE SPECIFIED DIMENSIONS ARE IN INCHES TOLERANCES ARE:
FRACTIONS ± DECIMALS .XX ± .XXX ± ANGLES

MATERIAL

FINISH

DO NOT SCALE DRAWING

CONTRACT NO.

APPROVALS	DATE
DRAWN	
CHECKED	
ISSUED	

SIZE A	FSCM NO.	DWG. NO.	REV.
SCALE		SHEET	

Ch17 Assembly and Detail Drawings for Design

Section-Chapter-Problem
3 17 5D

Detail of Vertical Vane
ECO Paravane Stabilizer

Course__________ Section_____
Student________________ Date______

Problem
3-17-5D

DWG. NO.	SH	REV.

REVISIONS

REV.	DESCRIPTION	DATE	APPROVED

	QTY REQD	FSCM NO.	PART OR IDENTIFYING NO.	NOMENCLATURE OR DESCRIPTION	MATERIAL SPECIFICATION	

PARTS LIST

UNLESS OTHERWISE SPECIFIED DIMENSIONS ARE IN INCHES TOLERANCES ARE:
FRACTIONS : DECIMALS .XX : .XXX : ANGLES

MATERIAL

FINISH

DO NOT SCALE DRAWING

CONTRACT NO.

APPROVALS	DATE
DRAWN	
CHECKED	
ISSUED	

SIZE	FSCM NO.	DWG. NO.	REV.
A			

SCALE SHEET

NEXT ASSY	USED ON
APPLICATION	

Ch17 Assembly and Detail Drawings for Design

Section-Chapter-Problem		Course______ Section_____	Problem
3 17 5E	Detail of the Weight ECO Paravane Stabilizer	Student______ Date______	3-17-5E

DWG. NO.	SH	REV.

REVISIONS

REV.	DESCRIPTION	DATE	APPROVED

	QTY REQD	FSCM NO.	PART OR IDENTIFYING NO.	NOMENCLATURE OR DESCRIPTION	MATERIAL SPECIFICATION	

PARTS LIST

UNLESS OTHERWISE SPECIFIED DIMENSIONS ARE IN INCHES TOLERANCES ARE:
FRACTIONS : DECIMALS .XX : .XXX : ANGLES

MATERIAL

FINISH

DO NOT SCALE DRAWING

CONTRACT NO.

APPROVALS	DATE
DRAWN	
CHECKED	
ISSUED	

NEXT ASSY	USED ON
APPLICATION	

SIZE	FSCM NO.	DWG. NO.	REV.
A			
SCALE		SHEET	

Ch17 Assembly and Detail Drawings for Design

Section-Chapter-Problem
3 17 5F

Assembly Drawing
ECO Paravane Stabilizer

Course__________ Section_____
Student________________ Date______

Problem
3-17-5F

A A palette for stacking unassembled wooden nightstands is shown in file "17-1.dxf" as a multiview drawing. The manufacturer of this new palette is requesting an assembly drawing that will illustrate how each of the pieces is connected. This will improve the floor supervisor's ability to design an appropriate manufacturing sequence. Pay attention to the bill of materials to show the correct fasteners.

B A model rocket is shown in file "17-2.dxf". This rocket has just begun a revision cycle and the feedback received from the consumer is that it is too short and the three fins do not adequately stabilize it. Redesign it to incorporate these changes and any others that you feel are necessary. Create an assembly drawing so the consumer can understand how to correctly build the rocket.

<table>
<tr><td colspan="3">Ch17 Assembly and Detail Drawings for Design</td><td rowspan="2">Problem
3-17-6</td></tr>
<tr><td>Section-Chapter-Problem
3 17 6</td><td>Assembly Drawings:
Palette, Rocket</td><td>Course__________ Section_____
Student__________________ Date_______</td></tr>
</table>

Use an isometric "box" to start an isometric drawing of the object shown. Measure directly from the views. Full Size.

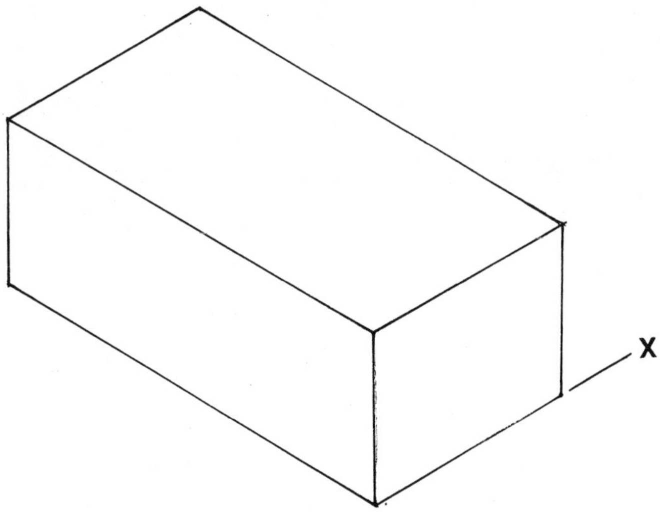

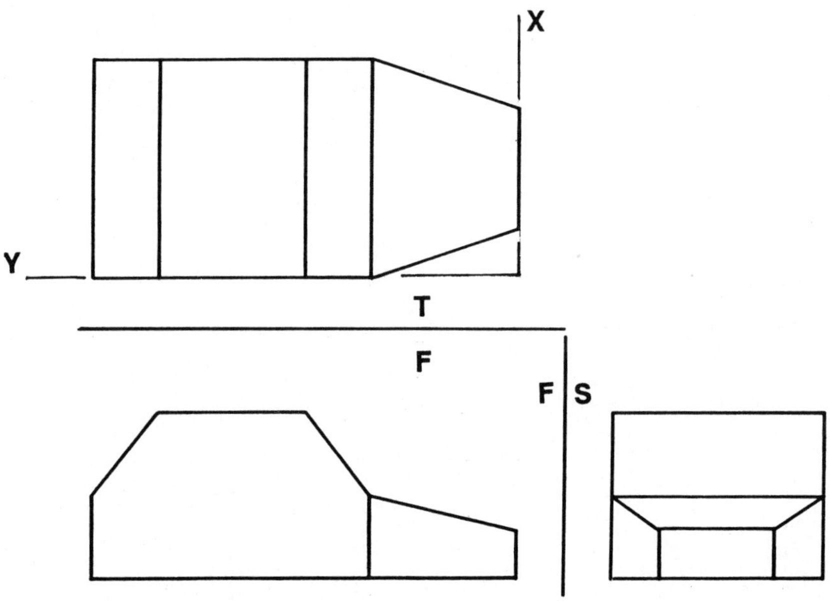

Ch18 Pictorial Drawings			Problem
Section-Chapter-Problem 3 18 1	Isometric Drawing Figure 18-4	Course__________ Section_____ **Student**________________ Date______	3-18-1

Use an isometric "box" to start an isometric drawing of the objects shown. Measure directly from the views. Half Size.

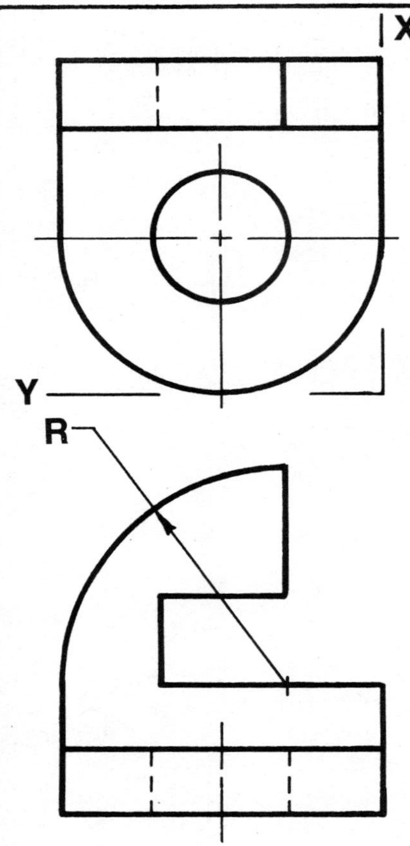

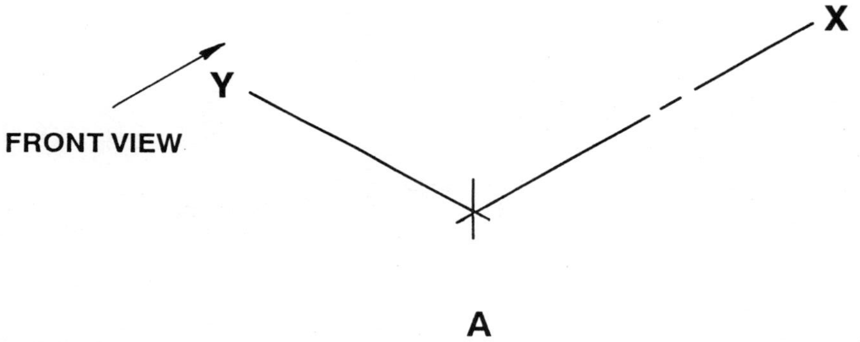

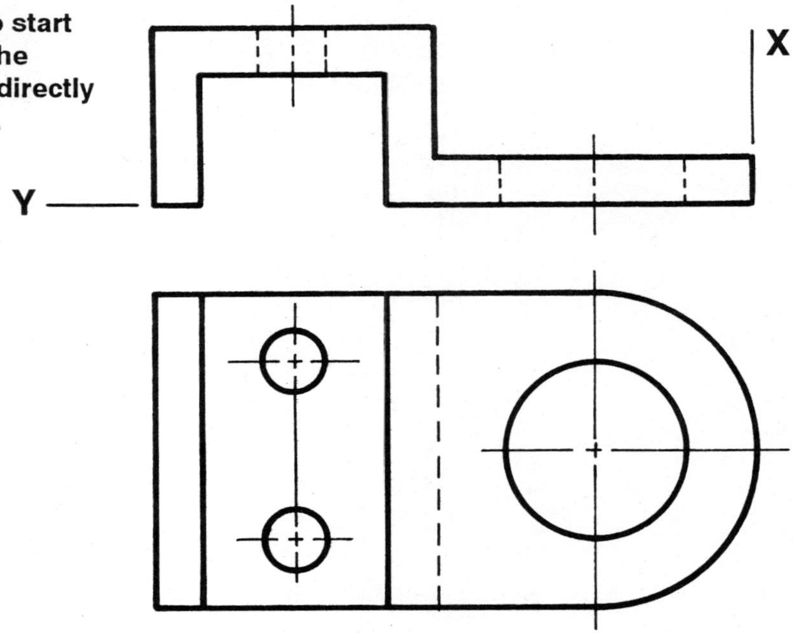

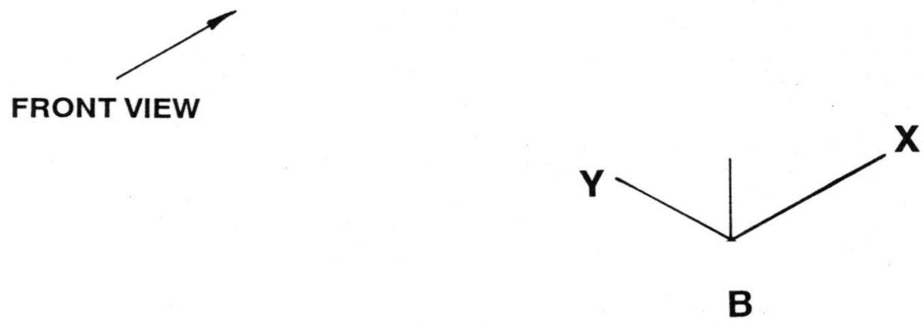

Ch18 Pictorial Drawings

Section-Chapter-Problem		Course_________ Section_____	Problem
3 18 2	Isometric Drawing Figure 18-4	**Student**____________ Date_____	3-18-2

A

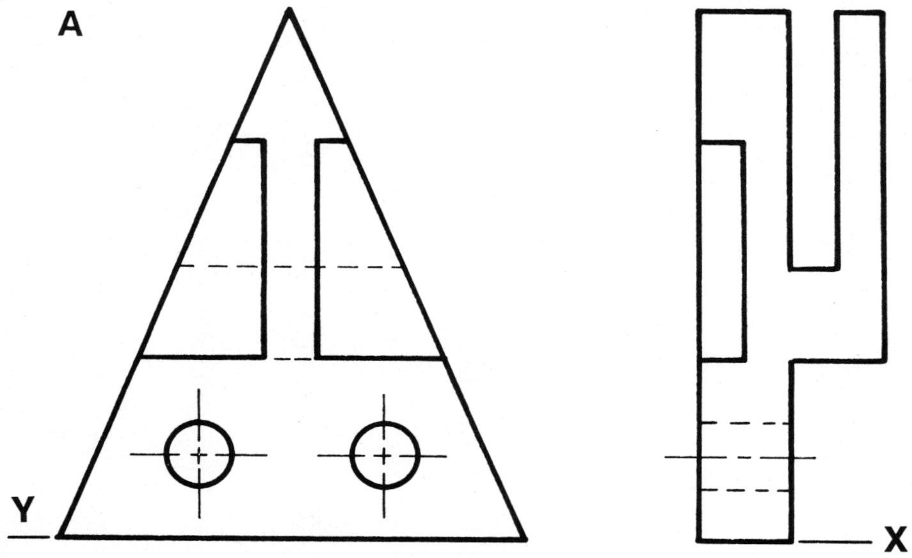

FRONT VIEW

Use a "box" to start the isometric drawing of the object shown. Measure directly from the views. Half Size.

B

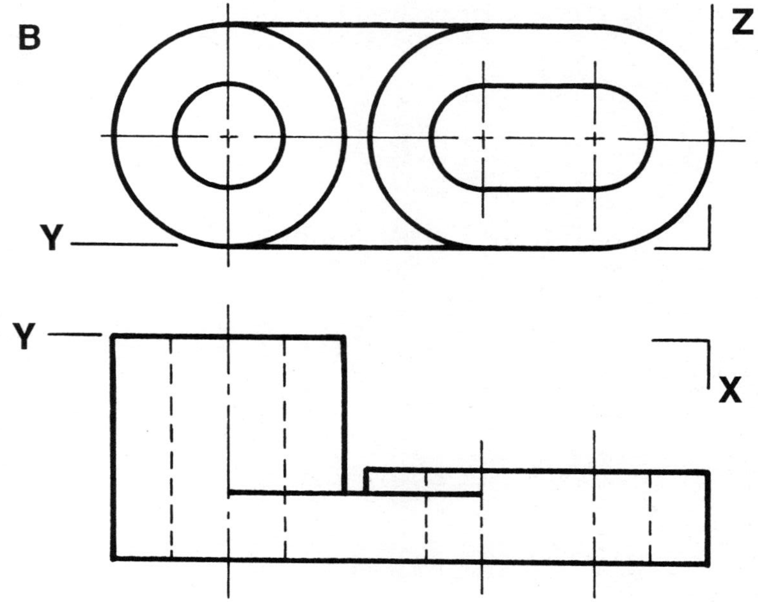

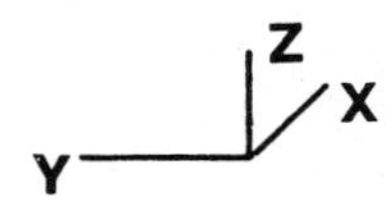

FRONT VIEW

Use a "box" to start the oblique drawing of the object shown. Measure directly from the views. Half Size.

Ch18 Pictorial Drawings			Problem
Section-Chapter-Problem 3 18 3	Isometric Drawing Figure 18-4 Oblique Drawing 18-40	Course________ Section_____ Student____________ Date______	3-18-3

The object shown is to be drawn with two point perspective on page 3-18-4B. Use a perspective "box" as illustrated.

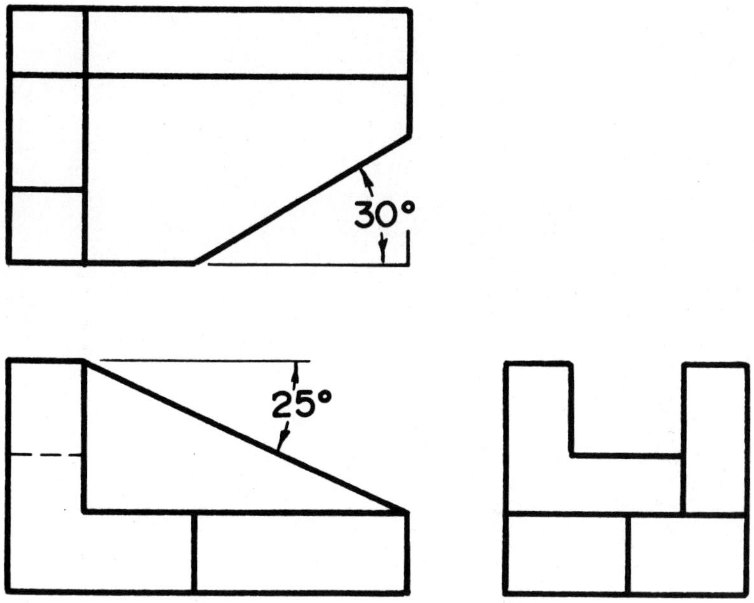

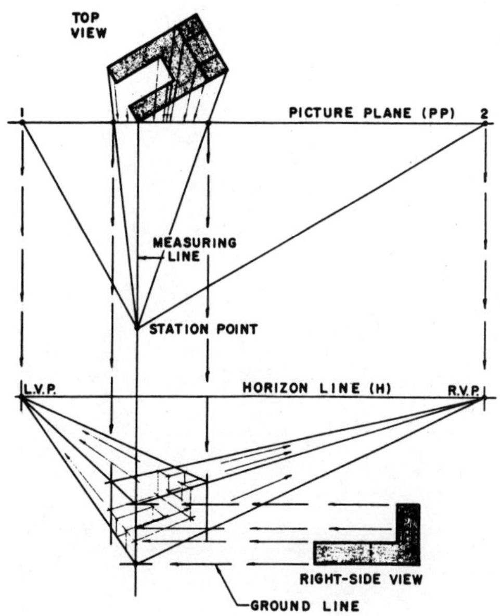

Ch18 Pictorial Drawings

Section-Chapter-Problem	Two Point Perspective	Course____________ Section_____	Problem
3 18 4A	Figure 18-33	**Student**____________________ Date______	**3-18-4A**

The object shown on page 3-18-4A is to be drawn with two point perspective on this page. Make all measurements from the views on page 3-18-4A. Full Size.

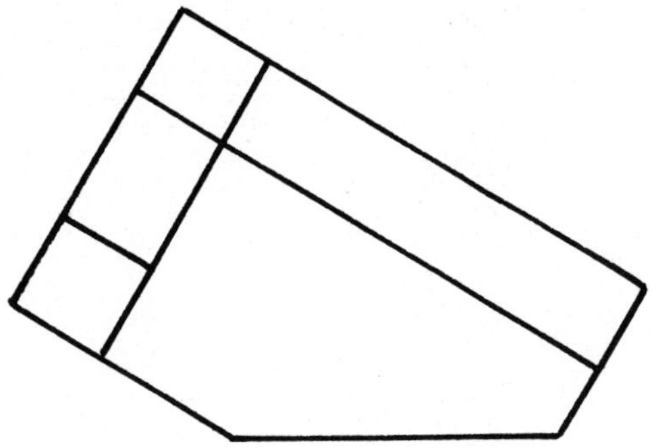

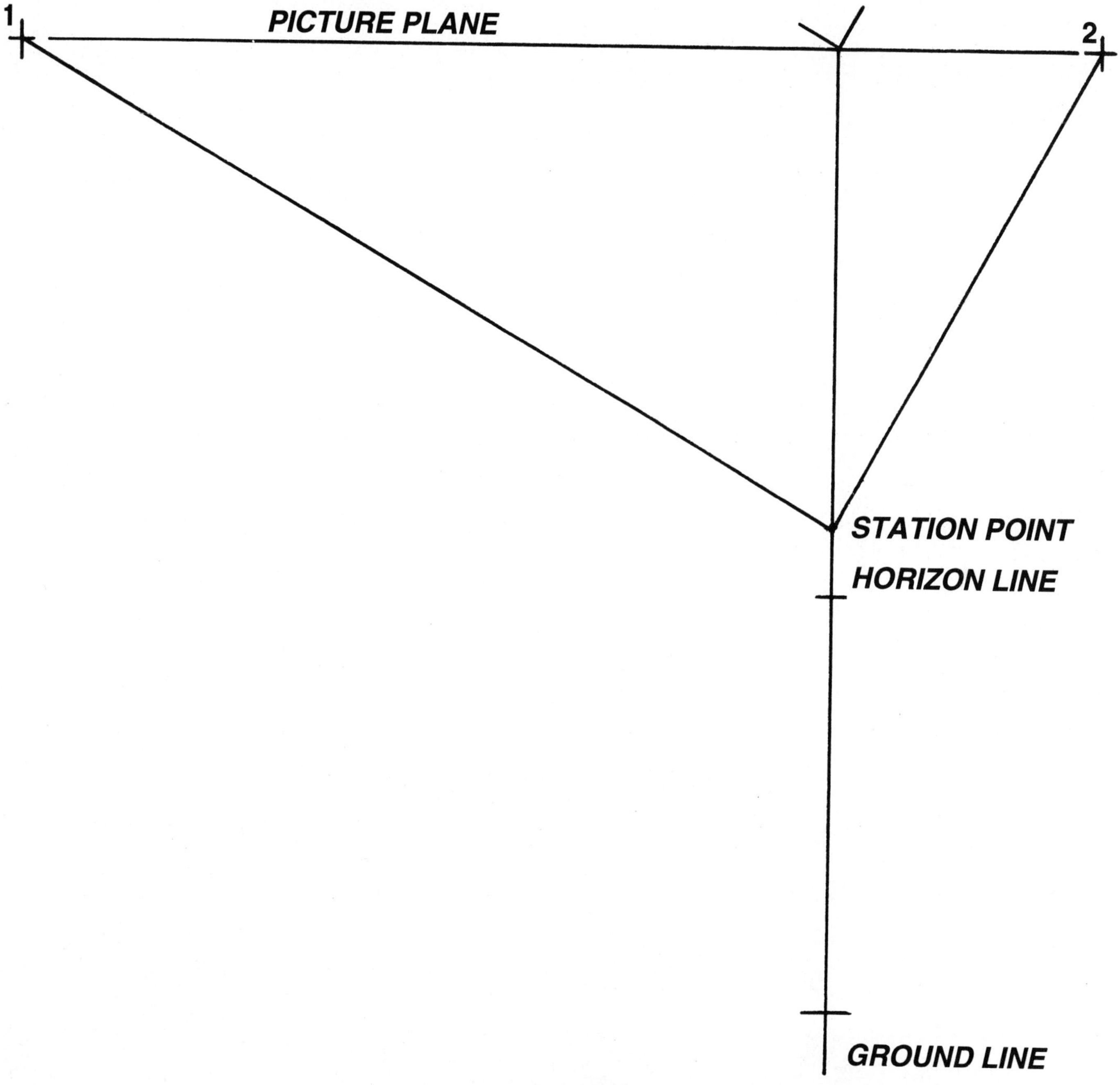

Ch18 Pictorial Drawings			Problem 3-18-4B
Section-Chapter-Problem 3 18 4B	Two Point Perspective Figure 18-33	Course____________Section_____ **Student**____________________Date______	

A1

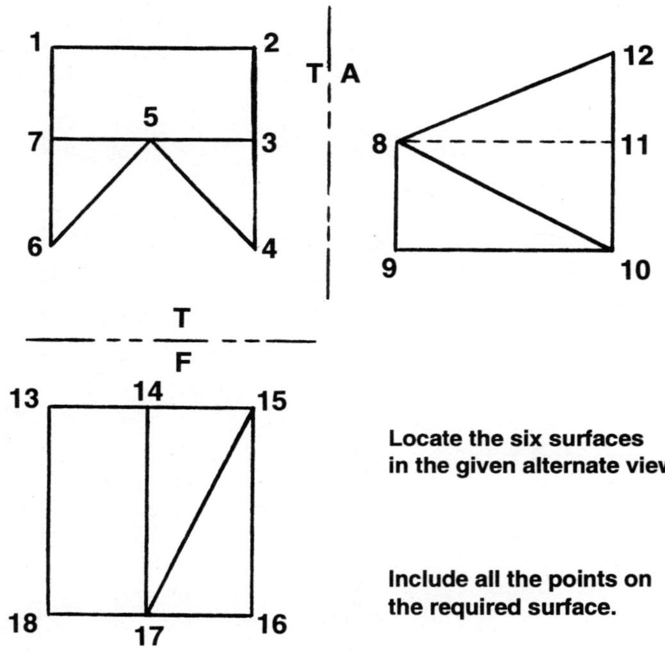

Locate the six surfaces
in the given alternate views.

Include all the points on
the required surface.

5, 6 in the A view ______________

14, 15, 17 in the A view ______________

8, 9 in the F view ______________

15, 16 in the A view ______________

3, 4, 5 in the F view ______________

2, 3, 4 in the A view ______________

On page 3-18-5B prepare a two point perspective of the object shown above. Full Size.

A2

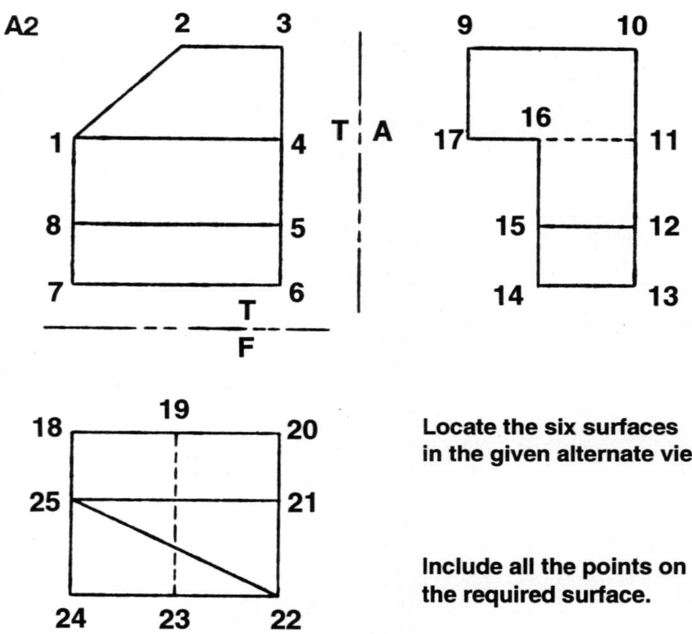

Locate the six surfaces
in the given alternate views.

Include all the points on
the required surface.

1, 2 in the A view ______________

16, 17 in the F view ______________

22, 23, 24 in the T view ______________

21, 22, 25 in the A view ______________

15, 16 in the F view ______________

25, 24 in the A view ______________

On page 3-18-5B prepare a two point perspective of the object shown above. Full Size.

Ch18 Pictorial Drawings

Section-Chapter-Problem	Analyzing orthographic views	Course ________ Section ______	Problem
3 18 5A	Two Point Perspective Figure 18-33	Student ______________ Date ______	3-18-5A

A1

A2

Ch18 Pictorial Drawings			Problem 3-18-5B
Section-Chapter-Problem 3 18 5B	Two Point Perspective Figure 18-33	Course____________Section_____ **Student**__________________Date______	

A Redraw the object shown in file "18-1.dxf" as an isometric and oblique drawing. It is currently shown as an orthographic projection.

B The top and front view of a child's tool bench top is shown in drawing "18-2.dxf". It requires the child to insert various shaped screw heads into their appropriate shaped hole. The drawing in this file only incorporates three shapes. Redesign this tool bench top to accommodate six different shapes. Construct this as an isometric or perspective drawing.

Ch18 Pictorial Drawings			**Problem**
Section-Chapter-Problem 3 18 6	Isometric Drawing Oblique Drawing	Course__________ Section_____ **Student**______________ Date______	**3-18-6**

Section 4
Objectives:

After studying and doing the exercises and problems in Section 4 of the workbook you should be able to:

- apply standard welding symbols to specified welds, and prepare basic weldment drawings.
- appreciate the function of template jigs in manufacturing processes.
- lay out a bolted or riveted truss made up of angles and gusset plates.
- draw several types of schematics from freehand sketches: piping, hydraulic circuits, and electronic/electrical.
- employ the five basics of curve plotting when plotting technical data.
- be able to visualize the power requirements for a specific mechanical system by observing the overall picture of moving a weight at a constant speed.

Supplemental Comments: Section 4

Chapter 19 Welding

Problems 4-19-1 and **4-19-2:** Welding symbols are an efficient way of communicating information to technical personnel.

Problems 4-19-3 and **4-19-4:** If a large number of units are to be manufactured, castings are less costly than weldments. If only a few units are required, weldments are probably less costly.

Problems 4-19-5A and **4-19-5B:** The picnic table in this problem shows an attractive and functional application of structural steel components and laminated wood. Weld symbols and hole callouts are required.

Problem 4-19-6: Use a CAD program to add the appropriate weld symbols to the drawing.

Chapter 20 Modern Manufacturing: Materials, Processes, and Automation

Problem 4-20-1: A template jig with press-fit steel drill-bushings is just one of many manufacturing aids used in industry.

Problem 4-20-2: Use a CAD program draw where relieving locators should be placed.

Chapter 21 Drafting Applications: Pipe, Structural, and Architecture

Problems 4-21-1A, 4-21-1B, and **4-21-1C:** A truss layout is an excellent introduction to structural detailing, gusset layouts, and parts lists.

Problems 4-21-2 and **4-21-3** are two single-line piping schematics.

Problem 4-21-4: Use a CAD program to add dimensions and bearings to the drawing and add the in-ground swimming pool.

Chapter 22 Drafting Applications: Electronics and Printed Circuit Boards

Problem 4-22-1: General practices for drawing electronic/electrical schematics are illustrated in this problem.

Problem 4-22-2: The basic operation of a motor control circuit is described. The circuit needs to be redrawn in a schematic.

Problem 4-22-3: Use a CAD program to draw a schematic that represents the circuit.

Chapter 23 Charts and Graphs

Problem 4-23-1: The Five Basic Components of plotting data are illustrated.

Problem 4-23-2: Use a CAD program to draw a multiple bar chart.

Chapter 24 The Design Process

Problem 4-24-1A: Calculations for horsepower requirements for a motor-gear-train-wire-rope hoist are given as an example.

Problem 4-24-1B: A change in the parameters for the motor-gear-train-wire-rope hoist requires new calculations. A format for presenting calculations is shown.

Fill in all spaces indicating the meaning of each number.

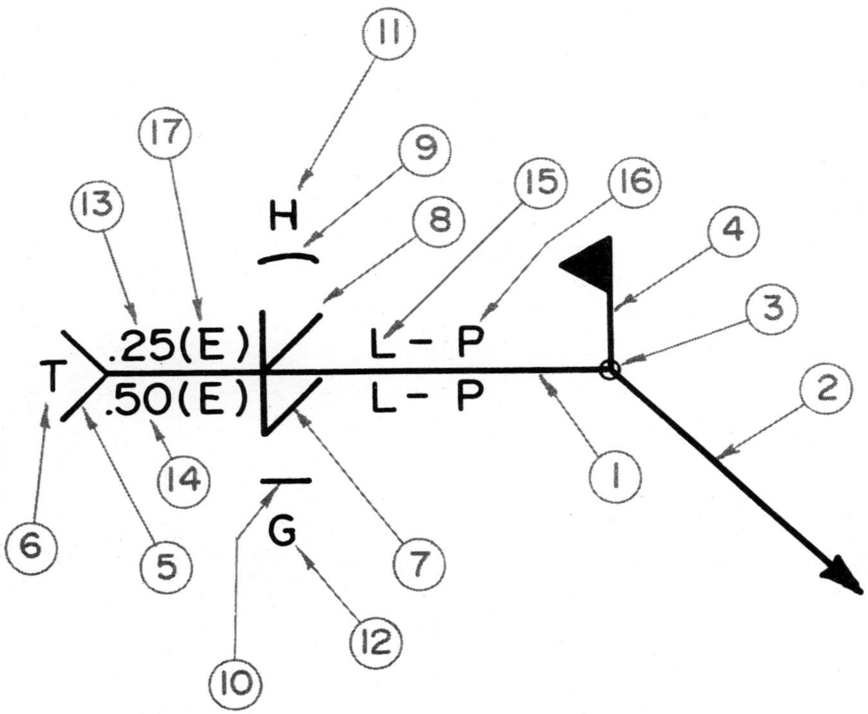

1	
2	
3	
4	
5	
6	
7	
8	
9	
10	
11	
12	
13	
14	
15	
16	
17	

Ch19 Welding

Section-Chapter-Problem	Welding symbols Figures 19-3,6,10,14,15,&20	Course________ Section____	Problem
4 19 1		Student________ Date____	4-19-1

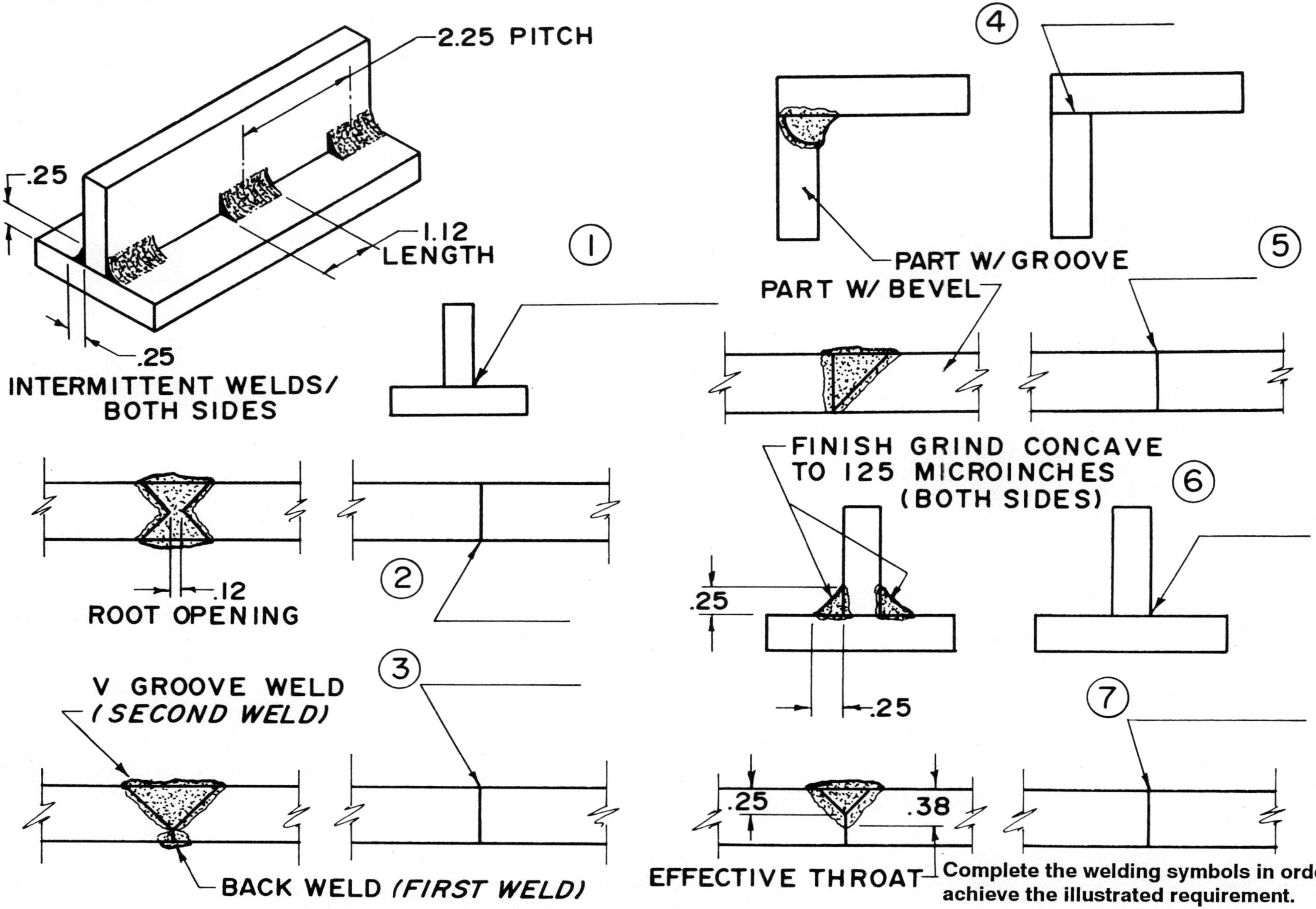

Complete the welding symbols in order to achieve the illustrated requirement.

Ch19 Welding			Problem
Section-Chapter-Problem 4 19 2	Welding symbols Figures 19-18,35,36,43, & 47	Course______ Section_____ Student__________ Date_____	4-19-2

Refer to the isometric drawing of a cast steel bearing and do the following:

1. change the casting into a weldment and prepare the front and side views in the space indicated. Half scale. Show weld symbols but do not dimension the object.

2. change the 38 and 35 dimensions each to 36.5mm.

3. add four φ 15mm holes in the base at the centers of the R12's.

4. Ignore Section A-A.

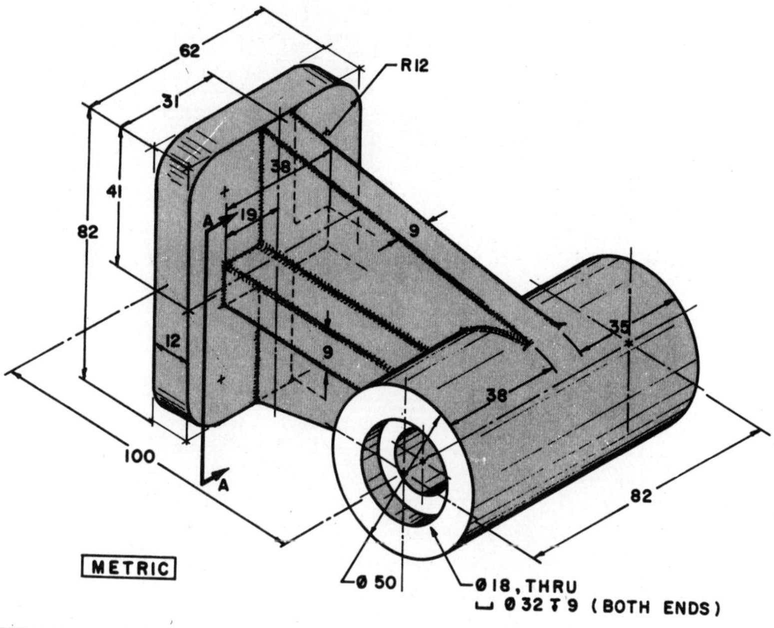

Ch19 Welding			Problem
Section-Chapter-Problem 4 19 3	Welding symbols Figures 19-3,6,10,14, & 15	Course____________ Section______ **Student**____________________ Date_______	4-19-3

Refer to the isometric drawing of a cast steel pipe connector and do the following:

1. change the casting into a weldment and prepare the front and side views in the space indicated. Half scale.
2. show weld symbols but do not dimension the object.
3. Ignore Section A-A.

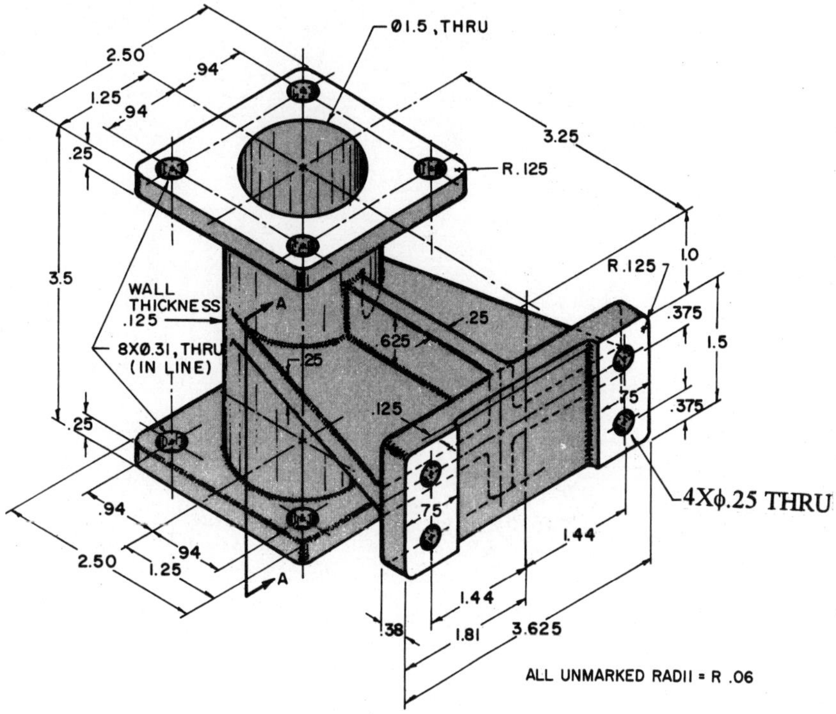

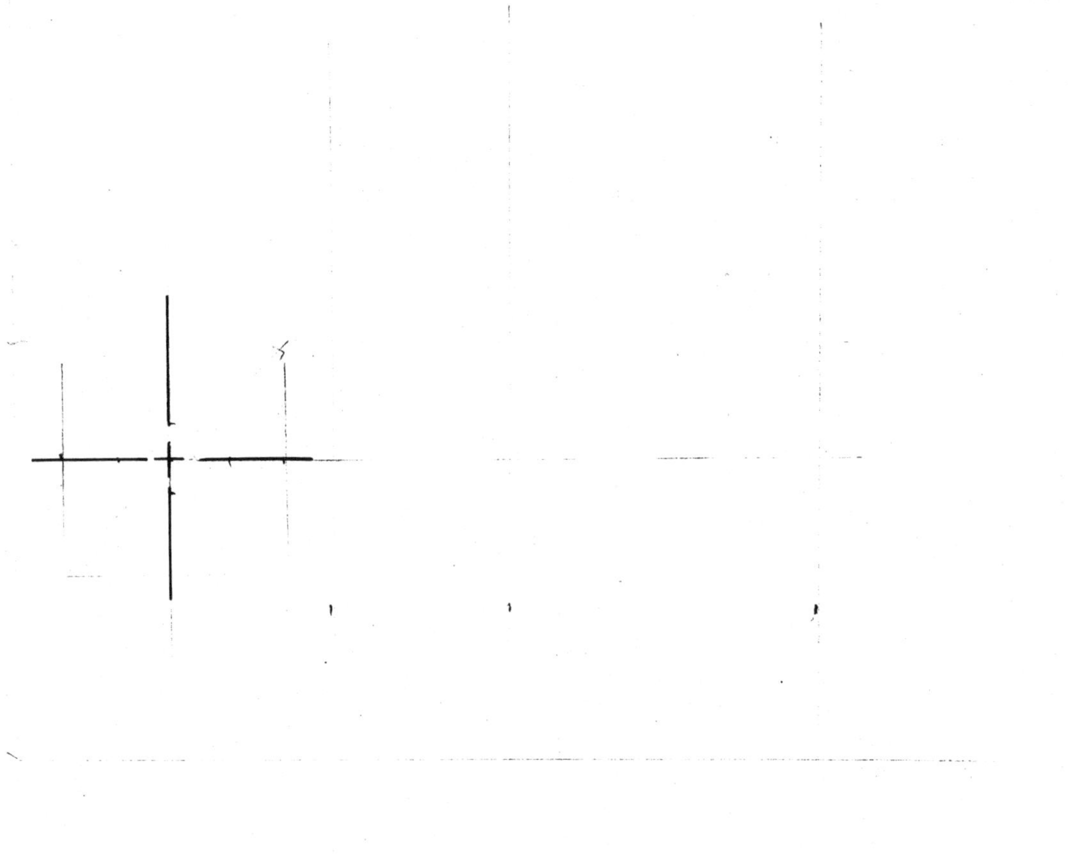

Ch19 Welding			Problem
Section-Chapter-Problem 4 19 4	Welding symbols Figures 19-3,6,10,14, & 15	Course__________ Section_____ Student_______________ Date_____	4-19-4

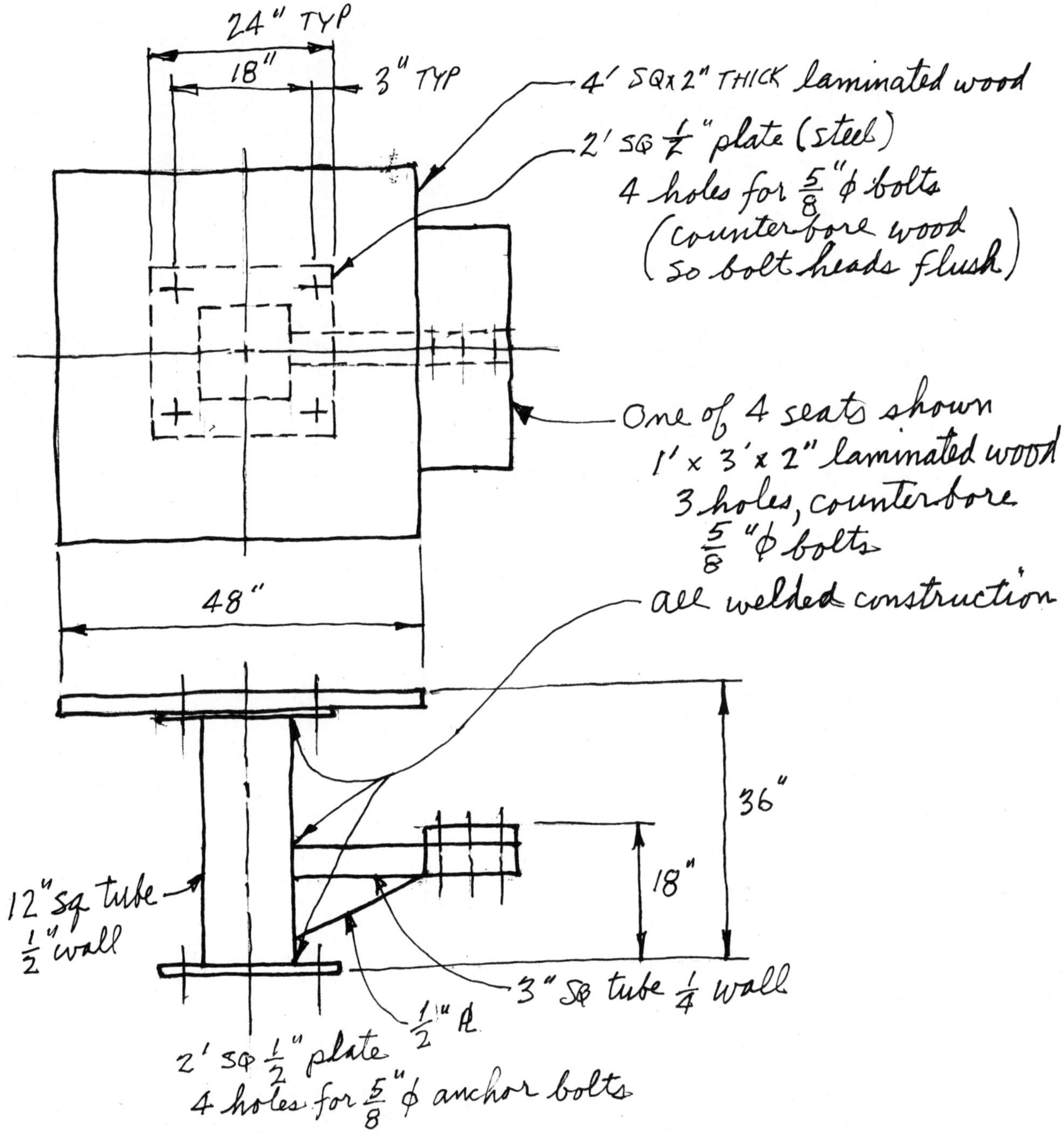

The basic structure of the picnic table shown is to be a weldment of steel plate and square tubing, and the seats and table top are to be 2" thick laminated wood.

1. On page 4-19-5B prepare a detail drawing of the steel weldment, complete with a parts list.

2. On page 4-19-5C prepare detail drawings of the table top and the four seats. Include a parts list.

3. On page 4-19-5D prepare an assembly drawing of the picnic table comlete with fasteners, and an assembly parts list.

Ch19 Welding			Problem 4-19-5A
Section-Chapter-Problem 4 19 5A	Welding practices Figures 19-13, & Coounterbored holes Figure 10-58	Course____________ Section______ Student____________________ Date_______	

DWG. NO.	SH	REV.

REVISIONS

REV.	DESCRIPTION	DATE	APPROVED

	QTY REQD	FSCM NO.	PART OR IDENTIFYING NO.	NOMENCLATURE OR DESCRIPTION	MATERIAL SPECIFICATION	

PARTS LIST

UNLESS OTHERWISE SPECIFIED
DIMENSIONS ARE IN INCHES
TOLERANCES ARE:
FRACTIONS :
DECIMALS .XX : .XXX :
ANGLES

MATERIAL

FINISH

DO NOT SCALE DRAWING

CONTRACT NO.

APPROVALS	DATE
DRAWN	
CHECKED	
ISSUED	

NEXT ASSY	USED ON

APPLICATION

SIZE A	FSCM NO.	DWG. NO.	REV.

SCALE		SHEET

Ch19 Welding

Section-Chapter-Problem
4 19 5B

Detail: Steel Picnic Table

Course__________ Section_____
Student________________ Date______

Problem 4-19-5B

DWG. NO.	SH	REV.

REVISIONS

REV.	DESCRIPTION	DATE	APPROVED

	QTY REQD	FSCM NO.	PART OR IDENTIFYING NO.	NOMENCLATURE OR DESCRIPTION	MATERIAL SPECIFICATION	

PARTS LIST

UNLESS OTHERWISE SPECIFIED DIMENSIONS ARE IN INCHES TOLERANCES ARE:
FRACTIONS ± DECIMALS .XX ± .XXX ± ANGLES

MATERIAL

FINISH

DO NOT SCALE DRAWING

CONTRACT NO.

APPROVALS	DATE
DRAWN	
CHECKED	
ISSUED	

NEXT ASSY	USED ON

APPLICATION

SIZE	FSCM NO.	DWG. NO.	REV.
A			

SCALE SHEET

Ch19 Welding

Section-Chapter-Problem	Detail: Wood Table Top and Seats	Course______ Section______	Problem
4 19 5C		Student______ Date______	4-19-5C

DWG. NO.	SH	REV.

REVISIONS

REV.	DESCRIPTION	DATE	APPROVED

	QTY REQD	FSCM NO.	PART OR IDENTIFYING NO.	NOMENCLATURE OR DESCRIPTION	MATERIAL SPECIFICATION	

PARTS LIST

UNLESS OTHERWISE SPECIFIED
DIMENSIONS ARE IN INCHES
TOLERANCES ARE:
FRACTIONS DECIMALS ANGLES
.XX :
.XXX :

MATERIAL

FINISH

DO NOT SCALE DRAWING

CONTRACT NO.

APPROVALS	DATE
DRAWN	
CHECKED	
ISSUED	

SIZE	FSCM NO.	DWG. NO.	REV.
A			

SCALE SHEET

NEXT ASSY	USED ON

APPLICATION

Ch19 Welding

Section-Chapter-Problem
4 19 5D

Assembly Drawing: Picnic Table

Course____________ Section_____
Student________________________ Date_______

Problem
4-19-5D

A The object given in file "19-1.dxf" requires some welds. The description of how to weld it is given above the title block. Add the appropriate symbols to the drawing.

B In file "19-1.dxf" are ten common welding symbols. Use an "A" size sheet of paper to design a reference sheet that will assist you in remembering welding symbols and terminology. Be sure to give an example drawing for each symbol and make any notes that will help you remember what the symbol means. You will need to rearrange them on the page to ensure that there is enough room between them.

<table>
<tr><td colspan="3">Ch19 Welding</td><td rowspan="2">Problem

4-19-6</td></tr>
<tr><td>Section-Chapter-Problem
4 19 6</td><td>Welding Symbols</td><td>Course__________ Section_____
Student__________________ Date______</td></tr>
</table>

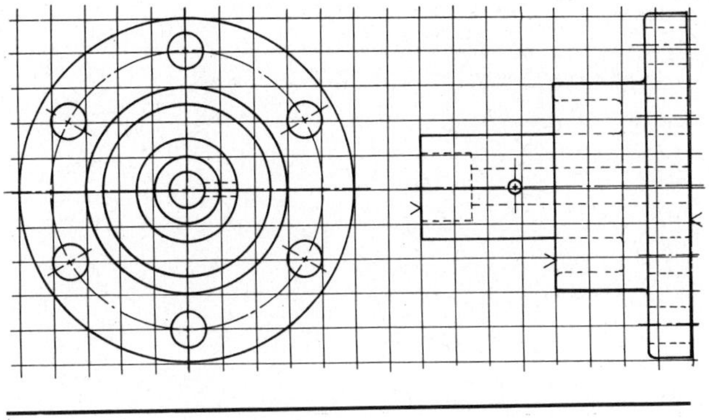

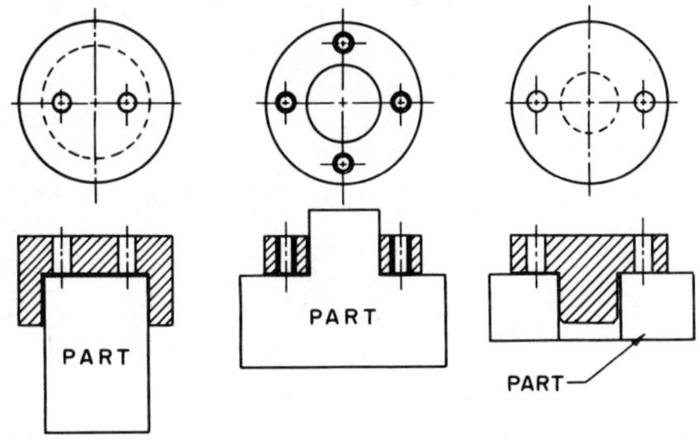

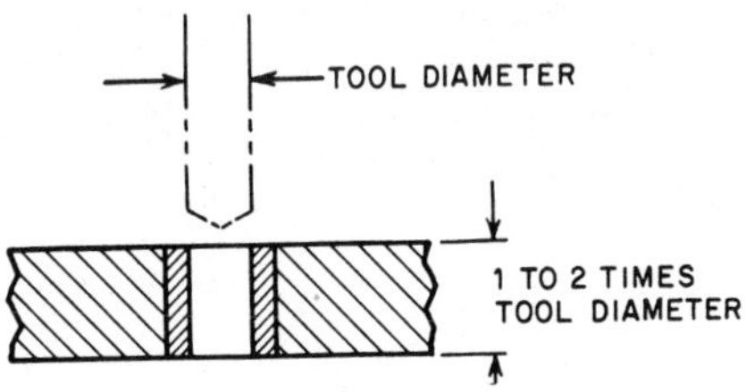

Press-fit drill bushings

A large number of the circular component shown, on the grid , are to be made in a production shop. The six holes need to be accurately located with respect to the 3" diameter cylindrical projection. On this sheet prepare a drawing of a template jig for drilling the six holes. Assume that the grid is on 0.5" spacing, and draw the jig full size in the space indicated. Use a press fit type drill bushing.

Ch20 Modern Manufacturing: Materials, Processes, and Automation			Problem
Section-Chapter-Problem 4 20 1	Drill bushings Figures 20-32, 55, & 58	Course________ Section_____ **Student**______________ Date______	**4-20-1**

A Shown in file "20-1.dxf" is a part that will be placed into a fixture so a groove can be milled into it for a pencil. The pencil groove is ½" wide and 9"long and the mill will travel from left to right. Draw where relieving locators should be placed and indicate two places where it should be clamped.
B Design a fixture that will be used to drill the holes for the object shown in file "20-2.dxf". The machinist will be using a drill press that has multiple chucks and therefore all three of the 3/8" holes are drilled at the same time. The other holes must be drilled at a later time in the operation, but the same fixture should be used.

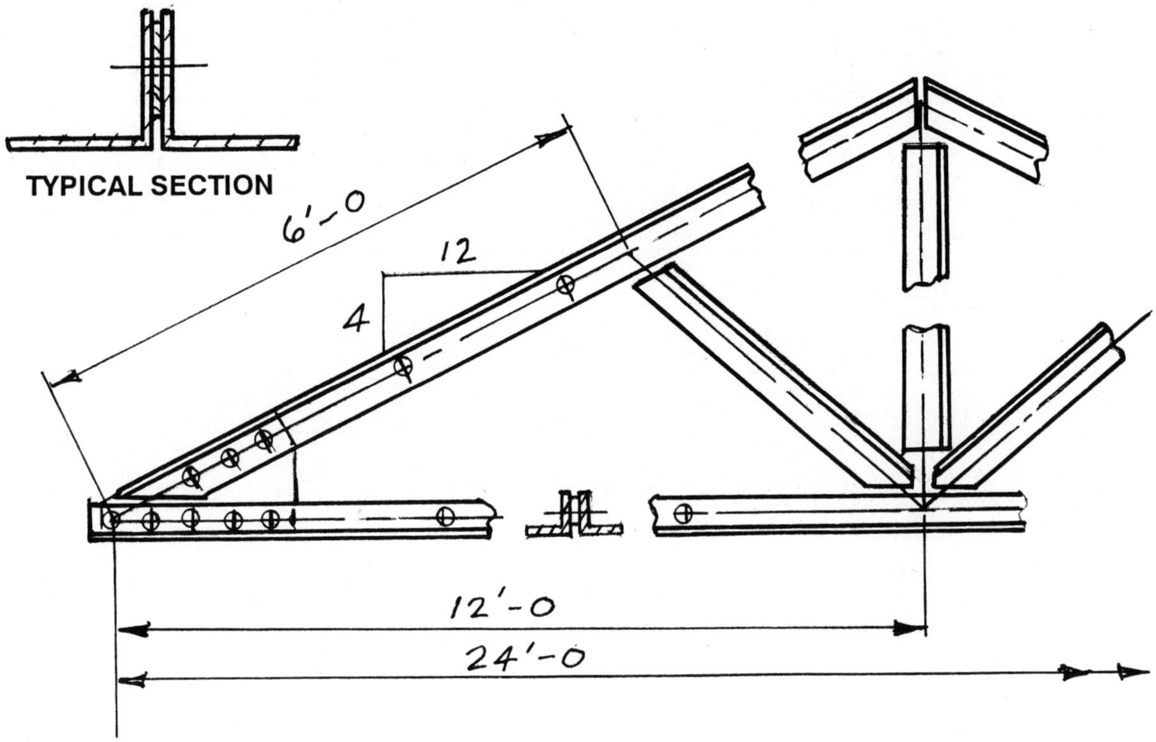

NOT TO SCALE

The truss shown was used in a commercial building, and had the following characteristics:

1. double steel angles ⊥ 3"X3"X1/4"
2. gusset plates 1/4" steel
3. hex head bolts ϕ1/2", hex nuts
4. spacers, ϕ2", located aproximately every 2 feet

On page 4-21-1B layout the truss with these characteristics (specifications), to a scale of 1/4" = 1'-0 to obtain centerline distances. Layout joints A, B, C, and D to a scale of 1-1/2" = 1'-0 to obtain lengths of angles, locate bolt holes, and establish sizes of the gussets. Assume that structural angles may be purchased in lengths of 30'-0.

General considerations:

A minumum of two bolts per member at a joint is recommended for safety. The locations of the bolts may be adjusted to improve the general appearance of the joint or to facilitate the production process, for example, simplify the shape of the gusset plate by making it rectangular, trapezoidal, symmetrical, etc.

On page 4-21-1C prepare a parts list of the angles, gussets, spacers, bolts, and nuts. Use numbers in circles to indicate parts on page 4-21-1B.

Joints in trusses may be bolted, riveted, welded, or combinatons of the three. An approach for laying out bolted joints for a truss using structural angles is to do the following:

a. construct the centerlines, to scale

b. locate the angles using recommended gage distances, G.D., and clearances, C. ; G.D. = approximately 0.6xleg of angle; clearance, C, = 1/4"+.

c. establish bolt locations using edge distance, E.D. = 1.5X bolt diameter (minimum), and bolt spacing, B.S. = 3Xbolt diameter (minimum).

d. layout gusset plates using E.D.'s .

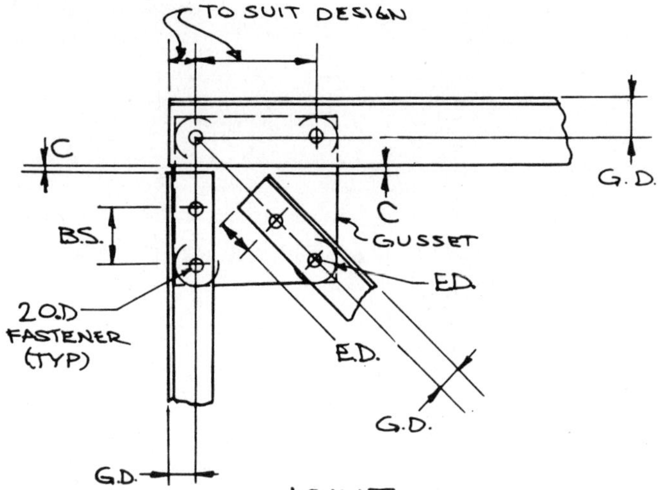

JOINT

An example joint layout is shown. Note the small radii used to locate edge distances

DWG. NO.	SH	REV.

REVISIONS

REV.	DESCRIPTION	DATE	APPROVED

	QTY REQD	FSCM NO.	PART OR IDENTIFYING NO.	NOMENCLATURE OR DESCRIPTION	MATERIAL SPECIFICATION	

PARTS LIST

UNLESS OTHERWISE SPECIFIED DIMENSIONS ARE IN INCHES TOLERANCES ARE:
FRACTIONS ± DECIMALS .XX ± .XXX ± ANGLES

MATERIAL

FINISH

CONTRACT NO.

APPROVALS	DATE
DRAWN	
CHECKED	
ISSUED	

NEXT ASSY	USED ON

APPLICATION

DO NOT SCALE DRAWING

SIZE	FSCM NO.	DWG. NO.	REV.
A			

SCALE SHEET

Ch21 Drafting Applications: Pipe, Structural, and Architectural

Section-Chapter-Problem	Truss Parts List	Course_____ Section_____	Problem
4 21 1C		Student_____ Date_____	4-21-1C

Refer to the double-line piping drawing shown and redo the drawing in a single-line orthographic.

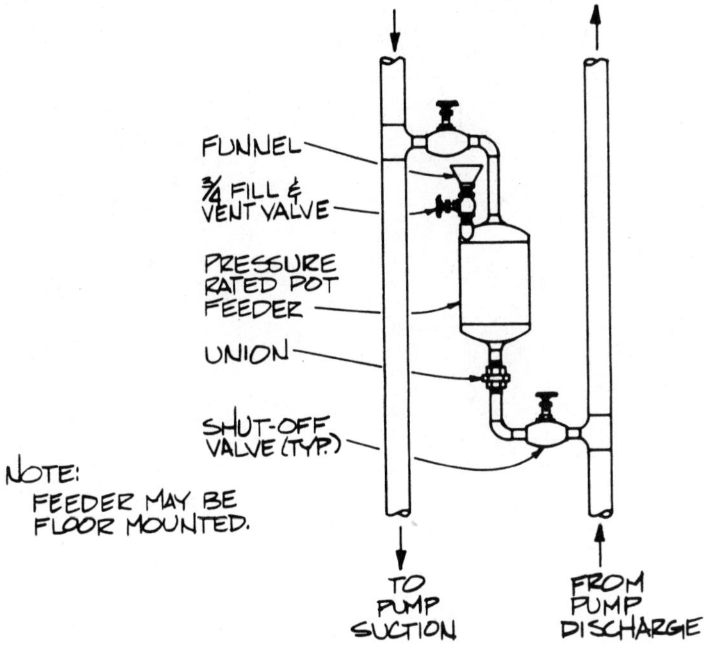

CHEMICAL POT FEEDER DETAIL
N.T.S.

Ch21 Drafting Applications: Pipe, Structural, and Architectural			Problem
Section-Chapter-Problem 4 21 2	Single-line pipe drawing Figure 21-20	Course__________ Section_____ **Student**________________ Date______	**4-21-2**

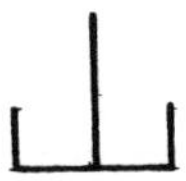
RESERVOIR

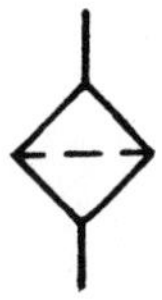
FILTER

PUMP

PRESSURE GAGE

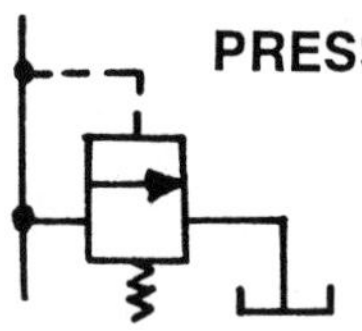
PRESSURE-RELIEF VALVE
SPRING-LOADED

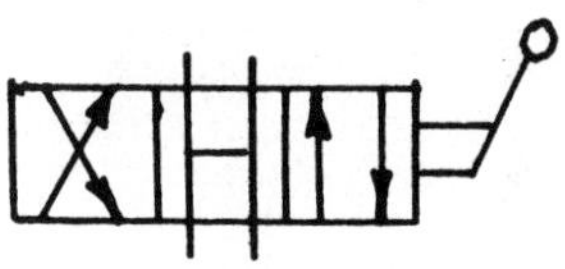
DIRECTIONAL VALVE
HAND CONTROLLED

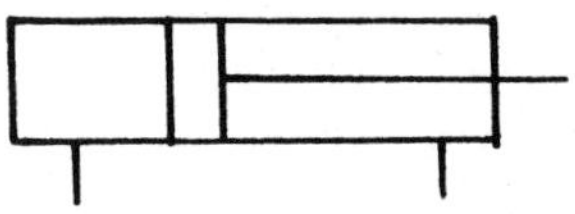
CYLINDER: DOUBLE ACTING

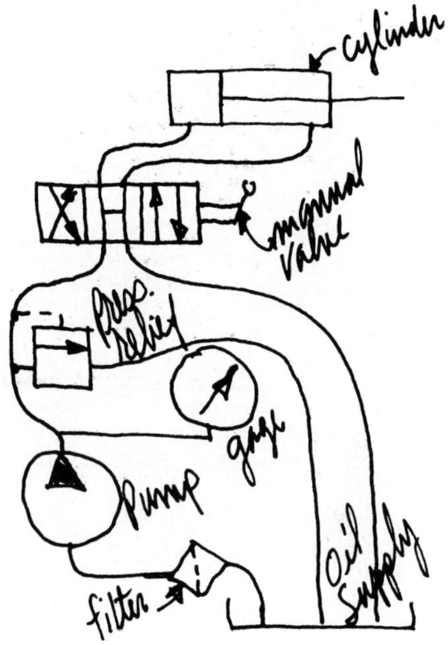

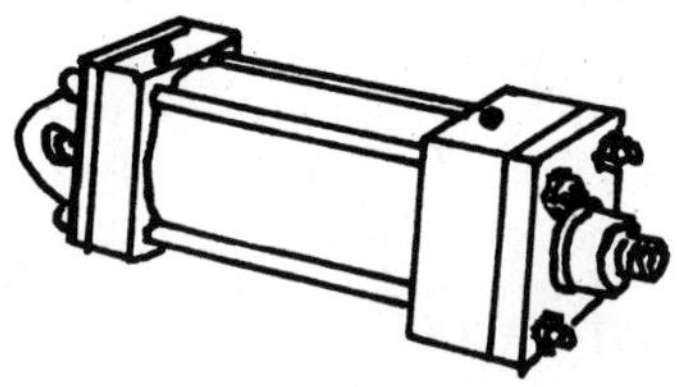

Using a straight edge and a circle templae redraw the freehand sketch of the hydraulic-power schematic shown and lable each component. Show the directional valve in the position for pushing the hydraulic piston to the right.

Ch21 Drafting Applications: Pipe, Structural, and Architectural			Problem
Section-Chapter-Problem 4 21 3	Sketch and finished schematic Figure 22-3	Course__________ Section_____ Student______________ Date_____	4-21-3

A Add dimensions and bearings to file "21-1.dxf." Add a 16' × 32' in-ground swimming pool at least 50' from the house.

B The owner of a single-story house wishes to enlarge it by adding a second floor. The first floor plan is shown in file "21-3.dxf". Duplicate this file and design the second floor plan. Keep piping considerations in mind so the pipe length is minimized.

Using a straight edge redraw the freehand sketch of the schematic of an overtone oscillator and use the general practices as guidelines.

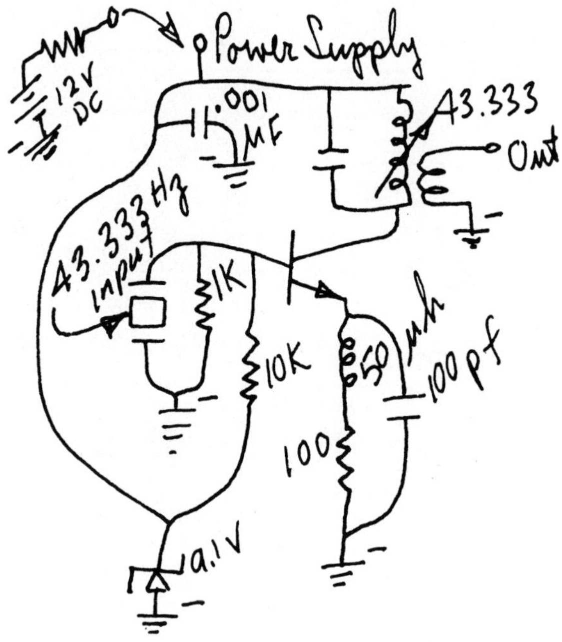

General practices for drawing electronic/electrical schematics

SIGNALS: Signals should flow from left to right; inputs on left margin, outputs on right margin.

LINES: Spacing between lines should be uniform. Cross over lines a litle as possible. If lines do cross, break one line preferably the one traveling in the short dimension of the paper. All lines should be horizontal or vertical.

CONNECTIONS: Positive connections generally go upward; negative, downward.

Ch22 Drafting Applications: Electronics and Printed Circuit Boards			Problem
Section-Chapter-Problem	Sketch and finished schematic	Course__________ Section_____	4-22-1
4 22 1	Figure 22-3	Student_______________ Date______	

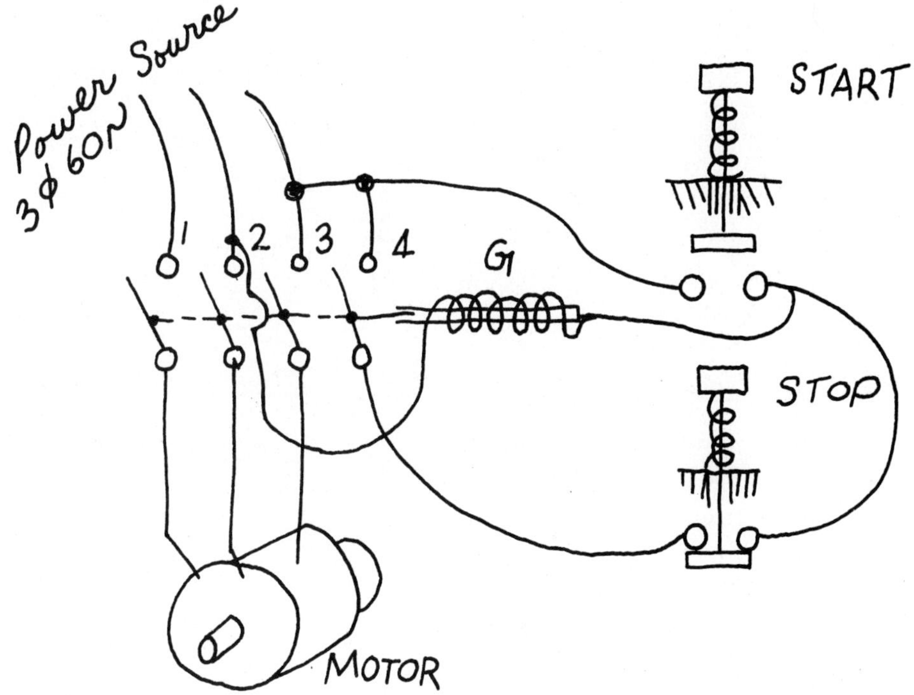

Use drawing instruments and redraw the freehand sketch of the schematic of a START-STOP control circuit for an a-c motor.

OPERATION: When the spring loaded, normally open START button is pressed, a path for the current is established from line 3, through the solenoid G to line 2.The current energizes the solenoid that causes all four, linked, switches to close.

Now switch 4 provides a different path for the solenoid current. The current flows through the normally closed STOP switch through the solenoid and finally to line 2.

The current will continue to flow until the STOP button is pressed. The STOP switch opens the current path through the solenoid so all four switches open quickly.

Ch22 Drafting Applications: Electronics and Printed Circuit Boards			Problem
Section-Chapter-Problem 4 22 2	Sketch and finished schematic Figure 22-3	Course____________Section_____ **Student**____________________Date_______	4-22-2

A Shown in file "22-1.dxf" are various electrical components. There is a "D" battery, two mini lamps, a fuse, a motor, and two switches. These are also connected correctly with wire. Draw a schematic diagram that would represent this circuit.

B A schematic electrical circuit is shown in file "22-2.dxf". Beside the circuit is a 4" × 5" rectangle that is the size of the printed circuit board. Draw the printed circuit board design for this schematic circuit. Connections must be accurate.

<table>
<tr><td colspan="3">Ch22 Drafting Applications: Electronics and Printed Circuit Boards</td><td>Problem</td></tr>
<tr><td>Section-Chapter-Problem
4 22 3</td><td>Schematic Circuit
PC Board</td><td>Course__________ Section_____
Student__________________ Date_______</td><td>4-22-3</td></tr>
</table>

Plot the data,BHP (brake horsepower) as a function of RPM, as given; RPM on the horizontal axis, BHP on the vertical. Include the Five Basic Components listed below. Use the graph at left as a guide. Suggestion: Use the axes shown, RPM from 0 to 5000, BHP from 0 to 150.

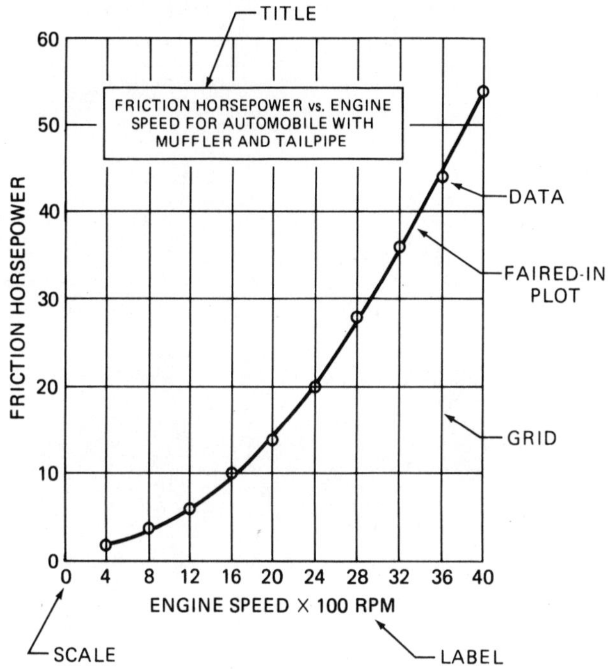

Basic components

Five Basic Components

1. The grid
2. The scales
3. The scales' labels
4. The plot
5. The title

RPM	BHP
500	15
1000	40
1500	65
2000	90
2200	
2500	115
3000	130
3500	145
4000	150
4500	145

Ch23 Charts and Graphs			Problem
Section-Chapter-Problem 4 23 1	Plotting data Figures 23-20 & 23-21	Course______ Section______ **Student**______ Date______	4-23-1

A A table of fictional data is given in file "23-1.dxf" about employment preference for beginning drafters. Draw a multiple bar chart that compares each drafting career preference for the given year.

B Develop a simple survey that is to be given to members in your drafting class. Use the data to design a variety of charts that are indicated in drawing file "23-2.dxf." Use colors or patterns to help define the charts.

Ch23 Charts and Graphs			Problem
Section-Chapter-Problem **4 23 2**	Bar Chart Survey Charts	Course__________ Section_____ **Student**______________ Date______	**4-23-2**

The hoist machinery sketched below includes the following:

- an electric motor with a reversing switch
- spur gear set (1)-(2)
- spur gear set (3)-(4)
- drum for wire rope (fastened to gear (4)
- wire rope supporing a 1,000 pound weight

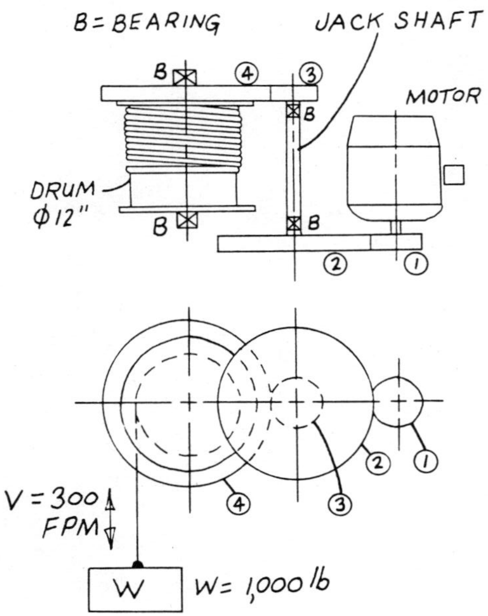

SPUR GEARS

	N	P
①	28	4
②	100	4
③	21	3
④	72	3

HORSEPOWER CALCULATIONS

$$HP = \frac{WV}{33{,}000}$$

Where W = pounds
V = FPM (feet per minute)
33,000 foot-pounds per minute per hosepower

$$HP = \frac{(1000)(300)}{33{,}000} = 9.09 @ thedrum$$

Add 5% to the horsepower (for bending the wire rope around the drum and for fiction in the gear sets .)

HP = (9.09)(1.05) = 9.55 (several commercial sizes of motors in this range are 5, 7, 10, 15, and 20.)
Therefore, select a 10 HP motor.

RPM required at motor:

$$DrumRPM = \frac{(300)(12)}{\pi(12)} = 95.5RPM$$

$$JackshaftRPM = \frac{(95.5)(72)}{(21)} = 327.4RPM$$

$$MotorRPM = \frac{(327.4)(100)}{(28)} = 1169RPM$$

Standard commercial motors are available at 1200 RPM and 1800 RPM; therefore,select a 1200 RPM.

PROBLEM: Note the following changes in parameters and calculate the horsepower required, the RPM's of the shafts needed, and the speed of the motor.

W = 1500 pounds; V = 250 FPM
Gear (1) = 28 teeth; Gear (2) = 96 teeth
Gear (3) = 35 teeth; Gear (4) = 125 teeth

Use the GIVEN, FIND, and SOLUTION format outlined in the sketch and do the problem.

Gearmotors may be used. Several available output RPM's are 750, 800, 850, 900, and 1000.

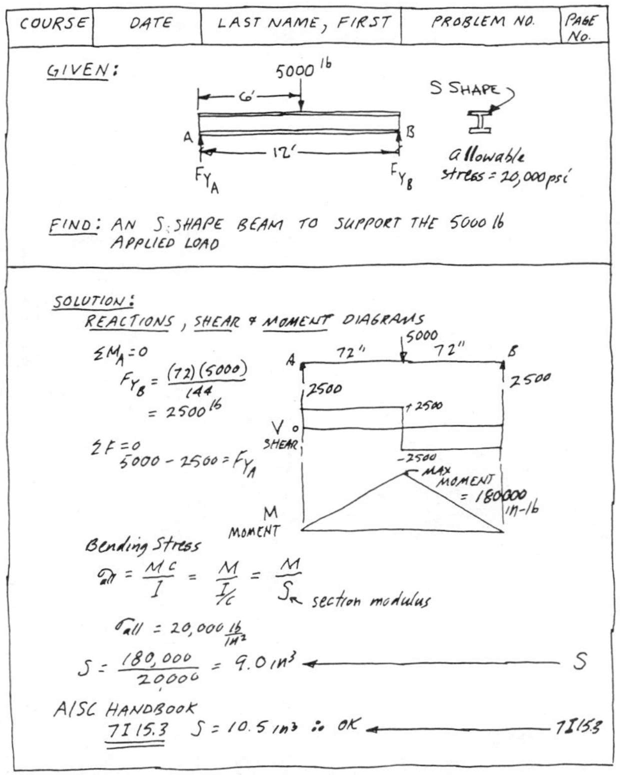

Ch24 The Design Process

Section-Chapter-Problem	Calculations: Motor, Gear Train, and Drum Hoist	Course________ Section_____	Problem
4 24 1B		Student____________ Date_____	4-24-1B

Ch24 The Design Process			Problem
Section-Chapter-Problem 4 24 2		Course____________Section_____ Student____________________Date______	4-24-2

Appendix A
Objectives:

After studying the discussions on mathematical topics, doing the exercises, and doing the problems, you should be able to:

- round off numbers to appropriate decimal places.
- calculate the decimal equivalents of fractions.
- evaluate formulas and set up ratios.
- add, subtract, and divide angles.
- employ the Law of Sines and the Law of Cosines.
- apply right-triangle equations to various geometrical problems.
- do dovetail calculations.
- calculate cross-sectional areas, theoretically and graphically.
- determine the volume of tapered components.
- perform triangulation calculations.
- find the mechanical advantage of crank-gear-drum systems.
- calculate bicycle speeds.
- know what angle to bank a curved roadway.
- determine the area of a flat pattern.

Supplemental Comments: Appendix A

The first 16 pages concentrate on basic mathematics and trigonometry for drafting. The remaining pages are applications.

Problem A-1 has you rounding off decimals, converting fractions to decimals, changing inches to millimeters, or millimeters to inches.

Problem A-2: When evaluating formulas, always check the units of the parameters for compatibility, and that the units on the left side of the equations equal the resulting units on the right side. For example, in Exercise C the units would be as follows:

$$(inches) = (inches) + (inches + inches) + \frac{(inches)^2}{(inches)}.$$

On the right side, inches are to the first power in all terms, so the equation is satisfactory. Ratios and proportions are on this page also.

Problems A-3, A-4, A-5, and **A-6** illustrate operations on angles, consider triangles (scalene, isosceles, and right) and demonstrate the use of the right triangle to solve graphical problems.

Problems A-7, A-8, and **A-9** consider various polygons. Note that the sum of the internal angles of a polygon of N sides is equal to (N–2)(180). Several properties of circles are noted and illustrated.

Problems A-9, A-10, and **A-11** focus on trigonometry basics and some applications.

Problems A-12 and **A-13** show the use of an internal dovetail.

Problems A-14, A-15, and **A-16** highlight two basic laws for solving triangle problems—the Law of Sines and the Law of Cosines.

Problem A-17: To facilitate calculating areas, divide odd areas into familiar shapes such as rectangles and triangles.

Problem A-18: Numerous fractions occur in architectural drafting.

Problem A-19: Typical questions from an ADDA (American Design Drafting Association) quiz are included here for your interest.

Problem A-20 illustrates the use of proportional equations to calculate heights.

Problem A-21 requires a volume calculation for concrete. At 145 pounds per cubic foot, what is the weight of the concrete?

Problem A-22 includes compass directions to determine angles.

Problems A-23A and **A-23B** contain directions for calculating an area graphically. An exercise is proposed for comparing the graphical technique with the theoretical answer. The theoretical area under the curve for y = (X**2)/15, between 0 and 5, is 8.33 square units.

Problem A-24 requires trigonometry to solve the problem.

Problems A-25A and **A-25B** utilize an equation for calculating the volume of a tapered component. Then, density times volume equals weight.

Problem A-26 needs the angles in decimal form. Six or more decimal places are recommended due to the large distances and angles close to 90 degrees.

Problem A-27 illustrates the application of leverage to assist boat owners.

Problem A-28 uses a typical gearing TRAIN VALUE (TV) equation, which generally is the product of the teeth in the driving gears divided by the product of the teeth in the driven gears. Finally, the output speed is the product of the TV and the input speed.

Problem A-29 demonstrates a typical machine shop calculation—cutting speed for a cutting tool. The basic equation is $V = \omega r$, where $\omega = 2\pi N$ or the number of radians per revolution, and V is the velocity of the cutting point in feet per minute if the radius is in feet.

Problem A-30 shows that a banked curve on a highway accommodates a certain speed.

Problem A-31: A bend allowance for the 90-degree bends of the base plate should be considered.

Rounding Decimal Fractions

Procedure: To round a decimal fraction

- Determine the number of decimal places required in an answer.
- If the digit directly following the last decimal place required is less than 5, drop all digits that follow the required number of decimal places.
- If the digit directly following the last decimal place required is 5 or larger, add one to the last required digit and drop all digits that follow the required number of decimal places.

Example 1: Round 0.68247 in to three decimal places.

The digit following the third decimal place is 4.

0.684④7 in.

Because 4 is less than 5, drop all digits after the third decimal place.

0.682 in Ans

Example 2: Round 12.3876 mm to two decimal places.

The digit following the second decimal place is 7.

12.38⑦6 mm

Because 7 is greater than 5, add 1 to the 8.

12.39 mm Ans

Expressing Common Fractions as Decimal Fractions

Procedure: To express a common fraction as a decimal fraction divide the numerator by the denominator.

Example: Express ⅞ in as a decimal.

Divide the numerator 7 by the denominator 8.

$$8\overline{)7.000}\quad = 0.875$$

DECIMAL EQUIVALENT TABLE

1/64	.015625	17/64	.265625	33/64	.515625	49/64	.765625
1/32	.03125	9/32	.28125	17/32	.53125	25/32	.78125
3/64	.046875	19/64	.296875	35/64	.546875	51/64	.796875
1/16	.0625	5/16	.3125	9/16	.5625	13/16	.8125
5/64	.078125	21/64	.328125	37/64	.578125	53/64	.828125
3/32	.09375	11/32	.34375	19/32	.59375	27/32	.84375
7/64	.109375	23/64	.359375	39/64	.609375	55/64	.859375
1/8	.125	3/8	.375	5/8	.625	7/8	.875
9/64	.140625	25/64	.390625	41/64	.640625	57/64	.890625
5/32	.15625	13/32	.40625	21/32	.65625	29/32	.90625
11/64	.171875	27/64	.421875	43/64	.671875	59/64	.921875
3/16	.1875	7/16	.4375	11/16	.6875	15/16	.9375
13/64	.203125	29/64	.453125	45/64	.703125	61/64	.953125
7/32	.21875	15/32	.46875	23/32	.71875	31/32	.96875
15/64	.234375	31/64	.48438	47/64	.734375	63/64	.984375
1/4	.250	1/2	.500	3/4	.750	1	1.000

Millimetre-Inch Equivalents (Conversion Factors)

1 inch (in) = 25.4 millimetres (mm)

1 millimetre (mm) = 0.03937 inch (in)

A Round 4.02463 inches to three decimal places.
Ans________________

B Round 59.6841 mm to two decimal places.
Ans________________

C Using the decimal equivalent table,

a. find the decimal equivalent of 57/64 in; Ans__________

b. find the fractional equivalent of 0.4375 in: Ans__________

c. find the nearer fractional equivalent of 0.706 in Ans__________

D Express 0.983 in in millimetres. Round the answer to two decimal places. Ans________________

E Express 70.15 mm in inches. Round the answer to three decimal places. Ans________________

Evaluating Formulas

Procedure: To solve (evaluate) a formula substitute the numerical values for the letter values and solve by following the order of operations of arithmetic. The order of operations follows:

- Do all operations within the grouping symbol first. Parentheses, the fraction bar, and the radical symbol are used to group numbers. If an expression contains parentheses within parentheses or brackets, do the work within the innermost parentheses first.
- Do powers and roots next. The operations are performed in the order in which they occur. If a root consists of two or more operations within the radical symbol, perform all the operations within the radical symbol, then extract the root.
- Do multiplication and division next in the order in which they occur.
- Do addition and subtraction last in the order in which they occur.

Example 1: Determine the outside diameter of a gear with 16 teeth and a circular pitch of 0.7854 in. Use the formula

$$D_o = \frac{P_c(N + 2)}{3.1416}$$

where D_o = outside diameter
P_c = circular pitch
N = number of teeth

$$D_o = \frac{0.7854\,(16 + 2)}{3.1416}$$

Do the work in parentheses.

$$D_o = \frac{0.7854\,(18)}{3.1416}$$

Multiply.

$$D_o = \frac{14.1372}{3.1416}$$

Divide.

D_o = 4.5000 in Ans

Mechanical Drafting Mathematics

Problem A-1

Section-Chapter-Problem: Appendix A 1

Rounding Off
Decimal Equivalents
Evaluating Formulas

Course__________ Section_____

Student__________________ Date______

A $A = \pi\frac{(R^2 - r^2)}{2}$. Solve for A when $\pi = 3.1416$, $R = 7.000$ in. and $r = 3.000$ in.

Ans ____________

B All dimensions are in millimetres.

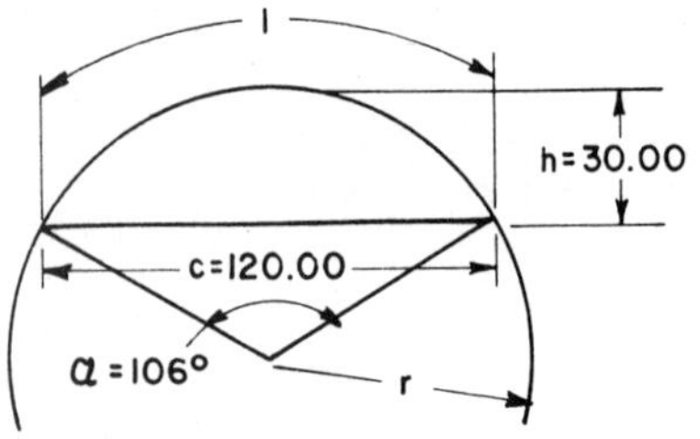

a. Find the radius of this circle.

$r = \frac{c^2 + 4h^2}{8h}$ Ans ____________

b. Find the length of the arc (l).

$l = 0.175ra$

Ans ____________

C All dimensions are in inches. Find the belt length on the pulleys.

Length of belt = $2C + \frac{11D + 11d}{7} + \frac{(D - d)^2}{4C}$

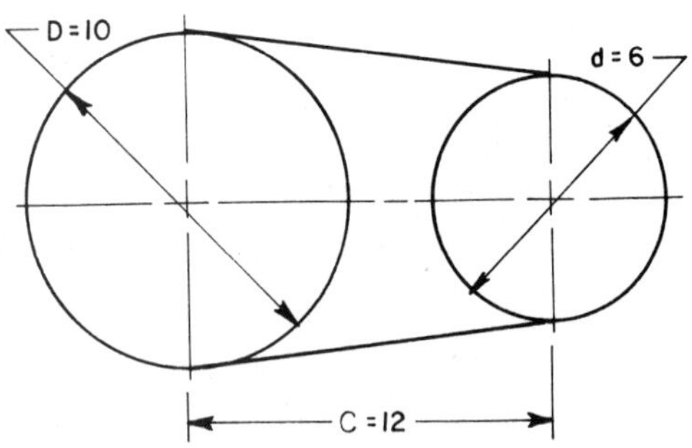

Ans ____________

D All dimensions in millimetres.

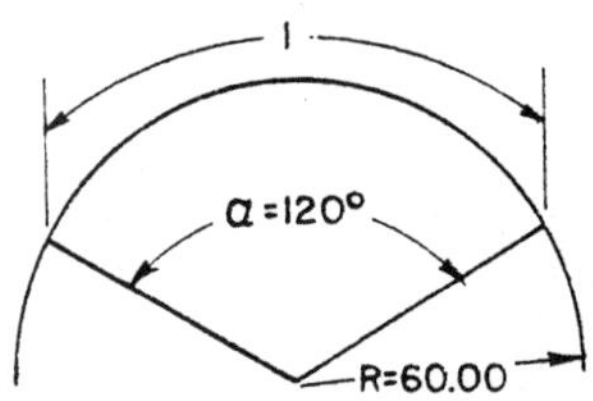

a. Find the length of this arc (l).

$l = \frac{3.1416Ra}{180°}$ Ans ____________

b. Find the area of this sector (A).

$A = \frac{1}{2}Rl$

Ans ____________

Ratio and Proportion

Ratios

Ratio and proportion are widely used in manufacturing applications, such as computing gear sizes and speeds.

Ratio is the comparison of two *like* quantities. The terms of a ratio are the two numbers that are compared. Both terms of a ratio must be expressed in the same units of measure. Ratios are generally expressed with a colon between the two terms, such as 3:8 (which is read 3 to 8) or as a fraction: ⅜.

The terms of a ratio must be compared in the order in which they are given. The first term is the numerator of a fraction, and the second is the denominator.

Examples:

1. The ratio 2 to 3 = 2 ÷ 3 = ⅔.
2. The ratio 3 to 2 = 3 ÷ 2 = 3/2.
   Generally, a ratio should be expressed in lowest fractional terms.

Proportions

A *proportion* is an expression that states the equality of two ratios. Proportions are expressed in the following two ways:

1. 3:4 :: 6:8, which is read: "3 is to 4 as 6 is to 8."
2. ¾ = 6/8. This form is the usual way to write a proportion.

A proportion consists of four terms. The first and the fourth terms are called *extremes*, and the second and third terms are called the *means*.

Example: 7/16 = 14/32. 7 and 32 are the extremes; 16 and 14 are the means.

In a proportion the product of the means equals the product of the extremes. If the terms are cross multiplied, their products are equal.

Example: ¾ = 6/8

Cross Multiply. ¾ x 6/8
3 x 8 = 4 x 6
24 = 24

The method of cross multiplying is used in solving proportions that have an unknown term. After the terms have been cross multiplied, the division principle of equality is applied.

$\frac{X}{7.5} = \frac{23.4}{20}$. Solve for the value of X.

Cross multiply.

20X = 7.5 (23.4)

20X = 175.5

Apply the division principle of equality. Divide both sides of the equation by 20.

$\frac{20X}{20} = \frac{175.5}{20}$

X = 8.775 Ans

Mechanical Drafting Mathematics			Problem A-2
Section-Chapter-Problem Appendix A 2	Evaluating Formulas Ratio and Proportion	Course________ Section______ Student____________ Date______	

A Of two gears that mesh, the one which has the greatest number of teeth is called the gear, and the one which has the fewer teeth is called the pinion. The proportion that expresses the relationship of gear teeth and speeds is

$$\frac{\text{Number of teeth on gear}}{\text{Number of teeth on pinion}} = \frac{\text{Speed of pinion}}{\text{Speed of gear}}$$

For each problem, a–e, determine the unknown value x. Where necessary, round answer to one decimal place.

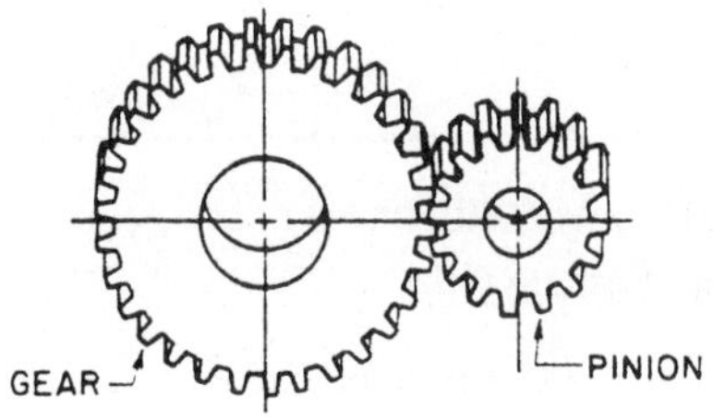

	Number of Teeth on Gear	Number of Teeth on Pinion	Speed of Gear (rpm)	Speed of Pinion (rpm)
a	48	20	120	x
b	32	24	x	210
c	35	x	160	200
d	x	15	150	250
e	54	28	80	x

Ans_____ Ans_____ Ans_____ Ans_____ Ans_____

Arithmetic Operations on Angles Expressed in Degrees, Minutes, and Seconds

The *degree* is the basic unit of angular measure. The symbol ° means degree. A radius rotated one revolution makes a complete circle or 360°. In the English system, computations and measurements are in degrees and minutes. Degrees, minutes, and seconds are used for applications requiring precise angular measurement.

1 circle = 360 degrees		1 degree = $\frac{1}{360}$ circle
1 degree = 60 minutes		1 minute = $\frac{1}{60}$ degree
1 minute = 60 seconds		1 second = $\frac{1}{60}$ minute

Relationship of degrees, minutes, and seconds

A degree is divided into 60 equal parts called *minutes*. The symbol for minute is ′. A minute is divided into 60 equal parts called *seconds*. The symbol for seconds is ″.

Adding Angles Expressed in Degrees, Minutes, and Seconds

Example Determine ∠2

$$\angle 2 = 78°43'27'' + 29°38'52''$$

$$\begin{array}{r} 78°43'27'' \\ +\ 29°38'52'' \\ \hline 107°81'79'' \end{array} = 107°82'19'' = 108°22'19'' \text{ Ans}$$

Note: 79″ = 60″ + 19″ = 1′19″; therefore, 107°81′79″ = 107°82′19″.
82′ = 60′ + 22′ = 1°22′; therefore, 107°82′19″ = 108°22′19″.

Subtracting Angles Expressed in Degrees, Minutes, and Seconds

Example Determine ∠2

$$\angle 2 = 57°13'28'' - 44°19'88''$$

$$\begin{array}{r} 57°13'28'' = 56°73'28'' = 56°72'88'' \\ -\ 44°19'42'' = 44°19'42'' = 44°19'42'' \\ \hline 12°53'46'' \text{ Ans} \end{array}$$

Note: Since 19′ cannot be subtracted from 13′, and 42″ cannot be subtracted from 28″, express 1° as 60′ and 1′ as 60″.
57°13′28″ = 56° + 1° + 13′ + 28″ = 56°73′28″ = 56°72′ + 1′ + 28″ = 56°72′88″.

Multiplying Angles Expressed in Degrees, Minutes, and Seconds

Example Determine ∠1

$$\angle 1 = 2(62°47')$$

$$\begin{array}{r} 62°47' \\ \times\ 2 \\ \hline 124°94' \end{array} = 125°34' \text{ Ans}$$

Note: 94′ = 1°34′.

B Add the angles in each of these problems.
a. 23°43′ + 71°12′ + 19°27′ Ans__________
b. 9°33′46″ + 37°12′19″ Ans__________

C Subtract the angles in each of these problems.
a. 63°23′ - 32°58′ Ans__________
b. 54°32′13″ - 19°13′42″ Ans__________

D Multiply the angles in each of these problems.
a. 2(43°43′) Ans__________
b. 5(22°10′13″) Ans__________

Dividing Angles Expressed in Degrees, Minutes, and Seconds

Example Determine ∠1, ∠2, and ∠3

∠1 = ∠2 = ∠3 = 128°37′21″ ÷ 3

Divide 128′ by 3.

$$3\overline{)128^\circ}$$ = 42°, 126°, remainder 2°

128 ÷ 3 = 42° plus a remainder of 2°.

Add the 2° (120′) to the 37′.

120′ + 37′ = 157′

Divide 157′ by 3.

$$3\overline{)157'}$$ = 52′, 156′, remainder 1′

157′ ÷ 3 = 52′ plus a remainder of 1′.

Add the 1′ (60″) to the 21″.

60″ + 21″ = 81″.

Divide 81″ by 3.

$$3\overline{)81''}$$ = 27″

81″ ÷ 3 = 27″

Combine.

42° + 52′ + 27″ = 42°52′27″ Ans

A Divide the angles in each of these problems.

a. 105°20′ ÷ 4

b. 110°51′5″ ÷ 5

In the figure,
∠1 = ∠2 = ∠3 = ∠4 = ∠5 = 53°41′
Determine ∠6.

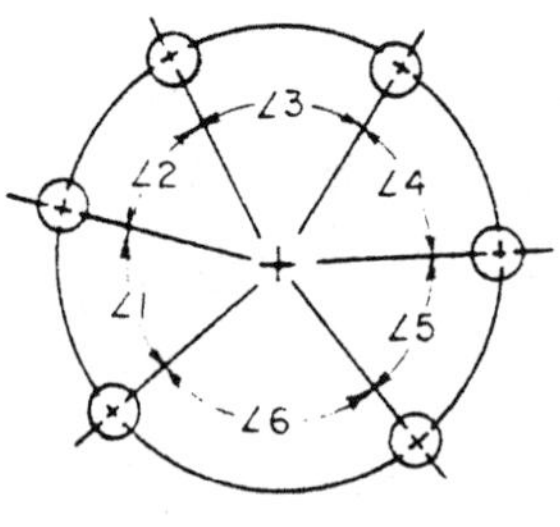

Ans________________

Ans________________

B In the figure,
∠1 = ∠2 = ∠3 = ∠4
Determine ∠1.

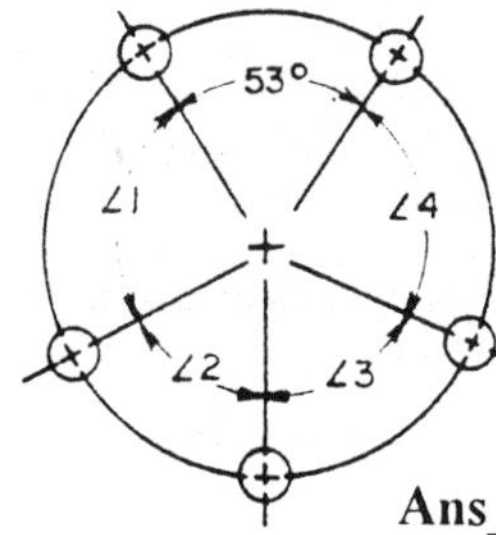

Ans________________

Degrees, Minutes, Seconds—Decimal Degree Conversion

In the metric system the preferred method of angular measure is decimal degrees. However, degrees, minutes, and seconds are also used. It is important to be able to compute and express angular measurements in both the English and the metric systems and to express angular measure between systems.

Expressing Decimal Degrees As Degrees, Minutes, and Seconds

The measure of an angle given in the form of decimal degrees, such as 41.1938°, must often be expressed as degrees, minutes, and seconds.

Procedure: To express decimal degrees as degrees, minutes, and seconds

- Multiply the decimal part of the degrees by 60′ in order to obtain minutes.
- If the number of minutes obtained is not a whole number, multiply the decimal part of the minutes by 60″ in order to obtain seconds. Round to the nearer whole second if necessary.
- Combine degrees, minutes, and seconds.

Example: Express 47.1938° as degrees, minutes, and seconds.

Multiply 0.1938 by 60′ to obtain minutes.
60′(0.1938) = 11.6280′
Multiply 0.6280 by 60″ to obtain seconds.
60″(0.6280) = 38″
Round to the nearer whole second.
Combine degrees, minutes, and seconds.
47° + 11′ + 38″ = 47°11′38″ Ans

C Express each of the following decimal-degrees as degrees and minutes.

a. 76.95°

b. 117.70°

Ans________________

Ans________________

Mechanical Drafting Mathematics			Problem A-4
Section-Chapter-Problem Appendix A 4	Dividing Angles Converting Angle Units	Course________ Section______ Student__________ Date______	

A Express the following decimal-degrees as degrees, minutes, and seconds.

a. 7.9250° Ans__________________

b. 37.9365°

Ans__________________

Expressing Degrees, Minutes, and Seconds As Decimal Degrees

Often an angle given in degrees and minutes is to be expressed as decimal degrees.

Procedure: To express degrees and minutes as decimal degrees

- Divide the minutes by 60.
- Combine whole degrees and decimal degrees. Round the answer to two places.

Example: Express 76°29′ as decimal degrees.

Divide 29 by 60 to obtain decimal degrees.

$29 \div 60 = 0.48°$

Combine whole degrees with decimal degrees.

$76° + 0.48 = 76.48°$ Ans

When working with English and metric units of measure, it may be necessary to express angles given in degrees, minutes, and seconds as angles in decimal degrees.

Procedure: To express degrees, minutes, and seconds as decimal degrees

- Divide the seconds by 60 in order to obtain decimal minutes.
- Combine whole minutes and decimal minutes.
- Divide the total minutes by 60 in order to obtain decimal degrees.
- Combine whole degrees and decimal degrees. Round the answer to four decimal places.

Example: Express 23°18′44″ as decimal degrees.

Divide 44 by 60 to obtain decimal minutes.

$44 \div 60 = 0.7333'$

Combine whole minutes and decimal minutes.

$18' + 0.7333' = 18.7333'$

Divide 18.7333 by 60 to obtain decimal degrees.

$18.7333 \div 60 = 0.3122°$

Combine whole degrees with decimal degrees.

$23° + 0.3122° = 23.3122°$ Ans

B Express each of the following degrees and minutes as decimal-degrees. Round the answer to two decimal places.

a. 93°18′ Ans__________________

b. 79°59′

Ans__________________

C Express each of the following degrees, minutes, and seconds as decimal-degrees. Round the answer to four decimal places.

a. 53°10′45″

b. 176°27′18″ Ans__________________

Types of Angles

An *acute angle* is an angle that is less than 90°.

A *right angle* is an angle of 90°.

An *obtuse angle* is an angle greater than 90° but less than 180°.

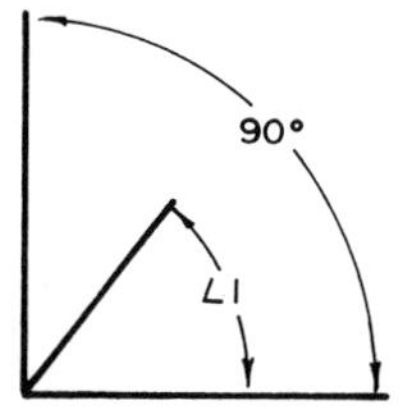

Acute angle

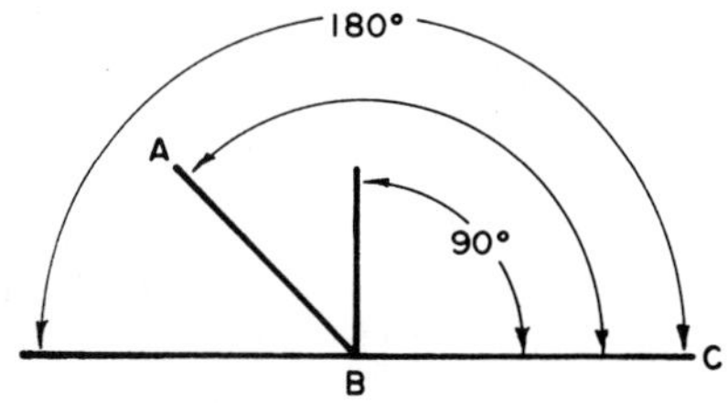

Obtuse angle

Two angles are *complementary* when their sum is 90°. For example, 38° + 52° = 90°. Therefore, 38° is the complement of 52°, and 52° is the complement of 38°.

Two angles are *supplementary* when their sum is 180°. For example, 84° + 96° = 180°. Therefore, 84° is the supplement of 96°, and 96° is the supplement of 84°.

Types of Triangles

A *polygon* is a closed plane figure formed by three or more straight-line segments.

A *triangle* is a three-sided polygon; it is the simplest kind of polygon.

The sum of the three angles of any triangle equals 180°.

Scalene Triangle

A *scalene triangle* has three unequal sides. It also has three unequal angles. Triangle ABC shown is scalene. Sides AB, AC, and BC are unequal, and angles A, B, and C are unequal.

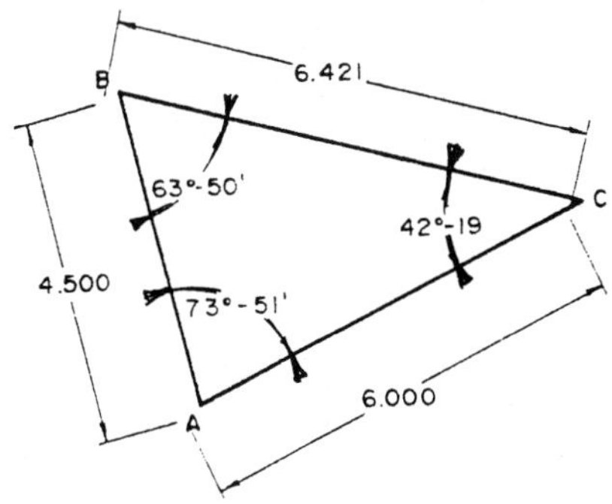

Scalene triangle

Isosceles Triangle

An *isosceles triangle* has two equal sides, called *legs*, and two equal *base angles* (the angles opposite the legs). In isosceles triangle EFG leg EF = leg FG = 27.60 mm and base ∠E base ∠G = 40°.

In an isosceles triangle an altitude to the base bisects the base and the vertex angle.

An *altitude* is a line drawn from a vertex perpendicular to the opposite side.

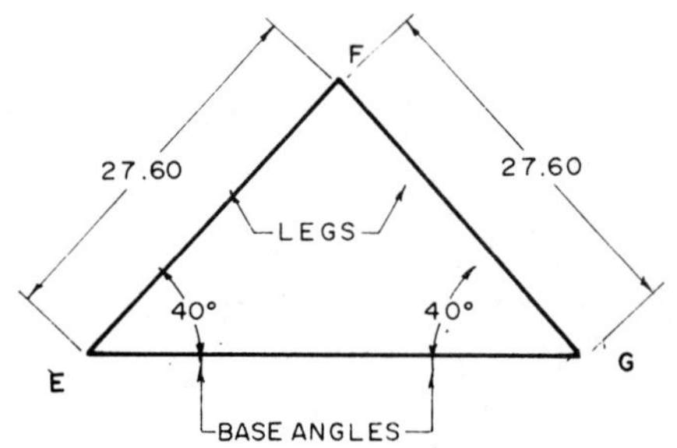

Isosceles triangle

Right Triangle

A *right triangle* has one right or 90° angle. The side opposite the right angle is called the *hypotenuse*. The other two sides are called *legs*.

In a right triangle the square of the hypotenuse is equal to the sum of the squares of the other two sides. This principle, called the *Pythagorean Theorem*, is often used for solving drafting problems. As applied to the right triangle with sides *a*, *b*, and *c* (hypotenuse) shown. The Pythagorean Theorem formula is expressed as $c^2 = a^2 + b^2$.

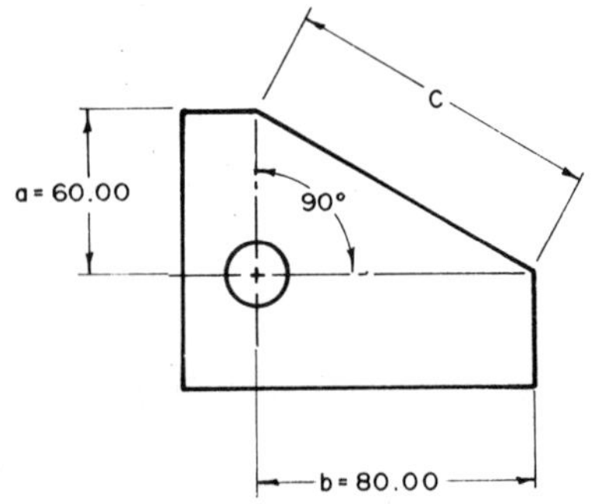

Right triangle

Example Solve for side c when side a = 60.00 mm and side b = 80.00 mm, as shown

$c^2 = a^2 + b^2$

Substitute the given values and solve for c.

$c^2 = (60.00\text{ mm})^2 + (80.00\text{ mm})^2$

$c^2 = 3600\text{ mm}^2 + 6400\text{ mm}^2$

$c^2 = 10{,}000\text{ mm}^2$

Add terms. Apply the root principle of equality. Extract the square root of both sides of the equation.

$\sqrt{c^2} = \sqrt{10{,}000}\text{ mm}^2$

$c = 100.00$ mm Ans

A Determine dimension x. All dimensions are in inches.

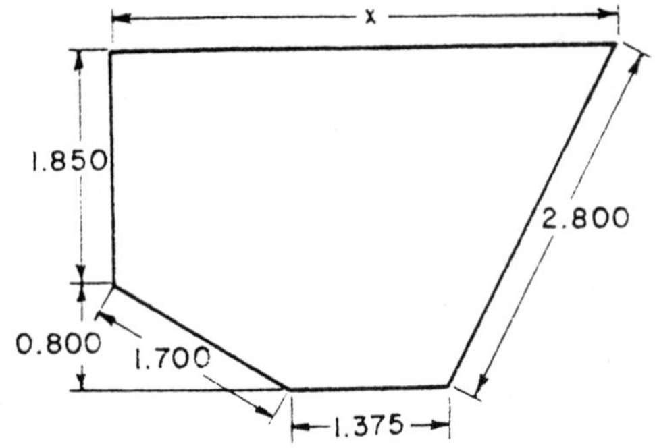

Ans____________________

B Point A is a tangent point. All dimensions are in inches. Determine dimension x.

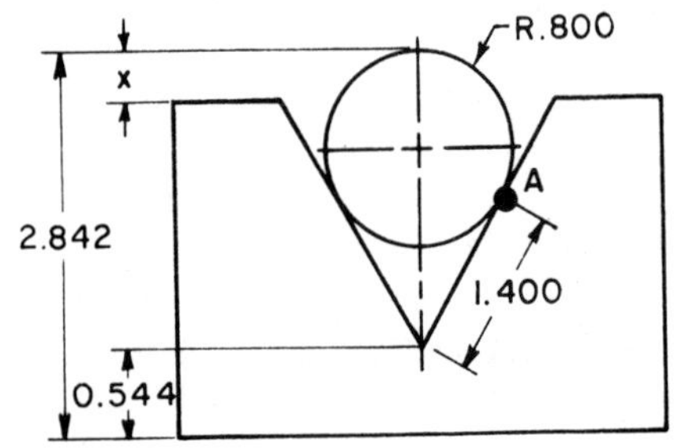

Ans____________________

Mechanical Drafting Mathematics			Problem A-6
Section-Chapter-Problem Appendix A 6	Right Triangle Applications	Course________ Section____ Student__________ Date_____	

Common Polygons

The types of polygons most widely used in drafting in addition to triangles are squares, rectangles, parallelograms, and regular hexagons. A *regular polygon* is one with equal sides and equal angles.

In addition to the description and properties of commonly used polygons, formulas are given that simplify computations when working with polygons.

Squares

A *square* is a regular four-sided polygon. Each angle equals 90°.

The following formulas are used with squares.

	where	
$a = 1.414s$		d = diagonal
$s = 0.7071\,d$		s = side
$A = s^2$ or $0.5d^2$		A = area

Rectangles

A rectangle is a four-sided polygon with opposite sides parallel and equal. Each angle is 90°.

The following formulas are used with rectangles.

	where	
$d = \sqrt{l^2 + w^2}$		d = diagonal
$l = \sqrt{d^2 - w^2}$		l = length
$w = \sqrt{d^2 - l^2}$		w = width
$A = lw$		A = area

A A rectangular sheet of aluminum is 30 in wide and 42 in long.

Determine:

a. the distance across opposite corners, diagonal d.

b. the area or the sheet in square feet.

Ans____________________

Ans____________________

Parallelograms

A *parallelogram* is a four-sided polygon with opposite sides parallel and equal and opposite angles equal.

The area of a parallelogram is equal to the product of the base and the height or altitude.

	where	
$A = bh$		A = area
		b = base
		h = height or altitude.

Note that h is perpendicular to b.

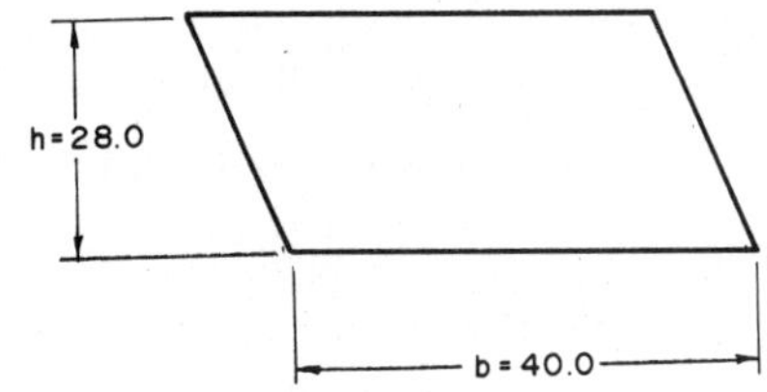

Example: A sheet-metal detail in the shape of a parallelogram is shown. Determine the area in square feet. Substitute values in the formula $A = bh$ and solve.

$A = bh$

$A = 28$ in (40 in)

$A = 1120$ sq in

Since 1 sq ft. = 144 sq in, 1120 sq in is divided by 144 to obtain square feet.

1120 sq in ÷ 144 = 7.78 sq ft Ans

Regular Hexagons

A *regular hexagon* is a six-sided figure with all sides equal and all angles equal. Each angle equals 120°.

The following formulas are used with regular hexagons.

	where	
$c = 2s$		s = side
$s = 0.5c$		c = distance across corners
$f = 1.732s$		f = distance across flats
$s = 0.577f$		A = area
$c = 1.55f$		
$f = 0.866c$		
$A = 2.598s^2 = 0.650c^2 = 0.866f^2$		

B

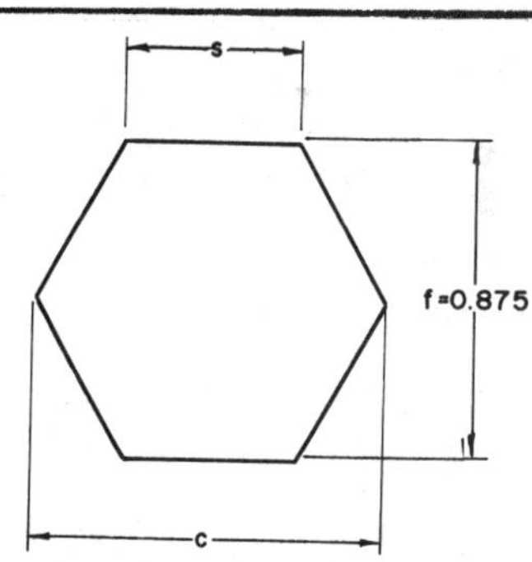

A hexagonal nut is shown. Given f (distance across flats) determine:

a. s (length of side)

b. c (distance across corners)

Ans____________________

Ans____________________

Mechanical Drafting Mathematics			Problem A-7
Section-Chapter-Problem Appendix A 7	Areas of Polygons	Course______ Section____ Student________ Date____	

A Determine the weight of a steel, hexagon head bolt, 3/4" diameter x 4" long. The head is 1.125" across the flats and is 1/2" thick. The 4" length dimension does not include the head. Deduct 4% of the weight to account for the cut threads. The density of steel is 0.28 lb/cubic inch.

Ans_______________

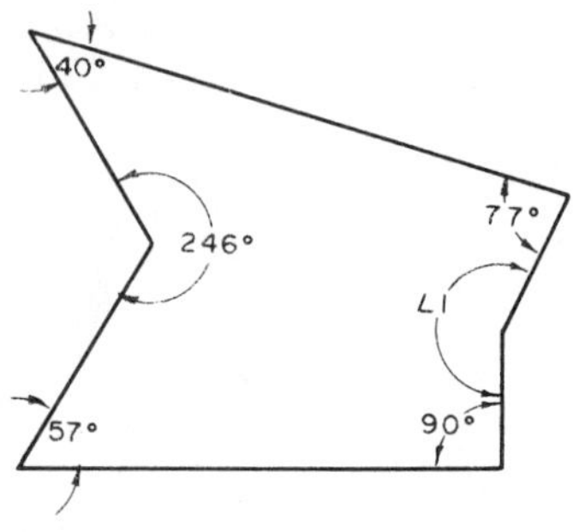

The following principle is widely applied when determining unknown angles in any type of polygon.

The sum of the interior angles of a polygon of N sides is equal to (N - 2) *times* 180°.

Example: Determine ∠1 in the polygon.

Count the number of sides. The polygon has six sides: N = 6. The sum of the six angles is (N - 2) 180° = (6 - 2) 180° = 4 (180°) = 720°.

Add the five known interior angles and subtract from 720° to find ∠1.

$\angle 1 = 720° - (57° + 246° + 40° + 77° + 90°)$

$\angle 1 = 720° - 510°$

$\angle 1 = 210°$ Ans

Determine the unknown angle ∠1 for the eight sided polygon.

∠A=138.5°
∠B=90°
∠C=315°
∠D=45°
∠E=58°
∠F=232°
∠G=85°

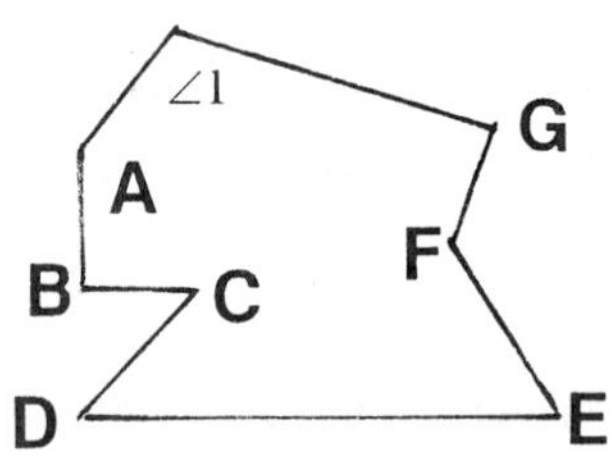

Ans_______________

Definitions of Properties of Circles

The following terms are commonly used to describe the properties of circles. These properties are often applied in mechanical drafting and design situations.

A *circle* is a closed curve of which every point on the curve is equally distant from a fixed point called the center.

A *circumference* is the length of the curved line which forms the circle.

A *chord* is a straight-line segment that joins two points on the circle. AB is a chord.

A *diameter* is a chord that passes through the center of a circle. CD is a diameter.

A *radius* (plural radii) is a straight-line segment that connects the center of the circle with a point on the circle. The radius is one-half the diameter.

An *arc* is that part of a circle between any two points on the circle. The symbol ⌒ written above the letters means arc. $\overset{\frown}{AB}$ is an arc.

A *tangent* to a circle is a straight line that touches the circle at one point only. The point on the circle touched by the tangent is called the *point of tangency* or *tangent point*.

A *segment* is that part of a circle bounded by a chord and its arc.

A *sector* is that part of a circle bounded by two radii and the arc intercepted by the radii.

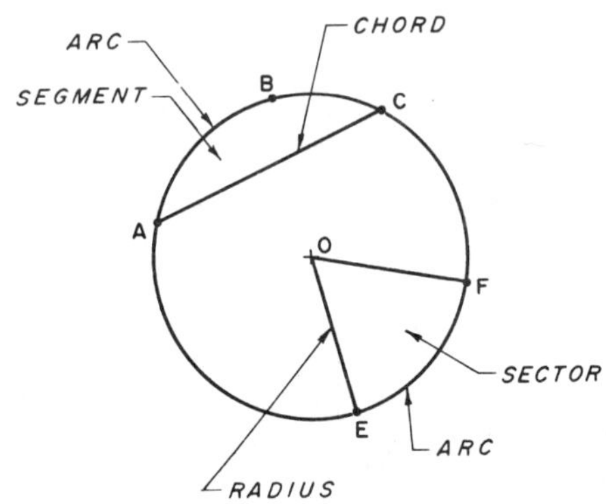

A *central angle* is an angle whose vertex is at the center of a circle and whose sides are radii.

An *inscribed angle* is an angle whose vertex is on the circle and whose sides are chords.

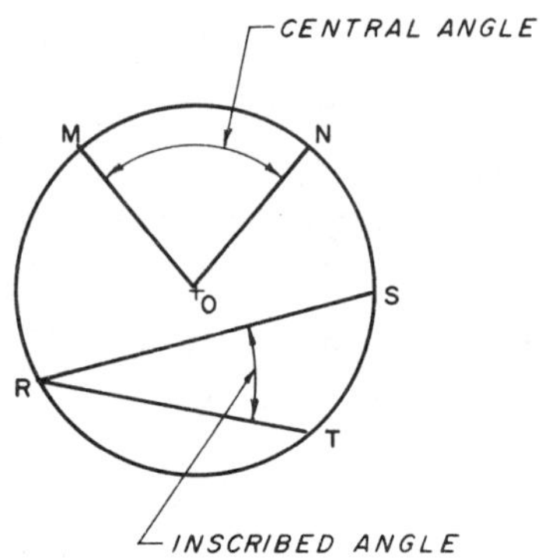

Mechanical Drafting Mathematics			Problem A-8
Section-Chapter-Problem Appendix A 8	Multi-sided Polygons Properties of Circles	Course_______ Section_____ Student_______ Date_____	

Geometric Principles of Circle Circumference, Central Angles, Arcs, and Tangents

The circumference of a circle is equal to pi (π) times the diameter. (Pi equals 3.1416 for precison machining operations.)

In the spaces provided construct 25 mm diameter circles and illustrate the statements **A, B, C**, and **D**.

A. *A line drawn from the center of a circle perpendicular to a chord bisects the chord and the arc cut off by the chord. Conversely, the perpendicular bisector of a chord passes through the center of the circle.*

B *A line perpendicular to a radius at its extremity is tangent to the circle. A tangent is perpendicular to a radius at its tangent point.*

C *Two tangents drawn to a circle from a point outside the circle are equal. The angle at the outside point is bisected by a line drawn from the point to the center of the circle.*

D *In the same or in equal circles, two central angles have the same ratio as the arcs that are cut off by the angles.*

Trigonometry: Trigonometric Functions

Trigonometry is the branch of mathematics used to compute unknown angles and sides of triangles. Many drafting problems that cannot be solved by the use of geometry alone are easily solved by trigonometry.

Ratio of Right Triangle Sides

In a right triangle the ratio of two sides of the triangle determines the sizes of the angles, and the angles determine the ratio of two sides.

Identifying Right Triangles by Name

The sides of a right triangle are named opposide side, adjacent side, and hypotenuse. The *hypotenuse* (hyp) is always the side opposite the right angle. It is always the longest side of a right triangle. The positions of the opposite and adjacent sides depend on the reference angle. The *opposite side* (opp side) is opposite the reference angle, and the *adjacent side* (adj side) is next to the reference angle.

Trigonometric Functions: Ratio Method

There are two methods of defining trigonometric functions: the unity or unit circle method, and the ratio method. Only the ratio method is presented here.

Since a triangle has three sides and a ratio is the comparison of any two sides, there are six different ratios. The names of the ratios are *sine, cosine, tangent, cotangent, secant*, and *cosecant.*

FUNCTION	SYMBOL	DEFINITION OF FUNCTION
sine of angle A	sine A	$\sin A = \frac{\text{opp side}}{\text{hyp}} = \frac{a}{c}$
cosine of angle A	cos A	$\cos A = \frac{\text{adj side}}{\text{hyp}} = \frac{b}{c}$
tangent of angle A	tan A	$\tan A = \frac{\text{opp side}}{\text{adj side}} = \frac{a}{b}$
cotangent of angle A	cot A	$\cot A = \frac{\text{adj side}}{\text{opp side}} = \frac{b}{a}$
secant of angle A	sec A	$\sec A = \frac{\text{hyp}}{\text{adj side}} = \frac{c}{b}$
cosecant of angle A	csc A	$\csc A = \frac{\text{hyp}}{\text{opp side}} = \frac{c}{a}$

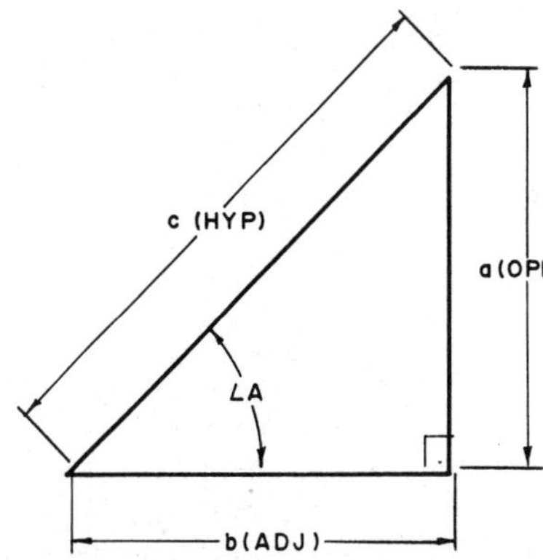

Mechanical Drafting Mathematics			Problem
Section-Chapter-Problem Appendix A 9	Properties of Circles Trigonometric Functions	Course______ Section______ Student______ Date______	A-9

Trigonometry: Basic Calculations of Angles

In order to solve for an unknown angle of a right triangle when neither acute angle is shown, at least two sides must be known. An understanding of the procedures required for solving for unknown angles is essential to the drafter.

Procedure: *To determine an unknown angle when two sides of a right triangle are known*

- Identify two given sides as adjacent, opposite, or hypotenuse in relation to the desired angle.
- Determine the functions that are ratios of the sides identified in relation to the desired angle.
- Choose one of the two functions, substitute the given sides in the ratio, and divide.
- Using either a calculator or a trigonometric function table, determine the angle that corresponds to the quotient obtained.

Example Determine $\angle B$ of the triangle. All dimensions are in millimetres.

In relation to $\angle B$, the 290.000-mm side is the adjacent side and the 575.00-mm side is the hypotenuse.

$$\text{cosine } \angle B = \frac{\text{adjacent side}}{\text{hypotenuse}}.$$

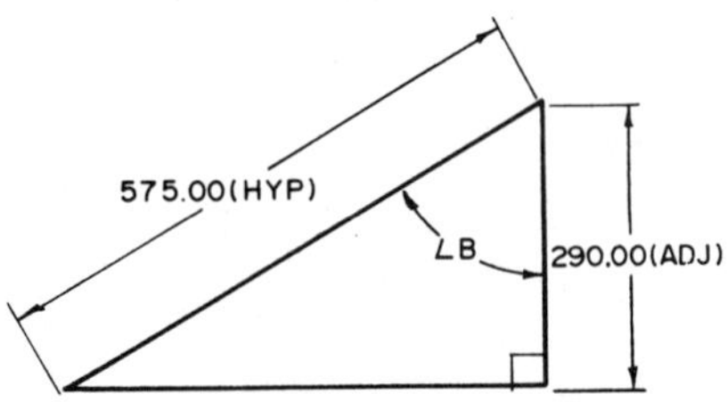

Choosing the cosine function:

$$\cos \angle B = \frac{\text{adj side}}{\text{hyp}}$$

$$\cos \angle B = \frac{290.00}{575.00}$$

$$\cos \angle B = 0.50435$$

$$\angle B = 59.72° \text{ or } 59°43' \text{ Ans}$$

A Determine $\angle 1$ and $\angle 2$ of the triangle. All dimensions are in inches.

Calculate either $\angle 1$ or $\angle 2$. Choose any two of the three given sides. In relation to $\angle 1$, the 3.420-in side is the opposite side. The 5.845-in side is the hypotenuse.

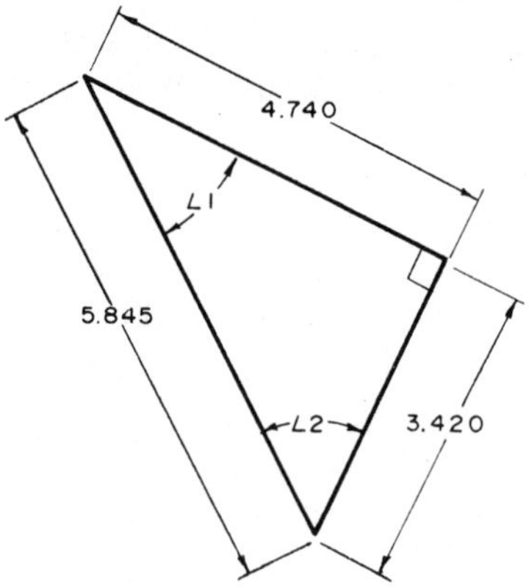

Ans__________________

Ans__________________

B Determine $\angle x$. All dimensions are in inches.

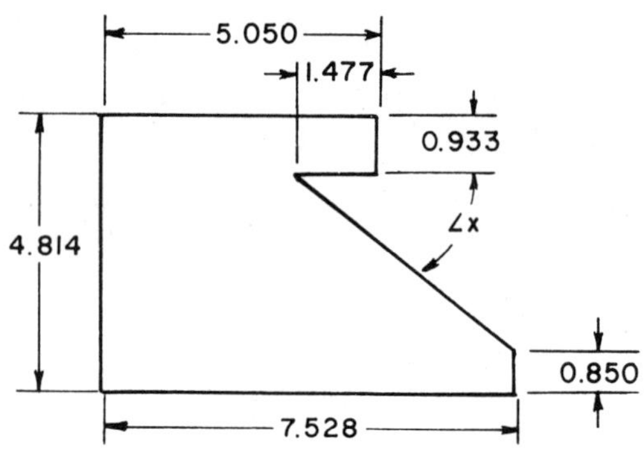

Ans__________________

Mechanical Drafting Mathematics

Section-Chapter-Problem	Trigonometry	Course______ Section____	Problem
Appendix A 10	Basic Calculations	Student__________ Date______	A-10

Trigonometry: Basic Calculations of Sides

Procedure: *To determine an unknown side when an acute angle and one side of a right triangle are known*

- Identify the given side and the unknown side as adjacent, opposite, or hypotenuse in relation to the given angle.
- Determine the trigonometric functions that are ratios of the sides identified in relation to the given angle.
- Using either a calculator or a trigonometric function table, find the function of the given angle and substitute this value.
- Solve as a proportion for the unknown side.

Example: Determine the side of the triangle All dimensions are in inches.

In relation to the 63°20′ angle, the 8.100-in side is the adjacent side and side x is the opposite side. Determine the two functions whose ratios consist of the adjacent and opposite sides:

$$\text{tangent } 63°20' = \frac{\text{opposite side}}{\text{adjacent side}},$$

Choosing the tangent

$$\tan 63°20' = \frac{\text{opp side}}{\text{adj side}}$$

$$\tan 63°20' = \frac{x}{8.100}$$

$$1.9912 = \frac{x}{8.100}$$

$$x = 1.9912\,(8.100)$$

$$x = 16.129 \text{ in Ans}$$

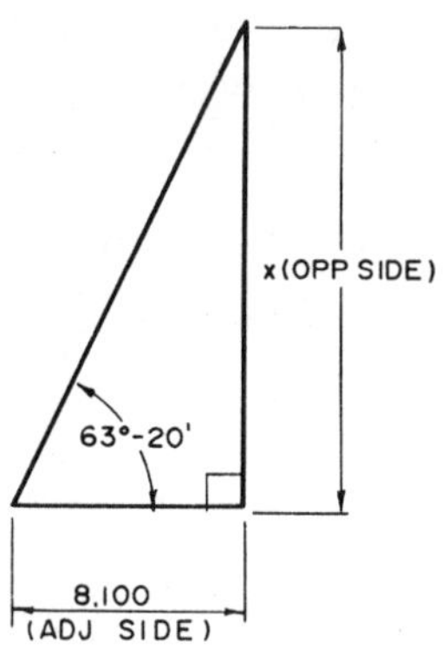

Trigonometry: Common Drafting Applications

Method of Solution

The following examples are simple practical applications of right-angle trigonometry, although they may not be given directly in the form of right triangles. To solve the examples it is necessary to project auxiliary lines to produce a right triangle. The unknown, or a dimension required to compute the unknown, is part of the triangle. The auxiliary lines may be projected between given points or from given points. The lines may be projected parallel or perpendicular to center lines, tangents, or other reference lines.

Tapers and Bevels

Example Determine the included taper angle of the shaft. All dimensions are in inches.

The problem must be solved by using a figure in the form of a right triangle. Therefore, project line AB from point A parallel to the center line. Right △ABC is formed in which ∠BAC is one-half the included taper angle.

$$\text{Side AB} = 10.500''$$

$$\text{Side BC} = \frac{1.800'' - 0.700''}{2} = 0.550''$$

Using sides AB and BC, solve for ∠BAC.

$$\tan \angle BAC = \frac{BC}{AB} = \frac{0.500}{10.500} = 0.05238$$

$$\angle BAC = 3°0'$$

The included taper angle = 2(3°0′) = 6°0′ Ans

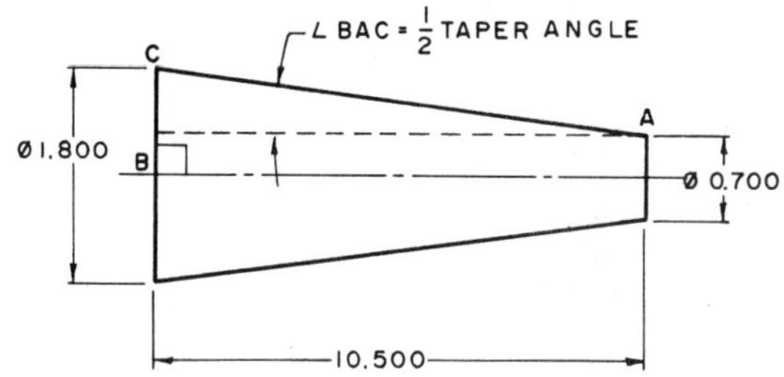

A Determine diameter x of the part shown All dimensions are in millimetres. Project line DE from point D parallel to the centerline in order to form right △DEF.

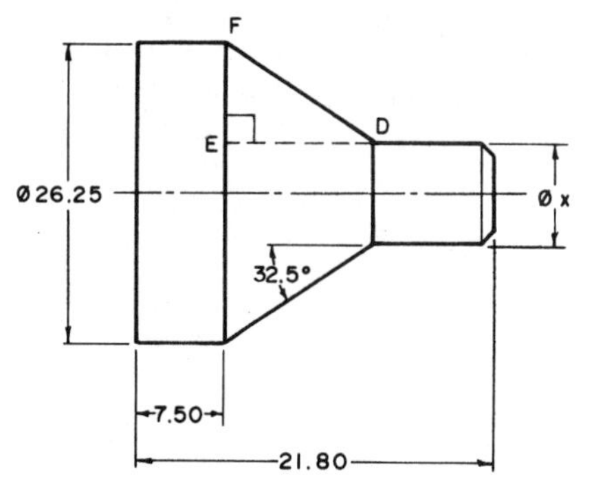

Ans__________

Mechanical Drafting Mathematics

Section-Chapter-Problem	Trigonometry	Course______ Section____	Problem
Appendix A 11	Drafting Applications	Student______ Date____	A-11

Isosceles Triangle Applications: Distance Between Holes and V-Slots

The solutions to many practical trigonometry problems are based on recognizing figures as isosceles triangles. In an isosceles triangle an altitude to the base bisects the base and the vertex angle.

Example five holes are equally spaced on a 5.200-in diameter circle. Determine the straight-line distance between centers of two consecutive holes.

Project radii from center O to hole centers A and B.

Project a line from A to B. $\angle AOB = \dfrac{360°}{5} = 72°$.

Since OA = OB, $\triangle AOB$ is isosceles. Project line OC perpendicular to AB from center O. Line OC bisects

$\angle AOB$ and side AB. In right $\triangle AOC$, $\angle AOC = \dfrac{72°}{2} = 36°$. $AO = \dfrac{5.200 \text{ in}}{2} = 2.600$ in.

Solve for side AC.

$$\sin \angle AOC = \frac{AC}{AO}$$

$$\sin 36° = \frac{AC}{2.600}$$

$$0.58779 = \frac{AC}{2.600}$$

$$AC = 0.58779(2.600)$$

$$AC = 1.528 \text{ in}$$

$$AB = 2(1.528 \text{ in}) = 3.056 \text{ in Ans}$$

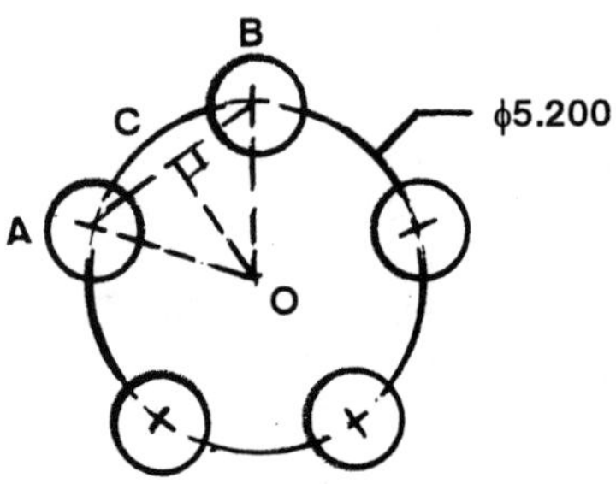

A Determine dimension x of the V-slot

All dimensions are in inches.

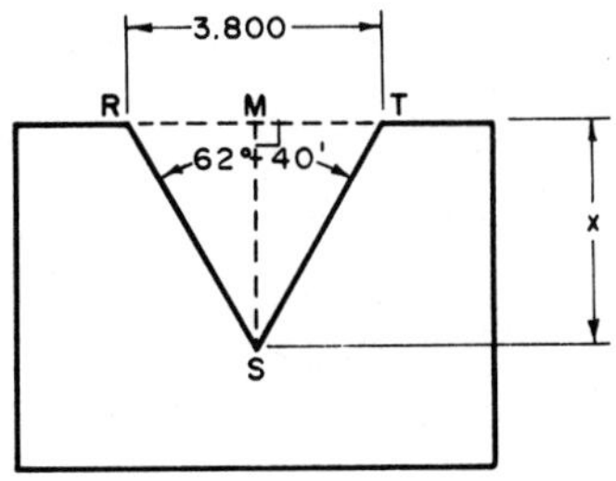

Ans________________

Tangents to Circles Applications: Angle Cuts and Dovetails

Example: An internal dovetail is shown

Two pins or balls are used to check the dovetail for both location and angular accuracy. Calculate check dimension x. All dimensions are in inches.

Project line HO from point H to the pin center O; HO bisects the 72°20′ angle. Project a radius from point O to the point of tangency K; $\angle HKO$ is a right angle, since a radius is perpendicular to a tangent at the point of tangency. In right $\triangle HOK$, $\angle KHO = \dfrac{72°20'}{2} = 36°10'$, $KO = \dfrac{1.000 \text{ in}}{2} = 0.500$ in

Solve for side HK.

$$\cot \angle KHO = \frac{HK}{KO}$$

$$\cot 36°10' = \frac{HK}{0.500}$$

$$1.3680 = \frac{HK}{0.500}$$

$$HK = 0.500(1.3680)$$

$$HK = 0.684 \text{ in}$$

KM = pin radius = 0.500 in

HM = HK + KM = 0.684 in + 0.500 in = 1.184. in

x = 6.312 in − 2(HM) = 6.312 in − 2(1.184 in) = 3.944 in Ans

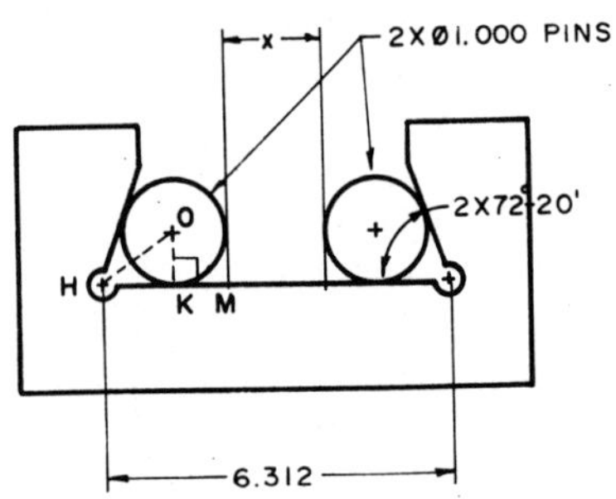

Mechanical Drafting Mathematics

Section-Chapter-Problem	Triangle Applications	Course______ Section____	Problem
Appendix A 12	Dovetail	Student______ Date____	A-12

A Determine gage dimension x. All dimensions are in inches.

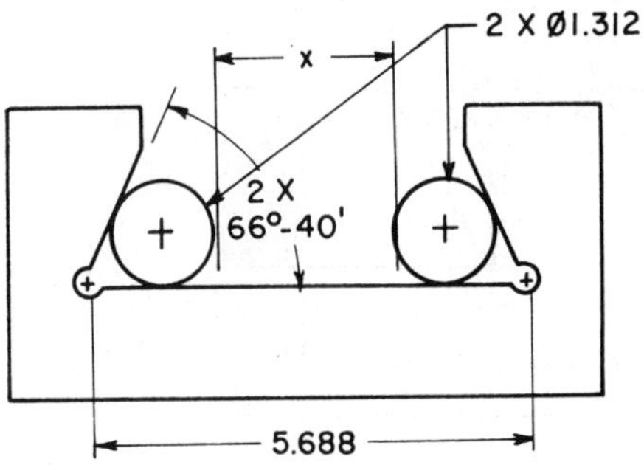

Ans____________________

Trigonometry: Oblique Triangles—Law of Sines and Law of Cosines

An *oblique triangle* is one that does not contain a right angle. Drafting and design problems often involve oblique triangles. These problems can be reduced to a series of right triangles, but the process can be cumbersome and time consuming. Two formulas, the *law of sines* and the *law of cosines*, can be used to simplify such computations. In order to use either formula, three parts of an oblique triangle must be known; at least one part must be a side.

An oblique triangle can contain an angle greater than 90°, such as oblique triangle ABC
Therefore, sine and cosine functions of angles between 90° and 180° must be determined.

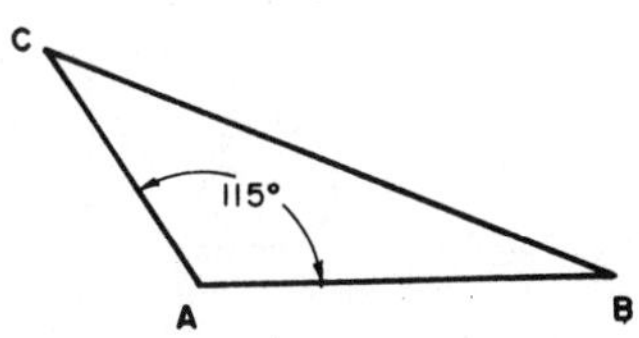

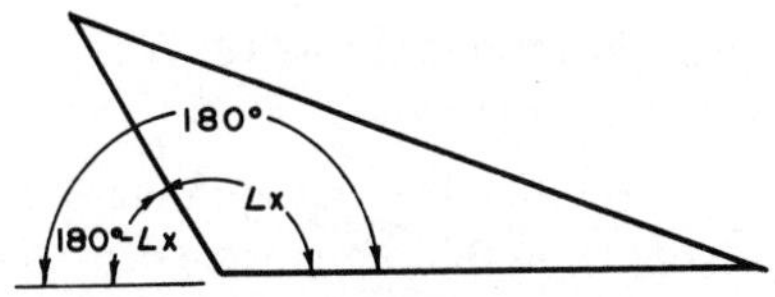

Sine Functions of Angles Between 90° and 180°

The sine of an angle greater than 90° and less than 180° equals the sine of the supplement of the angle. If the value of $\angle x$ is between 90° and 180°, as shown

$$\sin \angle x = \sin(180° - \angle x).$$

Examples:

1. $\sin 115° = \sin(180° - 115°) = \sin 65° =$ 0.90631 Ans
2. $\sin 90°42' = \sin(180° - 90°42') = \sin 89°18' =$ 0.99992 Ans
3. $\sin 176.25° = \sin(180° - 176.25°) = \sin 3.75° =$ 0.06540 Ans

Law of Sines

The law of sines states that, in any triangle, the sides are proportional to the sines of the opposite angles. That is

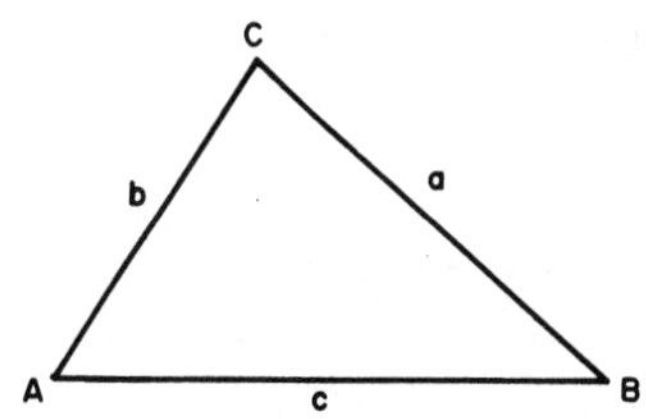

$$\frac{a}{\sin A} = \frac{b}{\sin B} = \frac{c}{\sin C}$$

The law of sines is used to solve the following kinds of problems:

- Problems where any two angles and any one side of an oblique triangle are known.
- Problems where any two sides and an angle opposite one of the given sides of an oblique triangle are known.

Solving Oblique Triangle Problems Given Two Angles and a Side

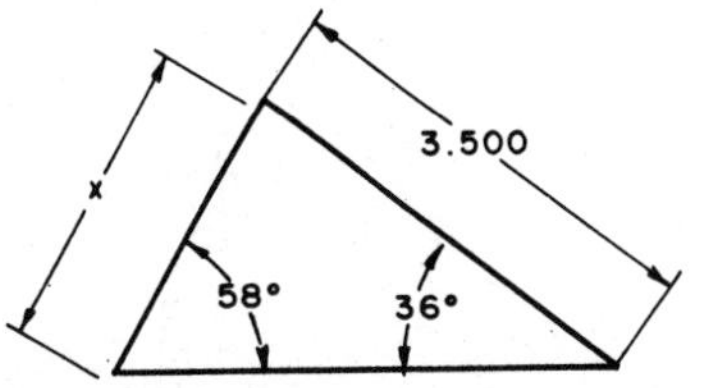

Mechanical Drafting Mathematics			Problem
Section-Chapter-Problem Appendix A 13	Law of Sines Applicatons	Course________ Section____ Student____________ Date_____	A-13

Example Given two angles and a side, determine side x of the oblique triangle. (All dimensions are in inches.)

Set up a proportion and solve for x.

$$\frac{x}{\sin 36^\circ} = \frac{3.500}{\sin 58^\circ}$$
$$\frac{x}{0.58779} = \frac{3.500}{0.84805}$$
$$0.84805x = 3.500(0.58779)$$
$$x = \frac{3.500(0.58779)}{0.84805}$$
$$x = 2.426 \text{ in Ans}$$

A Given two angles and a side of the oblique triangle formed by the intersection of hole centerlines shown All dimensions are in millimetres.

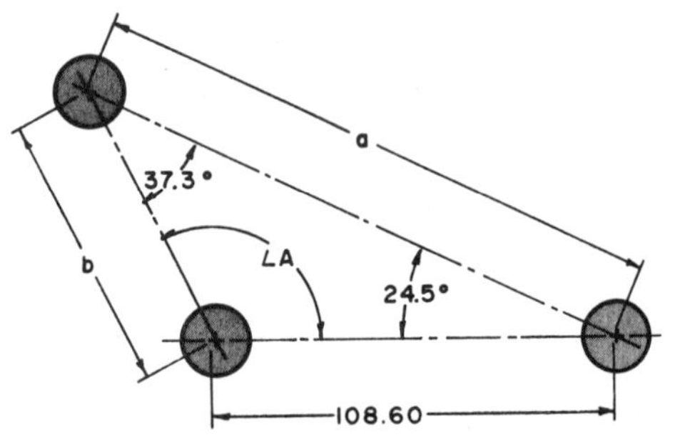

a. Determine ∠A.
b. Determine side a.

Ans____________________

Ans____________________

Solving Oblique Triangle Problems Given Two Sides and an Angle Opposite One of the Given Sides

Example Given two sides and an opposite angle of the oblique triangle formed by the intersection of

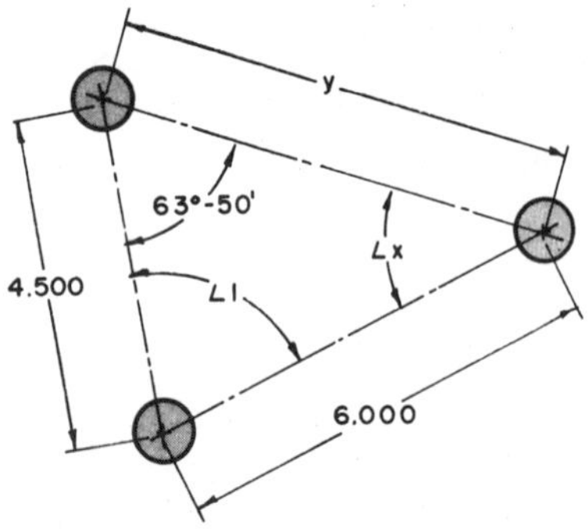

hole center lines shown ~~in Figure A-98~~. (All dimensions are in inches.)

a. Determine $\angle x$.
b. Determine side y.

a. Determine $\angle x$.

$$\frac{4.500}{\sin \angle x} = \frac{6.000}{\sin 63^\circ 50'}$$
$$\frac{4.500}{\sin \angle x} = \frac{6.000}{0.89752}$$
$$6.000(\sin\angle x) = 4.500(0.89752)$$
$$\sin\angle x = \frac{4.500(0.89752)}{6.000}$$
$$\sin \angle x = 0.67314$$
$$\angle x = 42^\circ 18' \text{ Ans}$$

b. Determine side y.

$$\angle 1 = 180^\circ - (63^\circ 50' + \angle x) =$$
$$180^\circ - (63^\circ 50' + 42^\circ 18') = 180^\circ - 106^\circ 8' = 73^\circ 52'$$
$$\frac{6.000}{\sin 63^\circ 50'} = \frac{y}{\sin 73^\circ 52'}$$
$$\frac{6.000}{0.89752} = \frac{y}{0.96062}$$
$$0.89752y = 6.000(0.96054)$$
$$y = \frac{6.000(0.96054)}{0.89752}$$
$$y = 6.422 \text{ in Ans}$$

B Given two sides and an opposite angle, determine $\angle x$ of the oblique triangle shown (All dimensions are in millimetres.)

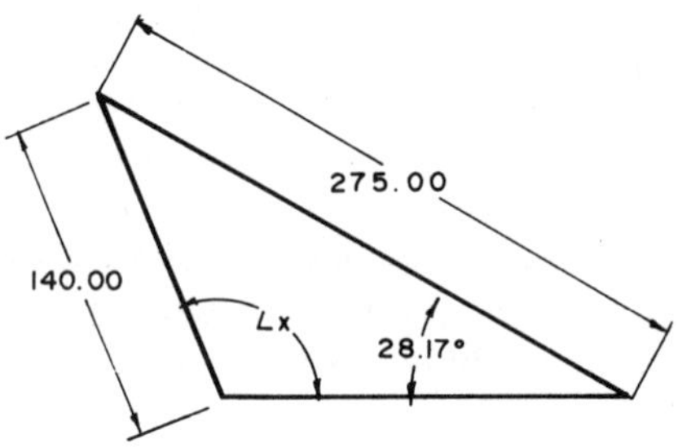

Ans____________________

Mechanical Drafting Mathematics

Problem A-14

Section-Chapter-Problem Appendix A 14

Law of Sines Applicatons

Course__________ Section_____

Student__________________ Date______

Cosine Functions of Angles Between 90° and 180°

The cosine of an angle greater than 90° and less than 180° equals the negative cosine of the supplement of the angle. If the value of $\angle x$ is between 90° and 180°, $\cos \angle x = -\cos(180° - \angle x)$. A negative function of an angle does *not* mean that the angle is negative; it is a negative function of a positive angle. For example, $-\cos 65°$ does *not* mean $\cos(-65°)$.

Examples:

1. $\cos 115° = -\cos(180° - 115°) = -\cos 65° = -0.42262$.
2. $\cos 90°42' = -\cos(180° - 90°42') = -\cos 89°18' = -0.01222$.
3. $\cos 176.25° = -\cos(180° - 176.25°) = -\cos 3.75° = -0.99786$.

Law of Cosines

The law of cosines states that, in any triangle, the square of any side is equal to the sum of the squares of the other two sides minus twice the product of these two sides multiplied by the cosine of their included angle.

$$a^2 = b^2 + c^2 - 2bc(\cos A)$$
$$b^2 = a^2 + c^2 - 2ac(\cos B)$$
$$c^2 = a^2 + b^2 - 2ab(\cos C)$$

The equations can be rearranged in this form:

$$\cos A = \frac{b^2 + c^2 - a^2}{2bc}$$
$$\cos B = \frac{a^2 + c^2 - b^2}{2ac}$$
$$\cos C = \frac{a^2 + b^2 - c^2}{2ab}$$

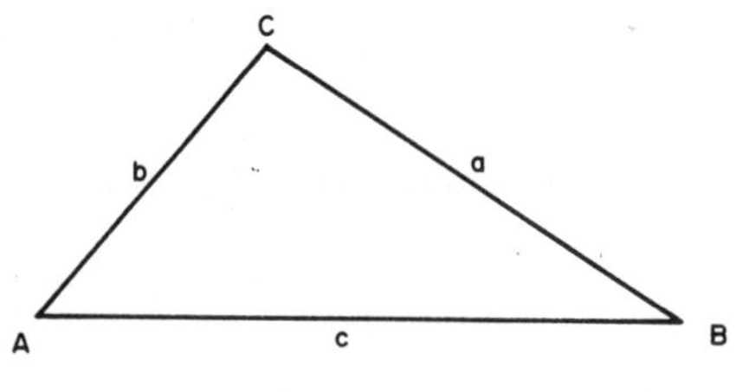

The law of cosines is used to solve the following kinds of problems:

- Problems where two sides and the included angle of an oblique triangle are known.
- Problems where three sides of an oblique triangle are known.

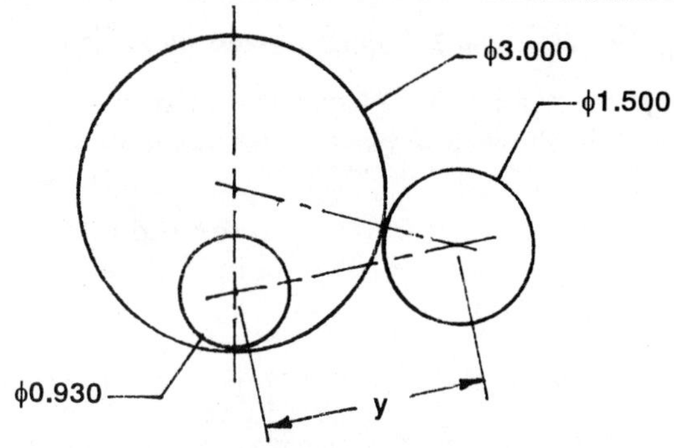

Solving Oblique Triangle Problems Given Two Sides and the Included Angle

Example Internally and externally tangent circles are shown. Determine dimension y, the distance between hole centers A and C. All dimensions are in inches.

In oblique $\triangle$ABC:

$$AB = \frac{3.000}{2} - \frac{0.930}{2} = 1.035$$
$$BC = \frac{3.000}{2} + \frac{1.500}{2} = 2.250$$

The 82° angle is the included angle between AB and BC. Substitute values in their appropriate places in the formula and solve for dimension y.

$$y^2 = 1.035^2 + 2.250^2 - 2(1.035)(2.250)(\cos 82°)$$
$$y^2 = 1.0712 + 5.0625 - 2(1.035)(2.250)(0.13917)$$
$$y^2 = 6.1337 - 0.6482$$
$$y^2 = 5.4855$$
$$y = 2.342 \text{ in Ans}$$

A Given two sides and the included angle, determine side x of the oblique triangle All dimensions are in inches.

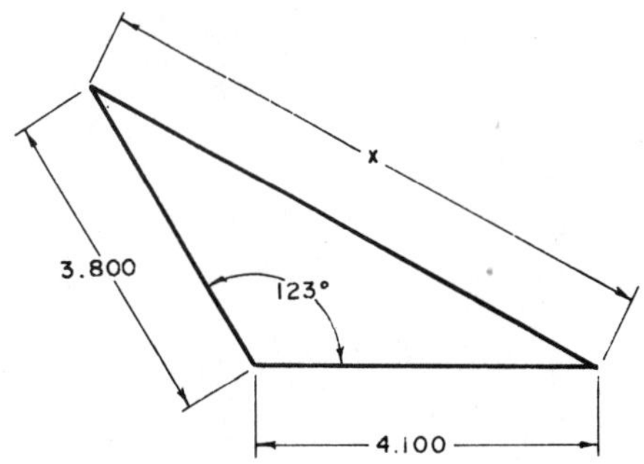

Ans________________

Solving Oblique Triangle Problems Given Three Sides

Example Four pins are located on a fixture at A, B, C, and D as shown ~~in Figure A-103~~. Compute ∠3. All dimensions are in millimetres.

∠3 = ∠1 + ∠2. Compute ∠1 in the right ΔADC. Compute ∠2 in oblique ΔABC. In right ΔADC, AC = 75.00, DC = 73.00.

$$\sin \angle 1 = \frac{DC}{AC}$$

$$\sin \angle 1 = \frac{73.00}{75.00} = 0.97333$$

$$\angle 1 = 76.73°$$

In oblique ΔABC, AC = 75.00, AB = 80.00, BC = 48.00. Substitute values in their appropriate places in the formula.

$$\cos \angle 2 = \frac{75.00^2 + 80.00^2 - 48.00^2}{2(75.00)(80.00)}$$

$$\cos \angle 2 = \frac{5625 + 6400 - 2304}{12{,}000} = 0.81008$$

$$\angle 2 = 35.90°$$

$$\angle 3 = \angle 1 + \angle 2 = 76.73° + 35.90° = 112.63° \text{ Ans}$$

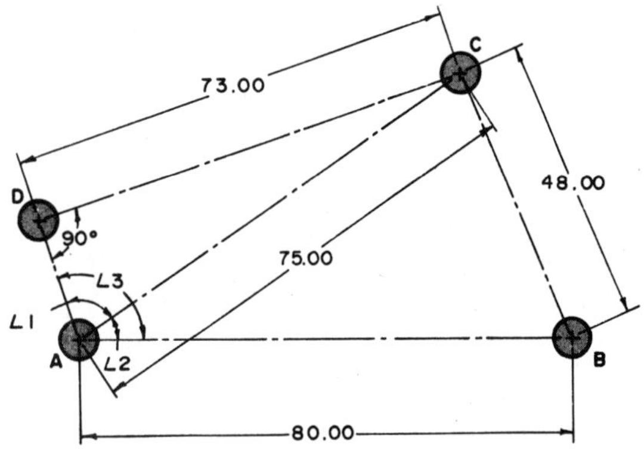

A Determine ∠x. All dimensions are in millimetres.

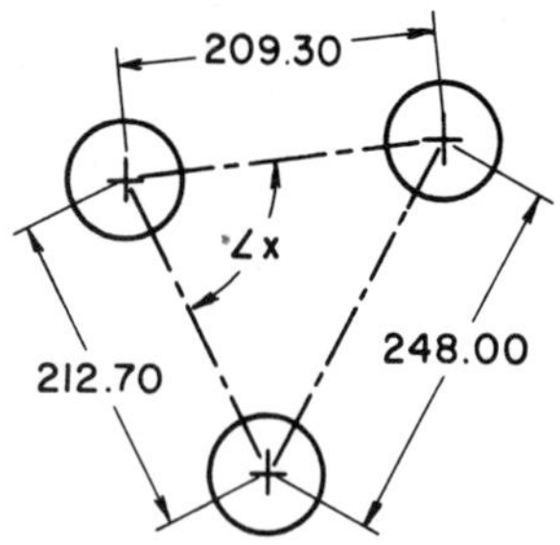

Ans________________

B

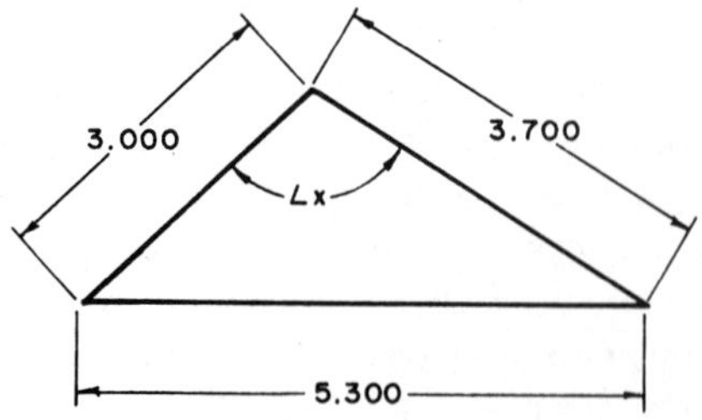

Given three sides, determine ∠x of the oblique triangle. All dimensions are in inches.

Ans________________

C Determine dimension *y*. All dimensions are in millimetres.

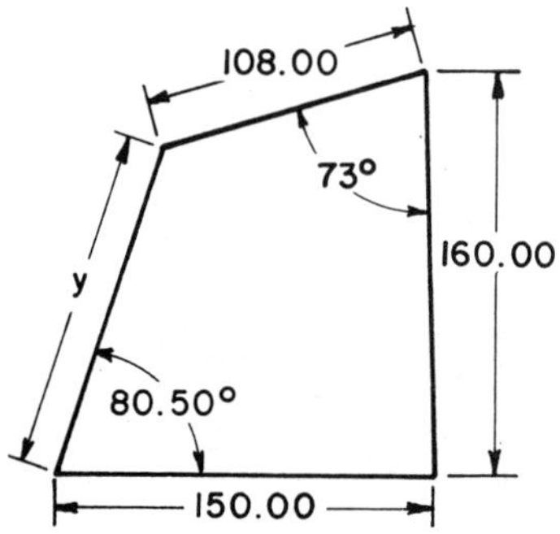

Ans________________

D Determine ∠x. All dimensions are in inches.

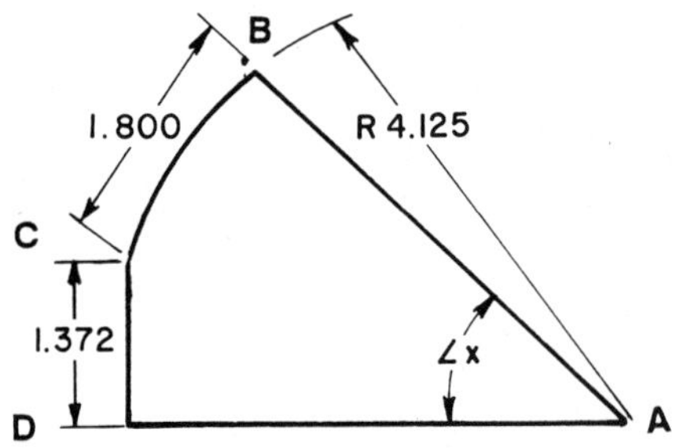

Ans________________

Mechanical Drafting Mathematics			**Problem**
Section-Chapter-Problem Appendix A 16	Oblique Triangle	Course__________ Section_____ Student__________________ Date______	**A-16**

The following problems beginning with this page, A-17, build on the mathematical operations, procedures, and principles presented in Problems A-1 through A-16. Some new mathematical topics and applications are introduced. Example solutions are presented where appropriate. Several design questions are included.

CUT and FILL cross sections used during highway construction are shown below in sketches made in the field by a surveyor. Determine angles A, B, C, and D (graphically and theoretically) based on the information given. Also, determine the cross sectonal areas of the cut and fill.

Assumed that the figures are drawn to scale, 1” = 40’.

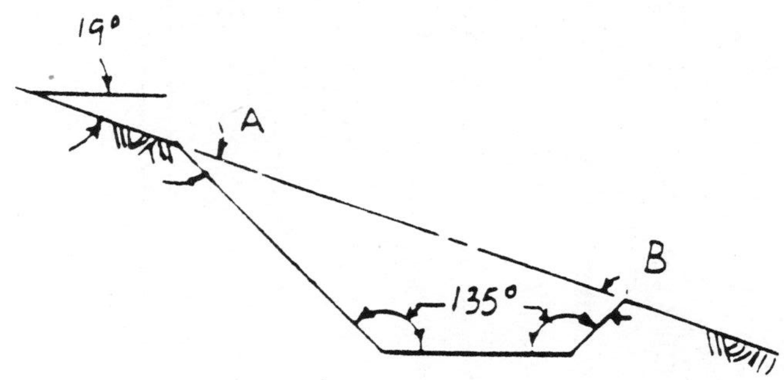

CUT

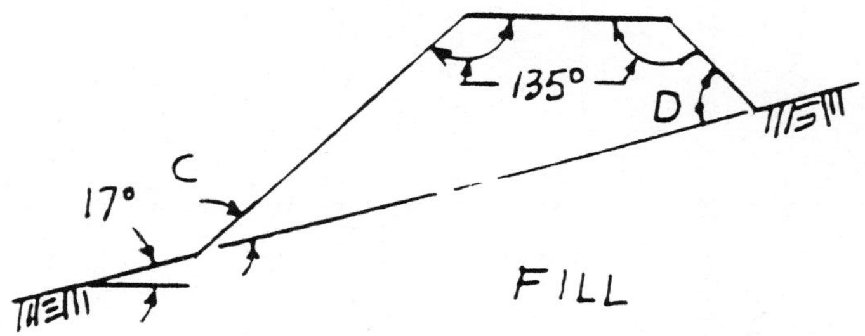

FILL

A = ________
B = ________
C = ________
D = ________

Area of CUT = ________
Area of FILL = ________

Mechanical Drafting Mathematics			Problem
Section-Chapter-Problem Appendix A 17	CUT and FILL Law of Cosines	Course________ Section_____ Student________ Date_____	A-17

The typical staircase dimensions shown are to be used for constructing a new stair case.

a. What will be the total "rise" and "run" of the 15 step staircase? Give your answer in fractional form.

b. If each rise is decreased by 1/8" and each run increased by 3/16" what would the total rise and run become?

a. Rise_________Run_________

b. Rise_________Run_________

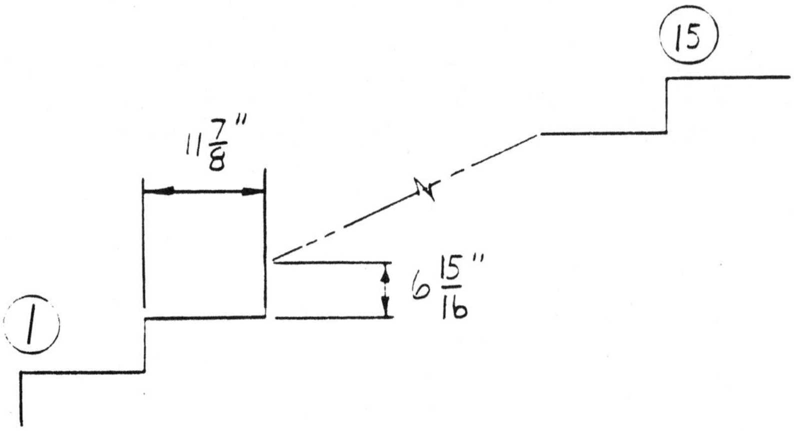

Mechanical Drafting Mathematics			Problem
Section-Chapter-Problem Appendix A 18	Rise and Run Stairway	Course______ Section____ Student______ Date____	A-18

The ADDA (American Design Drafting Asociation) offers a Drafter Certification Test to help drafters to:

1. enhance their credibility
2. improve their opportunities for promotion and pay increases
3. give them an edge in a highly competitive market

Three types of questions are used: Matching, true or false, and multiple choice.
No reference materials are allowed during a test.

The following questions (for your practice) are similar to those found in an ADDA flyer (6-17-97) describing the test:

1. Which of the following are not standard fits? Circle your answer(S).

   LT LF FN LC TF LN RC

2. True or False? Indicate your answers with an F or a T.

   a.____Apply dimensions in the views where the shape or part is shown most accurately.

   b.____Leader lines do not need to point to the center of a hole or circular feature.

   c. ____Dimensions that are not to scale should be underlined with a straight line.

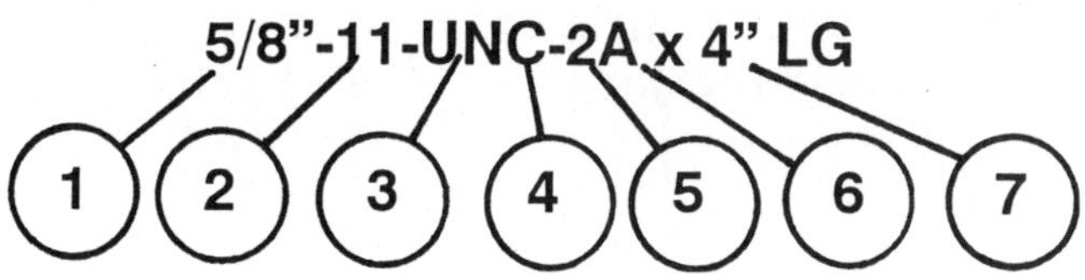

3. Match correct terms with the circled numbers by placing an appropriate letter after the number. The terms are: (a) length, (b) unified national series, (c) nominal size, (d) coarseness of threads, (e) class of fit, (f) external or internal thread, and (g) threads per inch.

Circled Numbers 1.____ 2.____3.____4.____5.____6.____7.____

For further information on the Drafter Certification Test call (301) 460 6875.

The following is an example of the use of proportional equations:
Example:

The shadow of a tall antenna measured 20 meters; at the same time the shadow of a 2 meter tall, vertical post measured 1.6 meters. How tall was the antenna?

$$\frac{1.6}{2.0} = \frac{20meters}{AntennaHeight}$$

$$AntennaHeight = \frac{(2.0)(20m)}{(1.6)} = 25meters$$

Answer the following:

a. Change the 1.6 meter in the example to 1.4 meters, and calculate the height of the antenna.

Antenna Height = ____________

b. A tall building casts a shadow 34 meters. Nearby a 54" tall Parking Meter casts a shadow 28" long. Determine the height of the building in feet.

Building Height = ____________

The concreat footing-retaining wall shown is 610 feet long. Determine how many truck loads of concrete were needed for pouring if the capacity of each truck was 12 yards.

(one yard or more accurately, “cubic yard” = 27 cubic feet)

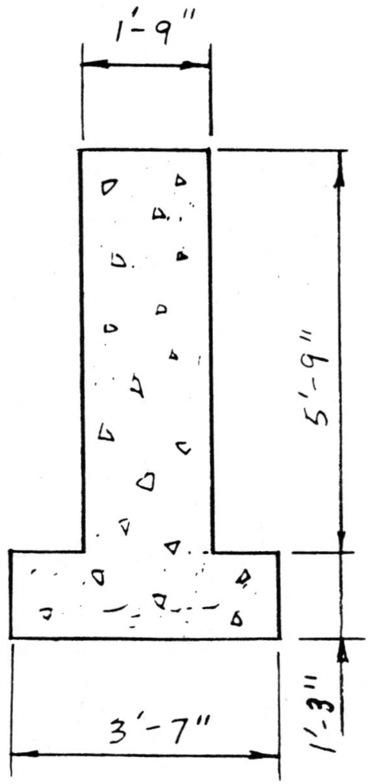

Number of truck loads @ 12 yards per load = ____________

Mechanical Drafting Mathematics			Problem A-21
Section-Chapter-Problem Appendix A 21	Concrete Bulkhead Volume	Course__________ Section_____ Student________________ Date______	

Use trigonometry to determine the distance D shown in the figure.

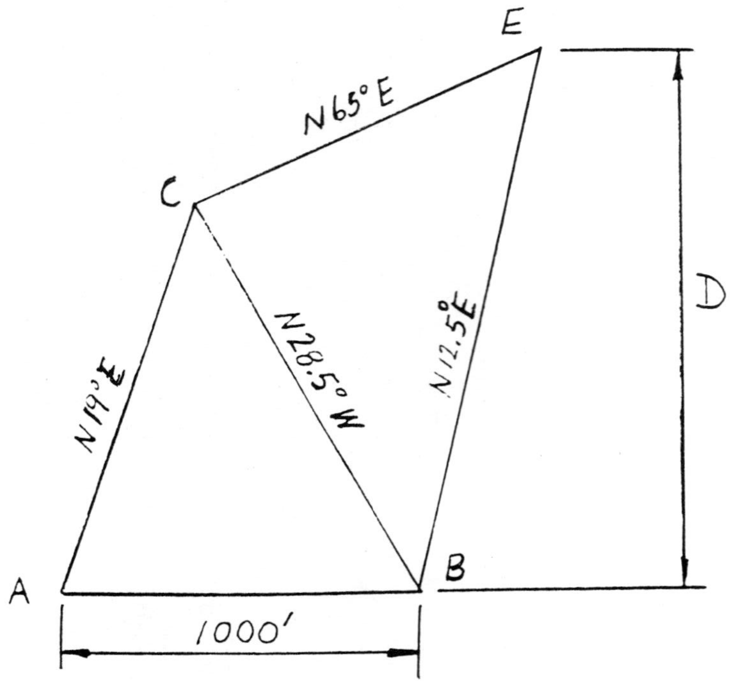

D = __________

Mechanical Drafting Mathematics			Problem
Section-Chapter-Problem Appendix A 22	Trigonometry Application	Course________ Section____ Student__________ Date____	A-22

One method of calculating the area between a curve and a horizontal axis is to do the following:

1. divide the area into convenient sections, the number of sections depending on the flatness or steepness of the the curve, more when the curve is steep.

2. estimate, by eye, the height of each individual section and draw a horizontal line the width of the section.

3. multiply the average height by the width of the section, and finally

4. add up all the calculated areas

Example:

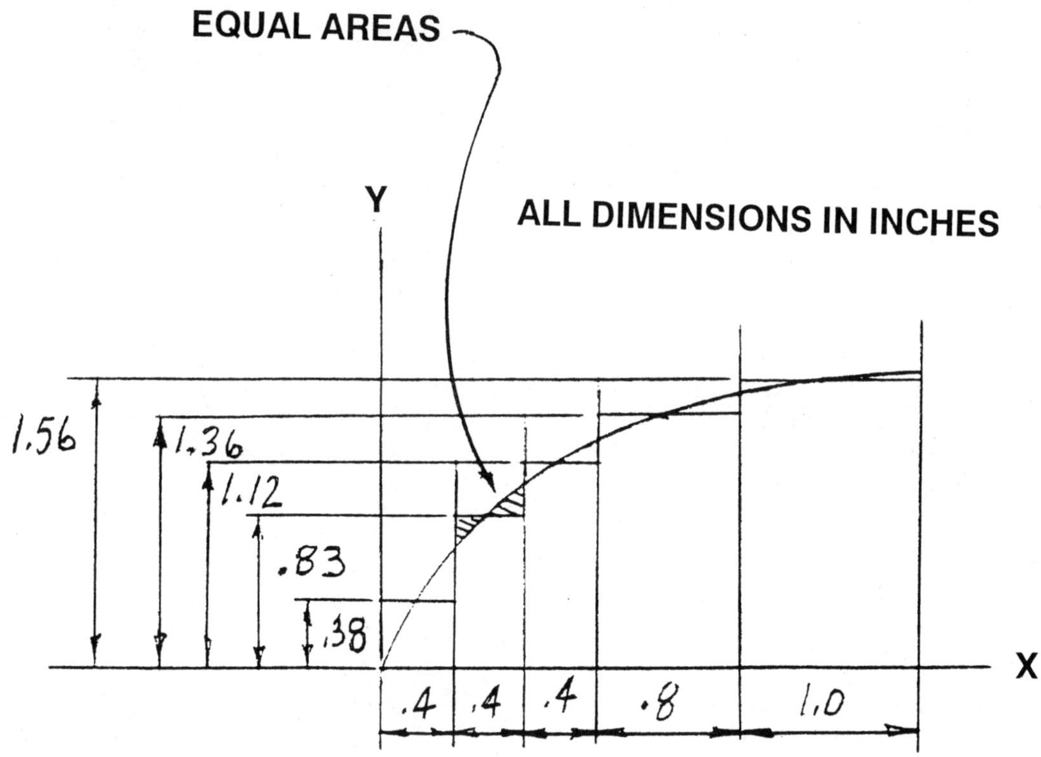

AREA = (.4)(.38) + (.4)(.83) + (.4)(1.12) + (.8)(1.36) + (1.0)(1.56)
= 3.58 square inches

Mechanical Drafting Mathematics			Problem
Section-Chapter-Problem **Appendix A 23A**	Graphical Technique Areas	Course______ Section______ Student______ Date______	**A-23A**

Do the following:

The curve shown is a plot of $x = \frac{x^2}{4}$ from x = 0 to x = 3.5. The theoretical area under the curve is equal to 3.57 in^2. Plot the curve $y = \frac{x^2}{5}$ from x = 0 to x = 5 to scale and determine the area graphically to compare the graphical method to the theoretical one; area = 10.4.

$$Y = \frac{X^2}{4}$$

Y	X
0	0
.25	1.0
.56	1.5
1.00	2.0
1.56	2.5
2.25	3.0
3.06	3.5

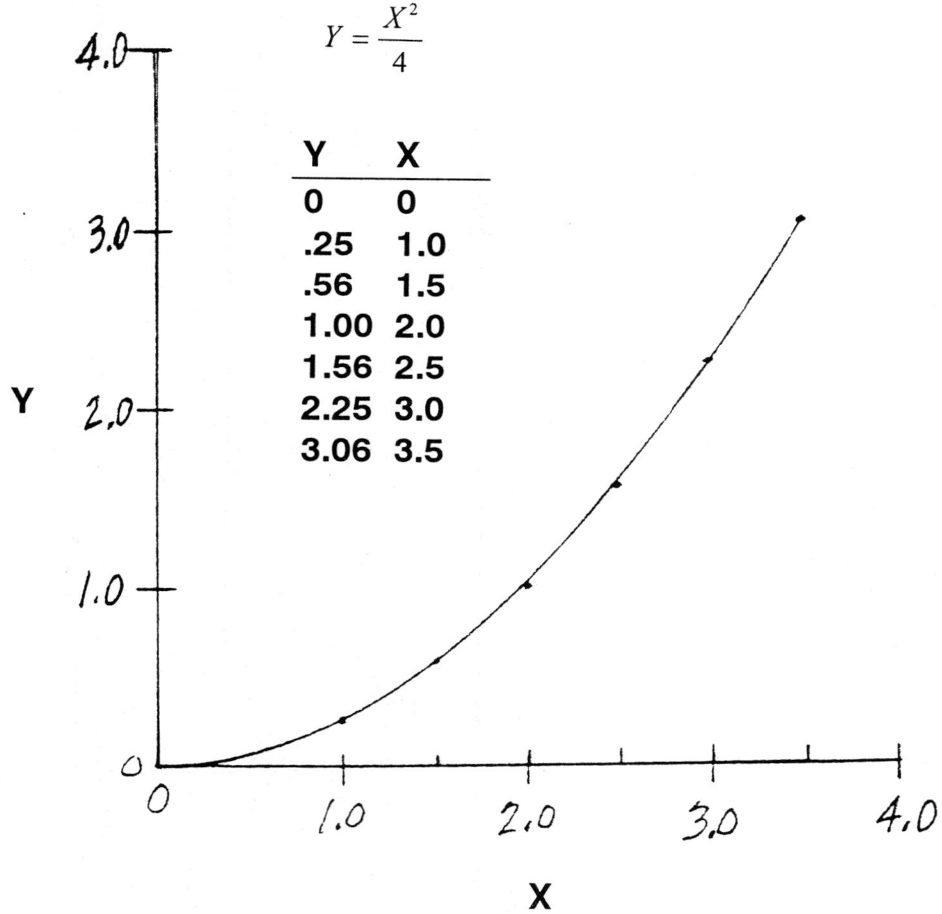

Area = ____________

Mechanical Drafting Mathematics			Problem
Section-Chapter-Problem Appendix A 23B	Graphical Technique Areas	Course________ Section____ Student______________ Date_____	A-23B

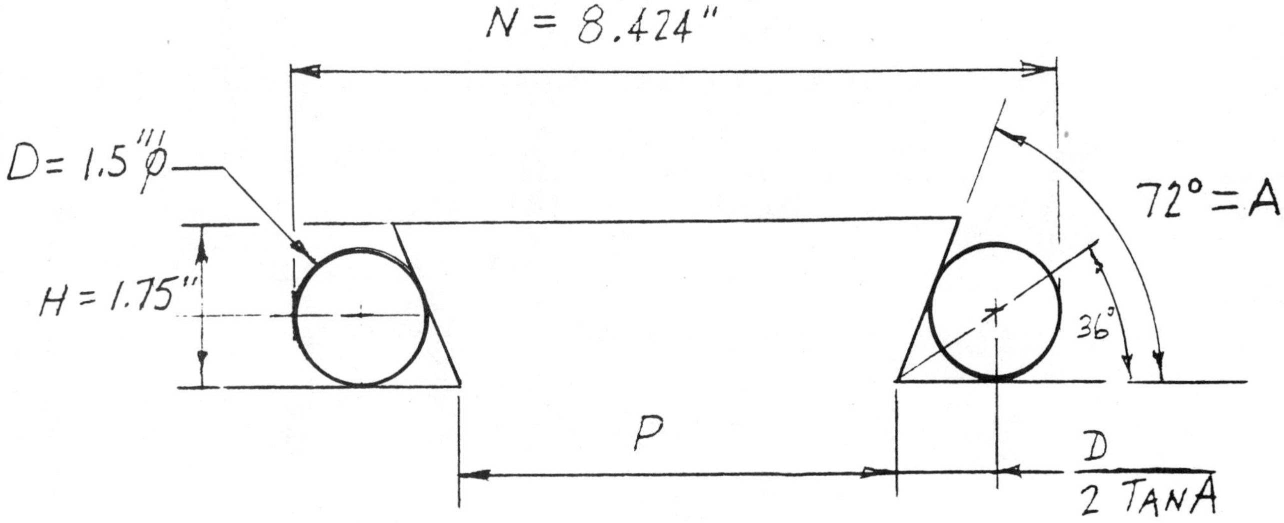

$$P = 8.424 - 2\left(\frac{0.75}{\tan 36^\circ} + .75\right) = 4.86''$$

For the given information determine the following:

a. Change N to 5.982", the angle to 60°, H to 1", and determind P.

P = __________

b. Change n to 3.9205", the angle to 50°, H = 1.1", the diameter to 0.875", and determine P.

P = __________

To determine the weight of a tapered member the following equations ae used:

$$V = \frac{\pi h}{3}(R^2 + Rr + r^2) \quad \text{and Weight = (volume)(density)}$$

Example:

Determine the volume and weight of a tapered member with the following dimensions: large diameter, D = 2", small diameter, d = 1", and h = 4". Density of steel = 0.28 lb/cubic inch.

$$V = \frac{3.14(4)}{3}[1^2 + (1)(0.5) + (0.5)^2] = 7.33in^2 \quad \text{Weight = (7.33)(0.28) = 2.05 pounds}$$

Do the following:

a. Determine the weight of the tapered steel-bar shown.

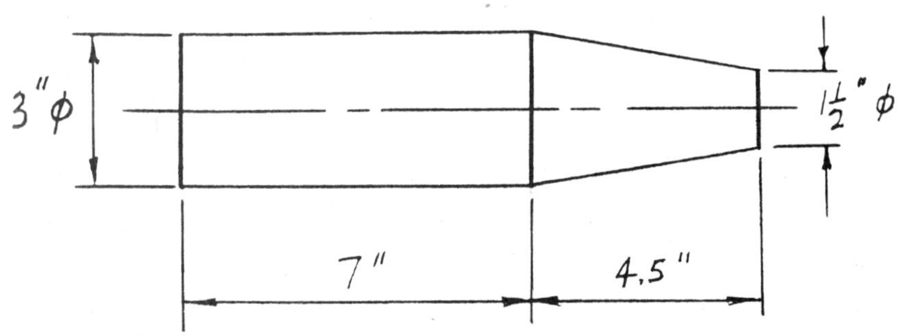

Weight = ____________

Mechanical Drafting Mathematics			Problem
Section-Chapter-Problem Appendix A 25A	Taper Calculations	Course________ Section_____ Student______________ Date______	A-25A

b. Determine the weight of the steel punch shown.

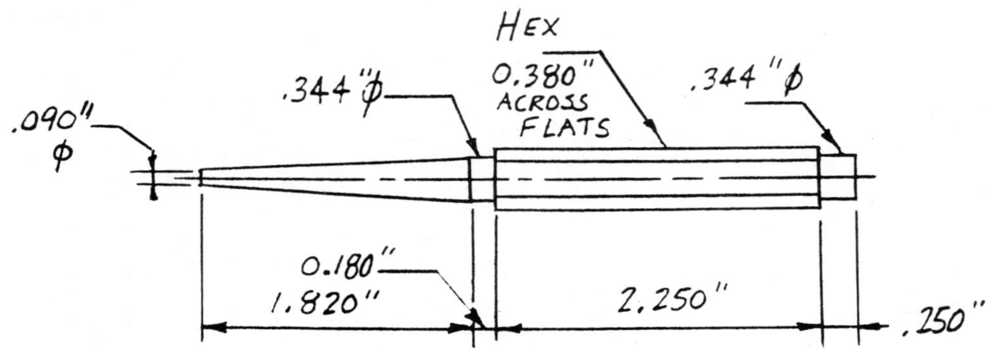

Weight = ______________

c. Determine the weight of the brass member shown. Brass density = 0.32 lb/cubic inch

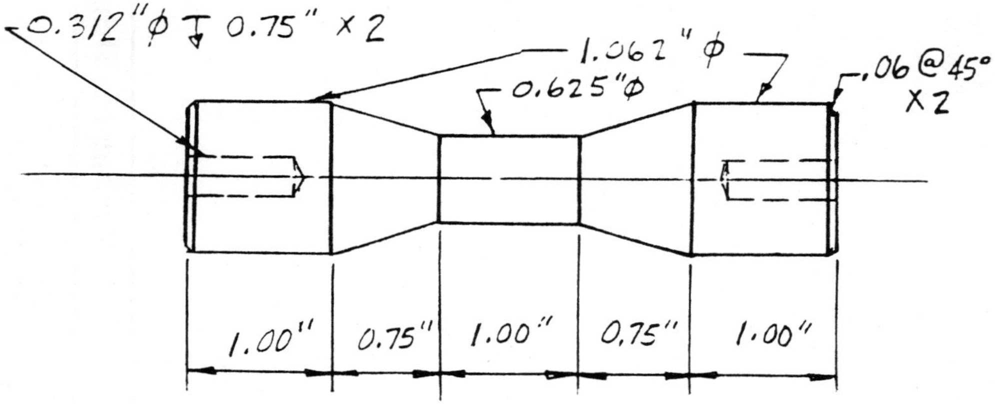

Weight = ______________

Mechanical Drafting Mathematics			Problem A-25B
Section-Chapter-Problem Appendix A 25B	Taper Calculations	Course ______ Section ____ Student ______ Date ____	

The following is an example triangulation problem and solution. A measured, horizontal-distance AB = 2,500 feet along one side of a gorge is used as a base line for determining the distance across the gorge. A prominent formation of rock, at the same elevation of point A, was sighted on the opposite side of the gorge from point A; the line of sight was at 84°17'22" with repect to line AB. A similar sighting from point B to the rock formation was at an angle of 81°21'9'. Determine the distance from the rock formation on a line perpendicular to line AB. (Suggestion: Use at least six decimal places in your calculations.)

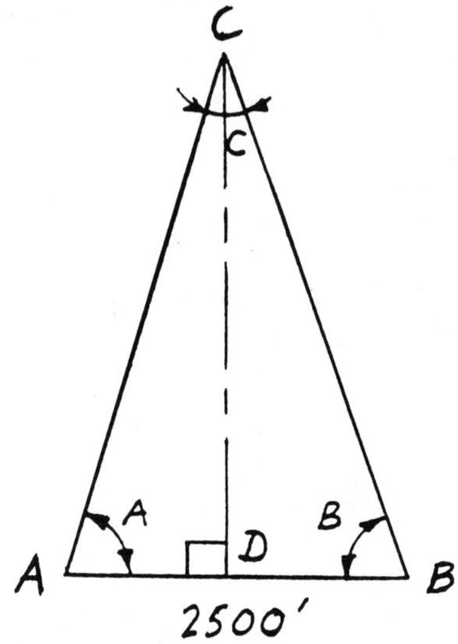

Angle A = 84°17'22", Angle B = 81°21'9"; therefore

Angle C = 14°21'29"

$$\frac{CB}{SinA} = \frac{2500}{SinC}; CB = \frac{SinA(2500)}{SinC}; CB = 10,031'$$

CD = SinB (10,031) = 9917.33'

Triangulation Problem:

Two students in a surveying class visitedthe Grand Canyon in Arizona. They took some preliminary measurements, sitings and angles for determining the distance across the canyon, but didn't have time to do the calculations. They were told that the distance across was about five miles. Use data from their log book and calculate the distance. Was their data accurate?

Log book data: Horizontal lLine GC, East-West direction, 4000 feet.

Angle at G at the West end of GC, measured from GC toward a large building P = 85°54'51"

Line of sight angle at C at the East end of GC = 84°33'21" to the same building. All lines were in the same horizontal plane.

DETERMINE the distance PD which is perpendicular to line GC and intersects GC at D.

PD = ____________________

Mechanical Drafting Mathematics			Problem
Section-Chapter-Problem Appendix A 26	Triangulation Law of Sines	Course________ Section_____ Student__________ Date_____	A-26

Determine the mechanical advantage of a typical boat trailer hoist. The sketch in the figure shows a 9" crank-arm connected to a 1.5" diameter pinion gear which drives an 8" diameter gear attached to a 6" diameter wire rope drum.The following calculations illustrate how to determine the mechanical advantage of boat hoist:

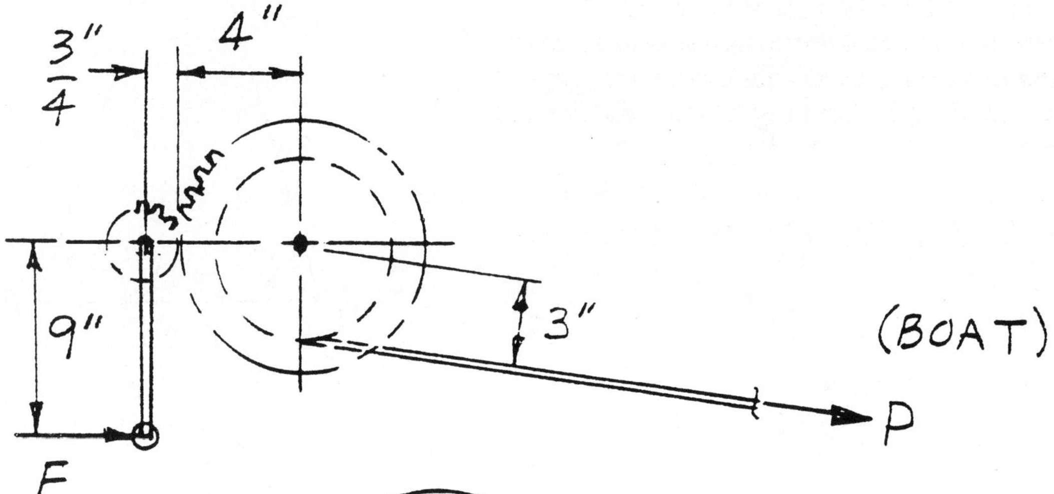

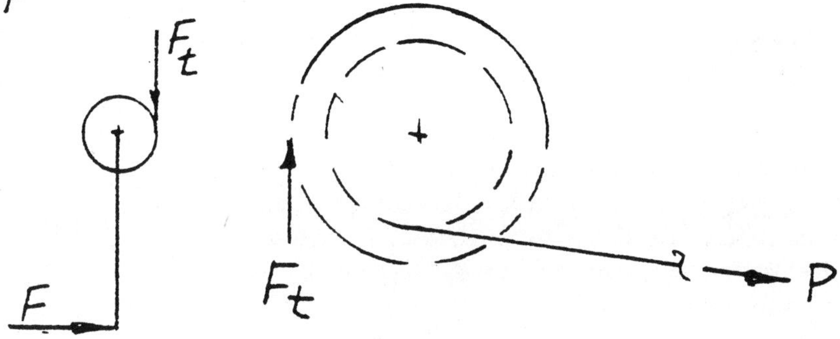

$$9F = \frac{3}{4}F_t \qquad F_t = \frac{4}{3}(9)(F) = 12F$$

$$F_t(4) = (3)P \qquad P = F_t\frac{4}{3} \qquad P = \frac{(12F)(4)}{3} = 16F$$

Therefore, the mechanical advantage = 16; that is, for example, a 50 pound force applied tby the boat operator produces 800 pounds of pull P to move the boat up the trailer.

Calculate the mechanical advantage of a trailer boat-hoist if the crank arm is 10", the pinion gear 2" in diameter, the gear 8.75" in diameter, and the drum 6.5" diameter.

Mechanical Advantage = ________________

The figure below shows a schematic version of an 18 speed bicycle which has 6 sprockets at the rear axle and 3 sprockets at the pedal axle. Assume that a rider pedals 65 complete pedal rotations per minute. Calculate the bicycle's speed if the rider uses a 26 tooth rear sprocket with a 28 tooth sprocket on the pedal axle.. The bicycle rides on 27" diameter tires.

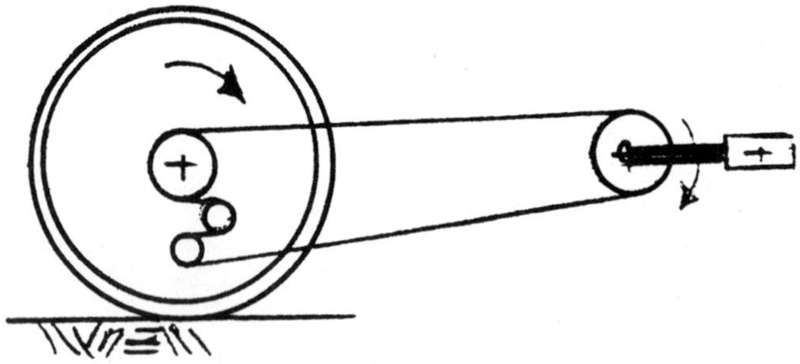

Example:

The distance T the bicycle travels for one revolution of the rear wheel = πD, where D is in feet; threfore,

T = (3.14)(27/12) = 7.07 feet.

$$\text{RPM of the rear wheel} = \frac{(driverteeth = 28)(driverRPM = 65)}{(driventeeth = 26)} = 70RPM$$

Therefore the speed of the bicycle in miles per hour is:

$$MPH = \frac{(70)(7.07)(60)}{(5280)} = 5.62$$

a. Determine the bicycle's speed for the following:
46 tooth front sprocket, 14 tooth rear sprocket
55 pedal shaft rotations per minute

b. Determine the bicycle's speed for the following:
42 tooth front sprocket, 20 tooth rear sprocket
60 pedal shaft rotations per minute

c. Determine the bicycle's speed for the following:
28 tooth front sprocket, 14 tooth rear sprocket
60 pedal shaft rotations per minute

a. Bicycle Speed = ____________

b. Bicycle Speed = ____________

c. Bicycle Speed = ____________

Mechanical Drafting Mathematics			Problem
Section-Chapter-Problem Appendix A 28	Bicycle Speed Gear Ratios	Course__________ Section_____ Student__________ Date_____	A-28

The relative cutting speed in FPM (feet per minute) of a cutting tool tip, used for machining objects in a lathe, is calculated with the following formula:

$$FPM = \frac{\pi DN}{12}$$

D = diameter of object in inches
N = revolutions per minute of the lathe.

Example:

A steel rod 1-5/8" diameter machined at 172 RPM

$$FPM = \frac{\pi(1.625)(172)}{12} = 73.2$$

a. Determine the cutting speed for a 2.50" diameter steel bar being turned in a lathe at 250 RPM

a. Cutting Speed in FPM = __________

b. The cutting speed in a lathe needs to be 24 FPM for a 7.2" diameter object. What rotational speed in RPM is required?

b. Rotational Speed in RPM = __________

An auto traveling at a velocity v, along a circlar road of radius r, will experience a centrifugal force in a radial direction outward from the center of the circular road. When the road is banked at a specific angle, no frictional force would be required to keep the auto on the road. The specific angle may be found from the following equation:

$$\tan\beta = \frac{v^2}{gr}$$

v = auto speed in feet per second
g = gravity constant 32.3 feet per second squared
r = radius of curve in feet

a. Determine the specific angle for 60 MPH and a 600 foot radius.

b. Determine the specific angle for 80 MPH and a 1000 foot radius.

c. If the correct angle is 10° for a 450 foot radius, what auto speed is required to go around the curve without slipping?

a. Specific Angle = __________

b. Specific Angle = __________

c. Auto Speed in FPS = __________

The steel base plate and the drive machinery shown in the figure consists of the following:

1. motor
2. coupling
3. gear reducer
4. the base plate
5, the fasteners

Determine the total weight of the base plate and the machinery. Include fasteners and lock washers (Note that lock washers are used whenever there is a possibility of vibrations.)

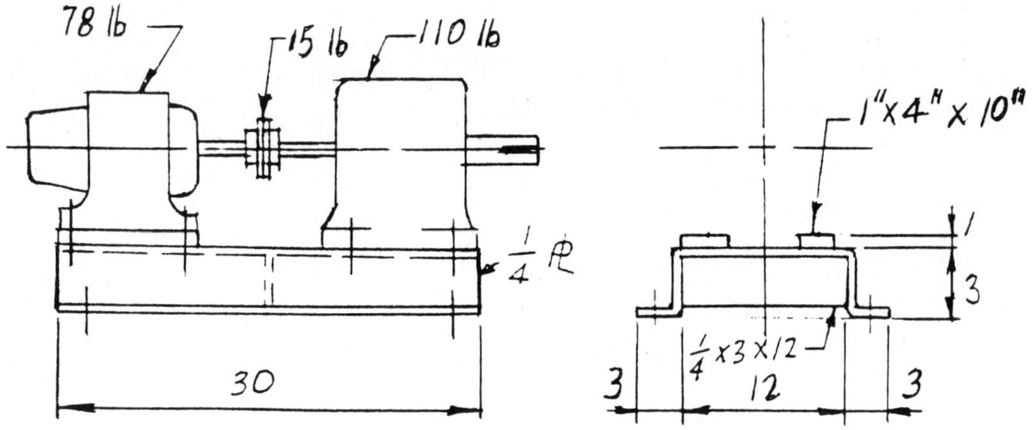

FASTENERS: 8--3/4" dia HEX BOLTS X 2-1/2" LG, HN, LOCK WASHERS

ALL DIMENSIONS IN INCHES

Total Weight=________________

Mechanical Drafting Mathematics			Problem
Section-Chapter-Problem Appendix A 31	Drive Base with Drive Weight	Course__________ Section_____ Student__________________ Date______	A-31

BEND ALLOWANCE FOR EACH 1° OF BEND (INCH)

Radii / Thickness	.031	.063	.094	.125	.156	.188	.219	.250	.281	.313	.344	.375	.438	.500	.531	.625
.013	.00064	.00120	.00174	.00228	.00282	.00338	.00392	.00446	.00500	.00556	.00610	.00664	.00774	.00883	.00937	.01101
.016	.00067	.00122	.00176	.00231	.00285	.00342	.00395	.00449	.00503	.00559	.00613	.00667	.00777	.00885	.00939	.01103
.020	.00069	.00125	.00179	.00233	.00287	.00343	.00397	.00452	.00506	.00561	.00616	.00670	.00780	.00888	.00942	.01106
.022	.00071	.00127	.00181	.00235	.00289	.00345	.00399	.00453	.00508	.00563	.00617	.00672	.00782	.00890	.00944	.01108
.025	.00074	.00129	.00184	.00238	.00292	.00348	.00402	.00456	.00510	.00566	.00610	.00674	.00784	.00892	.00946	.01110
.028	.00076	.00132	.00186	.00240	.00294	.00350	.00404	.00458	.00512	.00568	.00622	.00676	.00786	.00894	.00948	.01112
.032	.00079	.00134	.00189	.00243	.00297	.00353	.00407	.00461	.00515	.00571	.00625	.00679	.00789	.00897	.00951	.01115
.038	.00084	.00140	.00194	.00248	.00302	.00358	.00412	.00466	.00520	.00576	.00630	.00684	.00794	.00902	.00946	.01120
.040	.00085	.00141	.00195	.00249	.00303	.00359	.00413	.00468	.00522	.00577	.00632	.00686	.00796	.00904	.00958	.01122
.050		.00149	.00203	.00258	.00312	.00368	.00422	.00476	.00530	.00586	.00640	.00694	.00804	.00912	.00966	.01130
.064		.00160	.00214	.00268	.00322	.00378	.00432	.00486	.00540	.00596	.00650	.00704	.00814	.00922	.00976	.01140
.072			.00220	.00274	.00328	.00384	.00438	.00492	.00546	.00602	.00656	.00710	.00820	.00929	.00983	.01147
.078			.00225	.00279	.00333	.00389	.00443	.00497	.00551	.00607	.00661	.00715	.00825	.00933	.00987	.01152
.081			.00227	.00281	.00335	.00391	.00445	.00499	.00554	.00609	.00664	.00718	.00828	.00936	.00990	.01154
.091			.00235	.00289	.00343	.00399	.00453	.00507	.00561	.00617	.00671	.00725	.00835	.00944	.00998	.01162
.094			.00237	.00291	.00346	.00401	.00456	.00510	.00564	.00620	.00674	.00728	.00838	.00946	.00999	.01164
.102				.00298	.00352	.00408	.00462	.00516	.00570	.00626	.00680	.00734	.00844	.00952	.01006	.01170
.109				.00303	.00357	.00413	.00467	.00521	.00575	.00631	.00685	.00739	.00849	.00958	.01012	.01176
.125				.00316	.00370	.00426	.00480	.00534	.00588	.00644	.00698	.00752	.00862	.00970	.01024	.01188
.156					.00394	.00450	.00504	.00558	.00612	.00668	.00722	.00776	.00886	.00994	.01048	.01212
.188						.00475	.00529	.00583	.00637	.00693	.00747	.00802	.00911	.01019	.01073	.01237
.203								.00595	.00649	.00704	.00759	.00813	.00923	.01031	.01085	.01249
.218								.00606	.00660	.00716	.00770	.00824	.00934	.01042	.01097	.01261
.234								.00619	.00673	.00729	.00783	.00837	.00947	.01055	.01109	.01273
.250								.00631	.00685	.00741	.00795	.00849	.00959	.01068	.01122	.01286

EXAMPLE: MATERIAL THICKNESS = 1/8", INSIDE RADII = 1/4" R. WHERE THESE CROSS = .00534" IF THE INSIDE OF YOUR BEND IS 20° THE BEND ALLOWANCE (B/A) = .1068 (.00534 X 20°) BEND ALLOWANCE (B/A) PLUS TOTAL OF STRAIGHT LENGTHS = DEVELOPED LENGTH.

BEND ALLOWANCE FOR EACH 1° OF BEND (MILLIMETRE)

Radii / Thickness	0.80	1.58	2.38	3.18	3.96	4.76	5.56	6.35	7.15	7.94	8.74	9.52	11.12	12.70	13.50	15.85
.330	.0161	.0305	.0442	.0580	.0717	.0859	.0996	.1134	.1271	.1413	.1550	.1668	.1967	.2242	.2379	.2796
.406	.0169	.0311	.0448	.0586	.0723	.0865	.1002	.1140	.1277	.1419	.1556	.1694	.1973	.2248	.2385	.2802
.508	.0177	.0319	.0456	.0594	.0731	.0873	.1010	.1148	.1285	.1427	.1564	.1702	.1981	.2256	.2393	.2810
.559	.0181	.0323	.0460	.0598	.0735	.0877	.1014	.1152	.1289	.1431	.1568	.1706	.1985	.2260	.2397	.2814
.635	.0187	.0329	.0466	.0604	.0741	.0883	.1020	.1158	.1295	.1437	.1574	.1712	.1991	.2266	.2403	.2820
.711	.0193	.0335	.0472	.0610	.0747	.0889	.1026	.1164	.1301	.1443	.1580	.1718	.1997	.2272	.2409	.2826
.813	.0201	.0343	.0480	.0617	.0755	.0897	.1034	.1174	.1309	.1451	.1588	.1726	.2005	.2280	.2417	.2834
.965	.0213	.0355	.0492	.0629	.0767	.0909	.1046	.1183	.1321	.1463	.1600	.1737	.2017	.2291	.2429	.2845
1.016	.0217	.0358	.0496	.0633	.0771	.0913	.1050	.1187	.1325	.1467	.1604	.1741	.2021	.2295	.2433	.2849
1.270		.0378	.0516	.0653	.0790	.0932	.1070	.1207	.1345	.1486	.1624	.1761	.2040	.2315	.2453	.2869
1.625		.0406	.0543	.0681	.0818	.0960	.1097	.1235	.1372	.1514	.1652	.1789	.2066	.2343	.2480	.2897
1.829			.0559	.0697	.0834	.0976	.1113	.1251	.1388	.1530	.1667	.1805	.2084	.2359	.2496	.2913
1.981			.0571	.0709	.0846	.0988	.1125	.1263	.1400	.1542	.1679	.1817	.2096	.2371	.2508	.2925
2.058			.0577	.0715	.0852	.0994	.1131	.1269	.1406	.1548	.1685	.1823	.2102	.2377	.2514	.2931
2.311			.0597	.0734	.0872	.1014	.1151	.1288	.1426	.1568	.1705	.1842	.2122	.2396	.2534	.2950
2.388			.0603	.0740	.0878	.1020	.1157	.1294	.1432	.1574	.1711	.1848	.2128	.2402	.2540	.2956
2.591				.0756	.0894	.1035	.1173	.1310	.1448	.1589	.1727	.1864	.2143	.2418	.2556	.2972
2.769				.0770	.0907	.1049	.1187	.1324	.1461	.1603	.1741	.1878	.2157	.2432	.2570	.2986
3.175				.0802	.0939	.1081	.1218	.1356	.1493	.1635	.1772	.1910	.2189	.2464	.2601	.3018
3.962					.1001	.1142	.1280	.1417	.1555	.1696	.1834	.1971	.2250	.2525	.2663	.3079
4.775					.1064	.1206	.1343	.1481	.1618	.1760	.1897	.2035	.2314	.2589	.2726	.3143
5.156						.1235	.1373	.1510	.1648	.1789	.1927	.2064	.2344	.2618	.2756	.3172
5.537								.1540	.1677	.1819	.1957	.2094	.2373	.2648	.2785	.3202
5.944								.1572	.1709	.1851	.1988	.2126	.2405	.2680	.2817	.3234
6.350								.1603	.1741	.1883	.2020	.2157	.2437	.2711	.2849	.3265

EXAMPLE: MATERIAL THICKNESS = 3.175 MM, INSIDE RADII = 6.35 MM R. WHERE THESE CROSS = .1356 MM IF THE INSIDE OF YOUR BEND IS 20° THE BEND ALLOWANCE (B/A) = 2.712 MM (.1356 X 20°) BEND ALLOWANCE (B/A) PLUS TOTAL OF STRAIGHT LENGTHS = DEVELOPED LENGTH

From Drafting for Trades and Industry–Basic Skills, Nelson. Delmar Publishers Inc.

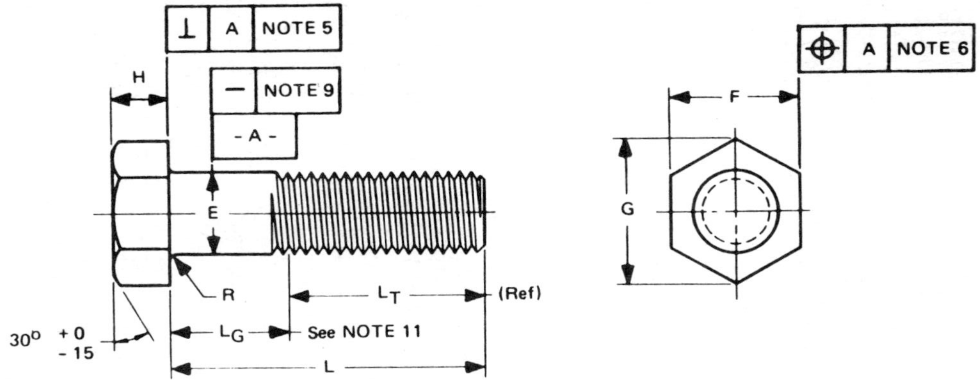

TABLE 19 DIMENSIONS OF HEX BOLTS

Nominal Size or Basic Product Dia (17)		E	F			G		H			R		L_T	
		Body Dia (7)	Width Across Flats (4)			Width Across Corners		Height			Radius of Fillet		Thread Length For Bolt Lengths (11) 6 in. and shorter	over 6 in.
		Max	Basic	Max	Min	Max	Min	Basic	Max	Min	Max	Min	Basic	Basic
1/4	0.2500	0.260	7/16	0.438	0.425	0.505	0.484	11/64	0.188	0.150	0.03	0.01	0.750	1.000
5/16	0.3125	0.324	1/2	0.500	0.484	0.577	0.552	7/32	0.235	0.195	0.03	0.01	0.875	1.125
3/8	0.3750	0.388	9/16	0.562	0.544	0.650	0.620	1/4	0.268	0.226	0.03	0.01	1.000	1.250
7/16	0.4375	0.452	5/8	0.625	0.603	0.722	0.687	19/64	0.316	0.272	0.03	0.01	1.125	1.375
1/2	0.5000	0.515	3/4	0.750	0.725	0.866	0.826	11/32	0.364	0.302	0.03	0.01	1.250	1.500
5/8	0.6250	0.642	15/16	0.928	0.906	1.083	1.033	27/64	0.444	0.378	0.06	0.02	1.500	1.750
3/4	0.7500	0.768	1 1/8	1.125	1.088	1.299	1.240	1/2	0.524	0.455	0.06	0.02	1.750	2.000
7/8	0.8750	0.895	1 5/16	1.312	1.269	1.516	1.447	37/64	0.604	0.531	0.06	0.02	2.000	2.250
1	1.0000	1.022	1 1/2	1.500	1.450	1.732	1.653	43/64	0.700	0.591	0.09	0.03	2.250	2.500
1 1/8	1.1250	1.149	1 11/16	1.688	1.631	1.949	1.859	3/4	0.780	0.658	0.09	0.03	2.500	2.750
1 1/4	1.2500	1.277	1 7/8	1.875	1.812	2.165	2.066	27/32	0.876	0.749	0.09	0.03	2.750	3.000
1 3/8	1.3750	1.404	2 1/16	2.062	1.994	2.382	2.273	29/32	0.940	0.810	0.09	0.03	3.000	3.250
1 1/2	1.5000	1.531	2 1/4	2.250	2.175	2.598	2.480	1	1.036	0.902	0.09	0.03	3.250	3.500
1 3/4	1.7500	1.785	2 5/8	2.625	2.538	3.031	2.893	1 5/32	1.196	1.054	0.12	0.04	3.750	4.000
2	2.0000	2.039	3	3.000	2.900	3.464	3.306	1 11/32	1.388	1.175	0.12	0.04	4.250	4.500
2 1/4	2.2500	2.305	3 3/8	3.375	3.262	3.897	3.719	1 1/2	1.548	1.327	0.19	0.06	4.750	5.000
2 1/2	2.5000	2.559	3 3/4	3.750	3.625	4.330	4.133	1 21/32	1.708	1.479	0.19	0.06	5.250	5.500
2 3/4	2.7500	2.827	4 1/8	4.125	3.988	4.763	4.546	1 13/16	1.869	1.632	0.19	0.06	5.750	6.000
3	3.0000	3.081	4 1/2	4.500	4.350	5.196	4.959	2	2.060	1.815	0.19	0.06	6.250	6.500
3 1/4	3.2500	3.335	4 7/8	4.875	4.712	5.629	5.372	2 3/16	2.251	1.936	0.19	0.06	6.750	7.000
3 1/2	3.5000	3.589	5 1/4	5.250	5.075	6.062	5.786	2 5/16	2.380	2.057	0.19	0.06	7.250	7.500
3 3/4	3.7500	3.858	5 5/8	5.625	5.437	6.495	6.198	2 1/2	2.572	2.241	0.19	0.06	7.750	8.000
4	4.0000	4.111	6	6.000	5.800	6.928	6.612	2 11/16	2.764	2.424	0.19	0.06	8.250	8.500

From The American Society of Mechanical Engineers - ANSI B18.2.1 - 1981